PHYSICS AND CHEMISTRY BASIS OF BIOTECHNOLOGY
VOLUME 7

FOCUS ON BIOTECHNOLOGY

Volume 7

Series Editors
MARCEL HOFMAN
Centre for Veterinary and Agrochemical Research, Tervuren, Belgium

JOZEF ANNÉ
Rega Institute, University of Leuven, Belgium

Volume Editors
MARCEL DE CUYPER
Katholieke Universiteit, Leuven
Interdisciplinaire Research Center, Kortrijk, Belgium

JEFF W.M. BULTE
National Institutes of Health, Bethesda, MD, U.S.A.

COLOPHON

Focus on Biotechnology is an open-ended series of reference volumes produced by Kluwer Academic Publishers BV in co-operation with the Branche Belge de la Société de Chimie Industrielle a.s.b.l.

The initiative has been taken in conjunction with the Ninth European Congress on Biotechnology. ECB9 has been supported by the Commission of the European Communities, the General Directorate for Technology, Research and Energy of the Wallonia Region, Belgium and J. Chabert, Minister for Economy of the Brussels Capital Region.

Physics and Chemistry Basis of Biotechnology
Volume 7

Edited by

MARCEL DE CUYPER
Katholieke Universiteit Leuven,
Interdisciplinaire Research Center, Kortrijk, Belgium

and

JEFF W.M. BULTE
National Institutes of Health, Bethesda, MD, U.S.A.

KLUWER ACADEMIC PUBLISHERS
DORDRECHT / BOSTON / LONDON

A C.I.P. Catalogue record for this book is available from the Library of Congress.

ISBN 0-7923-7091-0

Published by Kluwer Academic Publishers,
P.O. Box 17, 3300 AA Dordrecht, The Netherlands.

Sold and distributed in North, Central and South America
by Kluwer Academic Publishers,
101 Philip Drive, Norwell, MA 02061, U.S.A.

In all other countries, sold and distributed
by Kluwer Academic Publishers,
P.O. Box 322, 3300 AH Dordrecht, The Netherlands.

Printed on acid-free paper

Printed in the Netherlands.

EDITORS PREFACE

At the end of the 20th century, a tremendous progress was made in biotechnology in its widest sense. This progress was largely possible as a result of joint efforts of top academic researchers in both pure fundamental sciences and applied research. The surplus value of such interdisciplinary approaches was clearly highlighted during the 9th European Congress on Biotechnology that was held in Brussels, Belgium (11-15 July, 1999).

The present volume in the 'Focus on Biotechnology' series, entiteld 'Physics and Chemistry Basis for Biotechnology' contains selected presentations from this meeting. A collection of experts has made serious efforts to present some of the latest developments in various scientific fields and to unveil prospective evolutions on the threshold of the new millenium. In all contributions the emphasis is on emerging new areas of research in which physicochemical principles form the foundation.

In reading the different chapters, it appears that more than ever significant advances in biotechnology very often depend on breakthroughs in the biotechnology itself (e.g. new instruments, production devices, detection methods), which - in turn - can be realized by implementing the appropriate physical and chemical principles into the new application. This 'common' pattern is illustrated in the different chapters. Some highly relevant, next generation scientific topics that are treated deal with *de novo* synthesis of materials for gene transfection, imaging contrast agents, radiotherapy, aroma measurements, psychrophilic environments, biomimetic materials, bioradicals, biosensors, and more. Given the diversity of the selected topics, we are confident that scientists with an open mind, who are looking for new frontiers, will find several chapters of particular interest. Some of the topics will give useful, up-to-date information on scientific aspects that may be either right in, or at the interface of their own field of research.

We would like to thank the many authors who did such an excellent job in writing and submitting their papers to us. It was enjoyable to interact with them, and there is no question that the pressure we put on many of them was worthwhile.

Marcel De Cuyper Jeff W.M. Bulte

TABLE OF CONTENTS

BIOMIMETIC MATERIALS SYNTHESIS

ALEKSEY NEDOLUZHKO AND TREVOR DOUGLAS
Department of Chemistry, Temple University
Philadelphia, PA 19122-2585 USA

Abstract

Thc study of mincral formation in biological systems, biomineralisation, provides inspiration for novel approaches to the synthesis of new materials. Biomineralisation relies on extensive organic-inorganic interactions to induce and control the synthesis of inorganic solids. Living systems exploit these interactions and utilise organised organic scaffolds to direct the precise patterning of inorganic materials over a wide range of length scales. Fundamental studies of biomineral and model systems have revealed some of the key interactions which take place at the organic-inorganic interface. This has led to extensive use of the principles at work in biomineralisation for the creation of novel materials. A biomimetic approach to materials synthesis affords control over the size, morphology and polymorph of the mineral under mild synthetic conditions.

In this review, we present examples of organic-inorganic systems of different kinds, employed for the synthesis of inorganic structures with a controlled size and morphology, such as individual semiconductor and metal nanoparticles with a narrow size distribution, ordered assemblies of the nanoparticles, and materials possessing complex architectures resembling biominerals. Different synthetic strategies employing organic substances of various kinds to control crystal nucleation and growth and/or particle assembly into structures organised at a larger scale are reviewed. Topics covered include synthesis of solid nanoparticles in micelles, vesicles, protein shells, organisation of nanocrystals using biomolecular recognition, synthesis of nanoparticle arrays using ordered organic templates.

1. Introduction

The rapidly growing field of biomimetic materials chemistry has developed largely from the fundamental study of biomineralisation [1], the formation of mineral structures in biological systems. Many living organisms synthesise inorganic minerals and are able to tailor the choice of material and morphology to suit a particular function. In addition, the overall material is often faithfully reproduced from generation

M. De Cuyper and J.W.M. Bulte (eds.), Physics and Chemistry Basis of Biotechnology, 9–45.

to generation. The control exerted in the formation of these biominerals has captured the attention of materials scientists because of the degree of hierarchical order, from the nanometer to the meter length scale, present in most of these structures [2]. Biominerals are usually formed through complementary molecular interactions, the "organic-matrix mediated" mineralization proposed by Lowenstam [3]. These interactions between organic and inorganic phases are mediated by the organisms through the spatial localisation of the organic template, the availability of inorganic precursors, the control of local conditions such as pH and ionic strength, and a cellular processing which results in the assembly of complex structures. Our understanding of some of the fundamental principles at work in biomineralisation allows us to mimic these processes for the synthesis of inorganic materials of technological interest.

The biomimetic approach to materials chemistry follows two broad divisions that remain a challenge to the synthetic chemist. On the one hand mineral formation is dominated by molecular interactions leading to nucleation and crystal growth. On the other hand there is the assembly of mineral components into complex shapes and structures (tectonics [4]) which impart a new dimension to the properties of the material. So, interactions must be controlled at both the molecular length scale (Å) – to ensure crystal fidelity of the individual materials, as well as at organismal length scales (cm or m). The fidelity of materials over these dramatic length scales is not necessarily the same. An intense effort in biomimetic materials chemistry is focussed (often simultaneously) on these two length scales. There is not yet a generalised approach to the processing of materials from the molecular level into complex macroscopic forms that can be used for advanced materials with direct applications.

2. Principles

The processes of crystal nucleation and growth have been shown to be effectively influenced by using organised organic molecular assemblies as well as growth modifiers in solution. Langmuir monolayers, Langmuir-Blodgett films, phospholipid vesicles, water-in-oil microemulsions, proteins, protein–nucleic acid assemblies, nucleic acids, gels, and growth additives afford a degree of empirical control over the processes of crystal nucleation and growth. Stereochemical, electrostatic, geometric, and spatial interactions between the growing inorganic solid and organic molecules are important factors in controlled crystal formation [2, 5]. In addition, well defined, spatially constrained, reaction environments have been utilised for nanoscale inorganic material synthesis and in some instances these nanomaterials have been successfully assembled into extended materials.

Crystalline materials form from supersaturated solutions and their formation involves at least two stages i) nucleation and ii) growth. Controlled heterogeneous nucleation can determine the initial orientation of a crystal. Subsequent growth from solution can be modified by molecular interactions that inhibit specific crystal faces and thereby alter the macroscopic shape (morphology) of the crystal. While an ideal crystal is a tightly packed, homogeneous material that is a three dimensional extension of a basic building block, the unit cell, real crystals are often characterised by defects, and the surfaces are almost certainly different from the bulk lattice. These surfaces, in contact with the solution, grow from steps, kinks, and dislocations that interact with molecules and molecular arrays to give oriented nucleation and altered crystal morphology.

2.1. NUCLEATION

Crystals grow from supersaturated solutions. The kinetic barrier to crystallisation requires the formation of a stable cluster of ions/molecules (critical nucleus) before the energy to form a new surface (ΔG_S) becomes less than the energy released by the formation of new bonds (ΔG_B).

There are two generalised mechanisms for the nucleation of crystals; homogeneous nucleation and heterogeneous nucleation. Homogeneous nucleation requires that the critical crystal nucleus forms spontaneously, as a statistical fluctuation, from solution. A number of molecules/atoms come together forming a crystal nucleus that must reach a critical size before crystal growth occurs rather than the re-dissolution of the nucleus. Heterogeneous nucleation comes about through favourable interactions with a substrate that acts to reduce the surface free energy ΔG_S. The substrate, in the case of biomineralisation, is usually an organised molecular assembly designed specifically for the purpose of crystal nucleation. Much of this article will deal with the rational use of molecules and organised molecular assemblies that are designed to induce nucleation.

2.2. GROWTH

Once nucleation has taken place, and provided there is sufficient material, the nucleus eventually grows into a macroscopic crystal. The equilibrium morphology of a crystal, in a pure system, reflects the molecular symmetry and packing of the substituents of that crystal. The observed morphology of the crystal is strongly affected for example by pH, temperature, the degree of supersaturation, and the presence of surface active growth modifiers. These effects can change the equilibrium morphology quite dramatically.

Solutes must diffuse to, and be adsorbed onto, the growing crystal surface before being incorporated into the lattice. Crystal faces that are fast growing will diminish in relative size while those that are slow growing will dominate the final morphology. The growth rate of a particular face is affected by factors such as the interactions of molecules adsorbed onto crystal faces as well as by the charge of a given face, the extent to which it is hydrated, and the presence of growth sites. The growth sites on a crystal surface are usually steps or dislocation sites and the molecule (atom/ion) must be incorporated there for crystal growth to occur.

2.3. BIOMOLECULES AND SUPRAMOLECULAR ASSEMBLIES AS TEMPLATES FOR CRYSTAL GROWTH

Generally, an organic template performs many functions during a biomimetic synthesis. It provides selective uptake of inorganic ions, and stabilises a critical nucleus, which might define the crystal polymorph. It may direct crystal growth in a certain plane, and, finally, it may terminate the crystal growth. It is often the case that in biomimetic synthesis it is the template that defines the size, polymorph, and morphology of the inorganic crystal. It is not true, however, for natural biomineralisation processes, where crystal growth is often controlled in a more complex and precise methods, such as enzymatic reactions.

It follows that in order to provide an environment for the mineralization, the template must meet several requirements. First, the template surface must contain specific binding sites to bind solution components and to stabilise critical nucleus formed at the template-solution interface. In order to limit vectoral crystal growth at the specific point, the template should present a spatially constrained structure.

Biomimetic synthesis may be organised by either using biological macromolecules, such as proteins or DNA, or creating artificial supramolecular assemblies. Although the latter are very simple systems from the chemical point of view, they often help to imitate some of the essential principles of biomineralisation processes. In the following sections applications of various organic macromolecules and assemblies for the synthesis of inorganic materials are reviewed.

3. Examples

3.1. PROTEINS

The ability of proteins to direct mineral formation is clearly recognisable by the simple observation that the shape, size, and mineral composition of seashells are faithfully reproduced by a species from generation to generation. This implies that the control of mineral formation is under some form of genetic control – most importantly at the level of protein expression. Many biomineral systems require the orchestration of multiple protein partners which have proved difficult to isolate but some clear examples exist of proteins which control biomineral formation [6, 7]. There are also some compelling examples, in the literature, of naturally occurring protein systems whose functions have been subverted to the formation of novel new materials [2, 8]. Other protein systems such as collagen gels (gelatine) have been responsible for much of our photographic technology through the encapsulation of nanocrystals of silver salts. Additionally, the synthesis of polypeptides of non-biological origin for inorganic materials applications is a new and novel direction in biomimetic synthesis.

3.1.1. Ferritin

Ferritin, an iron storage protein, is found in almost all biological systems [9]. It has 24 subunits surrounding an inner cavity of 60-80Å diameter where the iron is sequestered as the mineral ferrihydrite ($Fe_2O_3.nH_2O$ - sometimes also containing phosphate). The subunits of mammalian ferritin are of two types, H (heavy) and L (light) chain which are present in varying proportions in the assembled protein. These join together forming channels into the interior cavity through which molecules can diffuse [10, 11]. The native ferrihydrite core is easily removed by reduction of the Fe(III) at low pH ~4.5, and subsequent chelation and removal of the Fe(II). It is also easily remineralised by the air oxidation of Fe(II) at pH> 6. The oxidation is facilitated by specific ferroxidase sites which have been identified by site directed mutagenesis studies, as E27, E62, His65, E107 and Q141, which proceed through a diferric-μ-peroxo intermediate [12, 13] on the pathway to the formation of Fe(III). These sites are conserved on H chain subunits but absent from L chain subunits. Similar studies have also identified mineral nucleation sites which are comprised of a cluster of glutamates (E57, E60, E61, E64, and 67) inside the cavity. These sites are conserved on both H and L chain subunits. The charged cluster of glutamates that is the nucleation site, probably serves to lower the activation energy of nucleation by strong electrostatic interaction with the incipient crystal nucleus.

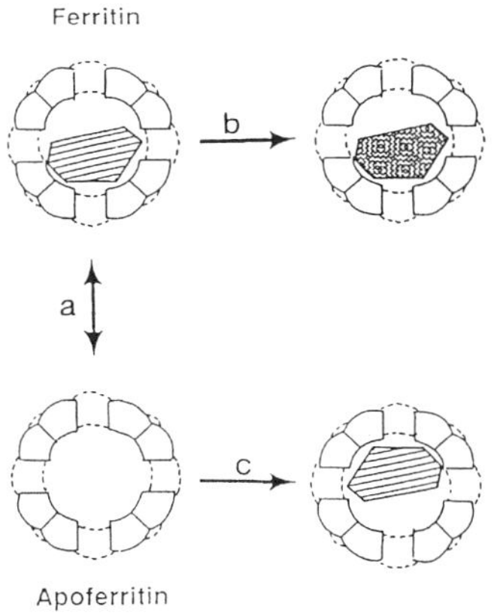

Figure 1. The reaction pathways for nanoparticle synthesis using ferritin. (a) Mineralization,/demineralisation, (b) metathesis mineralization, (c) hydrolysis polymerisation. Source: Reprinted by permission from Nature [17] copyright 1991 Macmillan Mgazines Ltd.

The intact, demineralised protein (apoferritin) provides a spatially constrained reaction environment for the formation of inorganic particles which are rendered stable to aggregation. Ferritin is able to withstand quite extreme conditions of pH (4.0-9.0) and temperature (up to 85°C) for limited periods of time and this has been used to advantage in the novel synthesis and entrapment of non-native minerals. Oxides of Fe(II/III) [14-16], Mn(III) [17, 18], Co(III) [19], a uranium oxy-hydroxide, an iron sulphide phase [20] prepared by treatment of the ferrihydrite core with H_2S (or Na_2S) as well as small semiconductor particles of CdS [21] have all been synthesised inside the constrained environment of the protein. The recent advances in site directed mutagenesis technology holds promise for the specific modification of the protein for the tailored formation of further novel materials.

3.1.2. Bacterial S-layers

The S-layer is a regularly ordered layer on the surface of prokaryotes comprising protein and glycoproteins. These layers can recrystalise as monolayers showing square, hexagonal or oblique symmetry on solid supports [22], with highly homogeneous and regular pore sizes in the range 2 to 8 nm. These proteins have also been implicated in biomineralisation of cell walls and their synthetic use is a great example of the biomimetic approach wherein an existing functionality is utilised for a nonbiological materials synthesis. The two-dimensional crystalline array of bacterial S-layers have been used as templates for ordered materials synthesis on the nanometer scale, both to initiate organised mineralization from solution [23, 24] as well as ordered templates for nanolithography [25]. Both techniques have produced ordered inorganic replicas of the organic (protein) structure.

Treatment of an ordered array of bacterial S-layers (having square, hexagonal or oblique geometry) to Cd^{2+} followed by exposure to H_2S results in the formation of nanocrystalline CdS particles aligned in register with the periodicity of the s-layer. Ordered domains of up to 1 μm were observed (Figure 2).

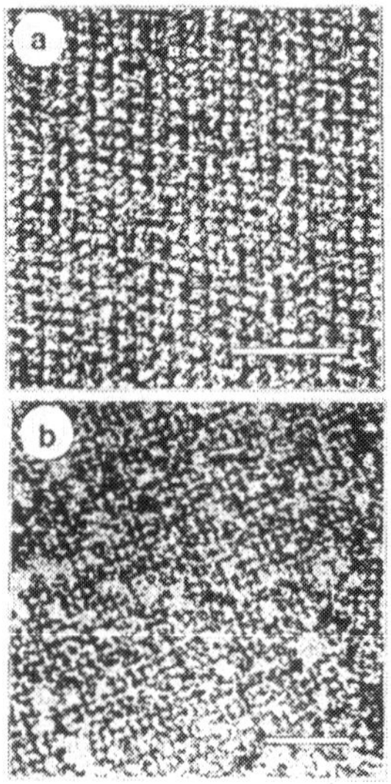

***Figure 2**. Transmission electron micrographs of self-assembled S-layers: (a) S-layer prior to mineralization (stained), (b) after CdS mineralization (unstained). Scale bars = 60 nm. Reprinted by permission from Nature [23] copyright 1997 Macmillan Mgazines Ltd.*

The interaction of S-layers with inorganic materials for the nanofabrication of a solid state heterostructure relies on the ability to crystallise these proteins into two-dimensional sheets. The crystallised protein was initially coated by a thin metal film of Ti which was allowed to oxidise to TiO_2. By ion milling, the TiO_2 was selectively removed from the sites adjacent to the protein leaving a hole with the underlying substrate exposed. Thus, the underlying hexagonal packing arrangement of the 2-d protein crystal layer has been used as a structural template for the synthesis of porous inorganic materials.

3.1.3. Anisotropic structures - Tobacco mosaic virus

It was recently reported that the protein shell of tobacco mosaic virus (TMV) could be used as a template for materials synthesis [26, 27]. The TMV assembly comprises approximately 2130 protein subunits arranged as a helical rod around a single strand of RNA to produce a hollow tube 300 nm x 18 nm with a central cavity 4 nm in diameter. The exterior protein assembly of TMV provides a highly polar surface, which has successfully been used to initiate mineralization of iron oxyhydroxides, CdS, PbS and silica (Figure 3). These materials form as thin coatings at the protein solution interface through processes such as oxidative hydrolysis, sol-gel condensation and so-crystallisation and result in formation of mineral fibres, having diameters in the 20-30 nm range. In addition, there is evidence for ordered end-to-end assembly of individual TMV particles to form mineralised fibres with very high aspect ratios, of iron oxide or silica, over 1μm long and 20-30 nm in diameter.

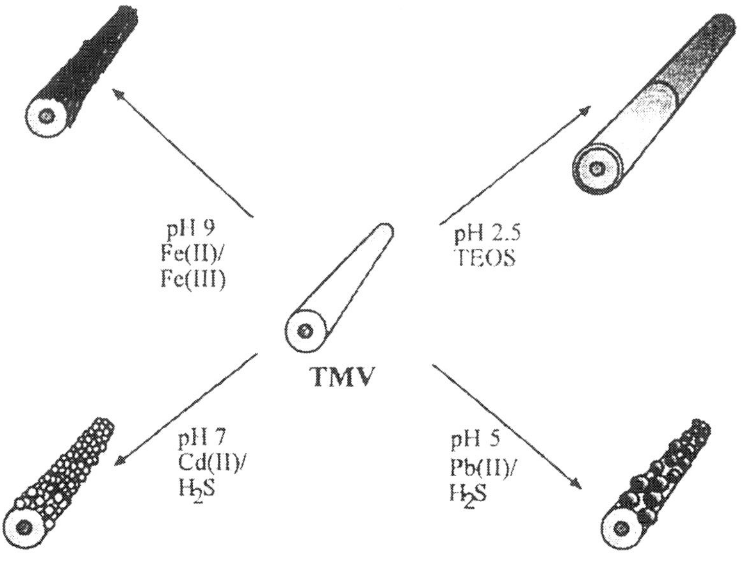

Figure 3. *Strategies for nanoparticle synthesis using tobacco mosaic virus. Reprinted by permission from Adv. Mater. [26] copyright 1999 Wiley -VCH Verlag.*

3.1.4. Spherical virus protein cages

Spherical viruses such as cowpea chlorotic mottle virus (CCMV) have cage structures reminiscent of ferritin and they have been used as constrained reaction vessels for biomimetic materials synthesis [8, 27]. CCMV capsids are 26 nm in diameter and the protein shell defines an inner cavity approximately 20 nm in diameter. CCMV is composed of 180 identical coat protein subunits that can be easily assembled *in vitro* into empty cage structures. Each coat protein subunit presents at least nine basic residues (arginine and lysine) to the interior of the cavity, which creates a positively charged interior interface that is the binding site of nucleic acid in the native virus. The outer surface of the capsid is not highly charged, thus the inner and outer surfaces of this molecular cage provide electrostatically dissimilar environments.

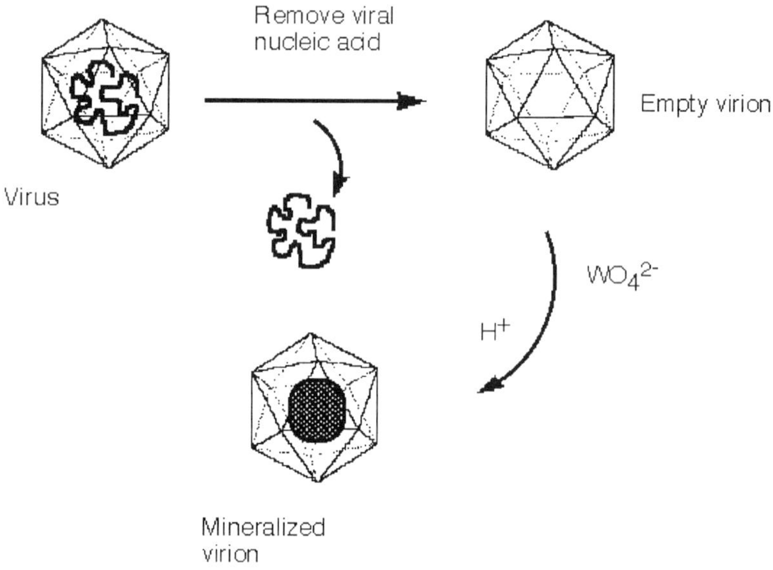

***Figure 4**. Strategy for biomimetic synthesis using cowpea chlorotic mottle virus. Adapted from [8].*

The protein cage of CCMV was used to mineralise polyoxometallate species such as $NH_4H_2W_{12}O_{42}$ at the interior protein-solution interface. It was suggested that mineralization was electrostatically induced at the basic interior surface of the protein where the negatively charged polyoxometalate ions aggregate, thus facilitating crystal nucleation. The protein shell therefore acts as a nucleation catalyst, similar to the biomineralisation reaction observed in ferritin, in addition to its role as a size constrained reaction vessel.

3.2. SYNTHETIC POLYAMIDES - DENDRIMERS

Some interesting synthetic polypeptides are emerging in the field of materials chemistry, in particular dendritic polymers based on poly(amidoamine) or PAMAM dendrimers. These polymers are protein mimics in that they too are polyamides, have fairly well defined structural characteristics (topology), and can accommodate a variety of surface functional groups. They are roughly spherical in shape and they can be terminated with amine, alcohol, carboxylate or ester functionalities. Two groups have demonstrated that pre treatment of either alcohol or amine terminated dendrimers with Pt(II), Pd(II), Cu(II) or $HAuCl_4$ followed by chemical reduction using hydrazine or borohydride resulted in the stabilisation of nanoparticles of the metals [28-31]. These were originally suggested to be stabilised within the matrix of the dendrimer sphere. In addition it has also been shown that dendrimers having different surface functionalities are able to stabilise nanoparticles of CdS (amine terminated [32]) and ferrimagnetic iron oxides (carboxyl terminated [33]). In this regard the functionalised dendrimer acts as a nucleation site by selective binding of the precursor ions and additionally passivates the nanoparticle by steric bulk to prevent extended solid formation.

3.3. GELS

A gel is a loosely cross-linked extended three dimensional polymer permeated by water through interconnecting pores. Gels are used as reaction media for crystal growth when especially big, defect free crystals are desired. Solutes are allowed to diffuse toward each other from opposite ends of a gel-filled tube. This creates a concentration gradient as the two fronts diffuse through each other, giving rise to conditions of local supersaturation. The gel additionally serves to suppress nucleation that allows fewer crystals to form, thus reducing the competition between crystallites for solute molecules, and the result is larger and more perfect crystals. It also acts to suppress particle growth that might otherwise occur by aggregation. Gels are easily deformed and so exert little force on the growing crystal [34].

Gelatine is used extensively in the photographic process for the immobilisation of silver and silver halide micro crystals. The most commonly used photographic emulsion comprises a gelatine matrix with microcrystals of silver halides distributed throughout. While gelatine is the most common matrix, albumen, casein, agar-agar, cellulose derivatives, and synthetic polymers have all been used as gel matrices. The silver halide crystals vary in size from 0.05μm to 1.7μm depending on the film type. Exposure of the film to light forms a "latent image" (a small critical nucleus of silver metal) that will catalyse the reduction (and growth of a silver crystal) of that particular grain when the film is developed. The development process is the chemical reduction of the silver halide grains and the growth, in its place, of a microcrystal of silver metal. The matrix serves to keep these microcrystals separate and prevent their aggregation that would result in loss of image resolution.

3.4. COMPOSITE MATERIALS

Proteins that have been isolated from biominerals exhibit a number of the properties mentioned in the preceding sections such as oriented nucleation, and confined reaction environments. The production of biocomposite ceramics is a low temperature route to strong, lightweight materials that has not yet been fully exploited. In bone, hydroxyapatite crystals are found in spaces within the collagen fibril. Purified collagen serves as a matrix for calcium phosphate growth in attempts to study that process and to create synthetic bone-like material. Matrix proteins isolated from bivalves have been shown to mediate nucleation and growth of calcium carbonate [35-38]. These materials are composites of microscopic crystals held together by a protein "glue" and have the advantages of both the hardness of the inorganic material, and the flexibility of the organic matrix. Composite materials such as these often have high fracture toughness thought to arise from interruption, by the protein, of the cleavage planes in the inorganic crystals. For example, the calcite crystal cleaves easily along the (104) planes. In the sea urchin skeleton the crystal fractures conchoidaly (like glass) and not cleanly along the (104) planes of calcite. It is suggested that this is due to the protein that is occluded within the crystal, preventing the cleavage along the (104) plane and thereby increasing the strength of the inorganic phase. These proteins have been isolated and shown to produce the same conchoidal fracture in synthetic calcite crystals grown in its presence. These materials are the inspiration for a new generation of materials incorporating both natural and synthetic polymers.

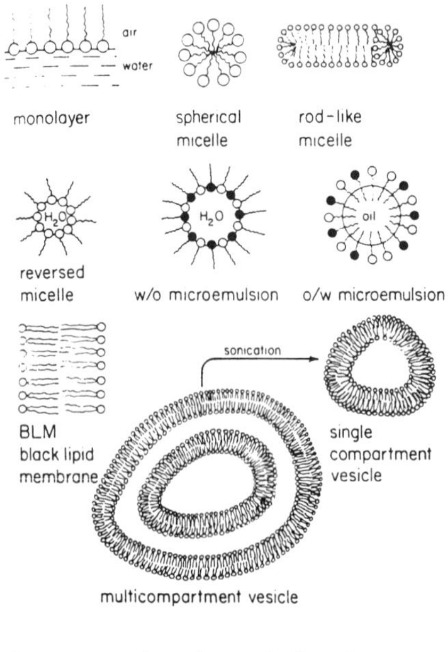

***Figure 5**. Schematic representation of organised surfactant assemblies. Reprinted by permission from Chem.Rev. [150] copyright 1987ACS Publications.*

3.5. ORGANIZED SURFACTANT ASSEMBLIES

Although surfactant assemblies are not 'biological' in the general sense of the word, they often give a good opportunity to mimic biomineralisation processes. These assemblies are schematically represented in Figure 5. With respect to the methods of their application in the synthesis of inorganic materials they may be separated into two groups on the basis of their geometry. Micelles, microemulsions and vesicles form one class of the assemblies, whose specific feature is maintenance of a confined environment for the crystal growth. Layered structures are the other class of assemblies, for which the periodicity of layered arrays is essential.

The phase behaviour of surfactant – water mixture depends on the water-to-surfactant ratio *w*. Another important parameter – cmc (critical micelle concentration) represents the constraint surfactant concentration for the formation of micelles.

3.5.1. Confined surfactant assemblies

3.5.1.1. Reverse micelles (water-in-oil microemulsions)

General principles for the synthesis of inorganic materials in these environments involve dissolution of reactants in the aqueous phase and the subsequent reactions, which occur due to micelle collisions, accompanied by the exchange of their aqueous phases.

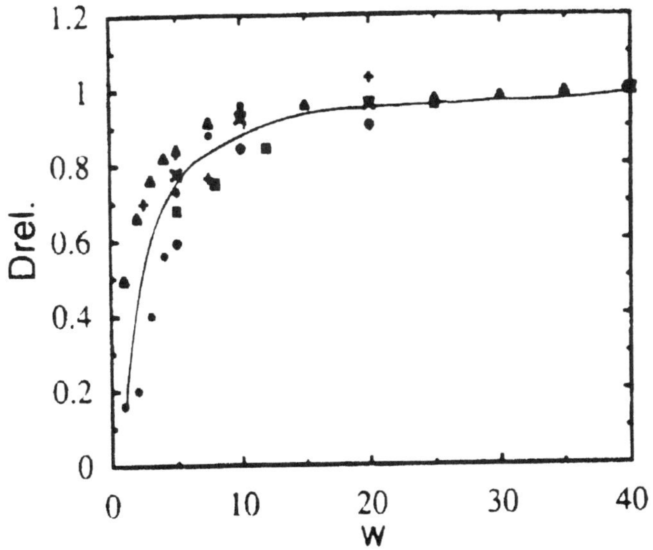

Figure 6. *The dependence of particle size on water-to-surfactant ratio. CdS (triangles), PbS (squares), $Cd_yZn_{1-y}S$ (circles), $Cd_yMn_{1-y}S$ (+), ZnS (X), Ag (octagons). Reprinted by permission from Langmuir [151] copyright 1997 ACS Publications.*

First reports on the application of reverse micelles for the synthesis of inorganic nanoparticles were devoted to the synthesis of precious metals. In the study of Boutonnet et al [39] metal cations dissolved in water pools of water/CTAB/octanol or water/ pentaethylene glycol dodecyl ether/hexane micellar solution, were reduced by hydrazine. The formed particles (Pt, Pd, Rh and Ir) were reported to have a narrow size distribution (standard deviation 10%) being in the size range 30 – 50 Å. The formation of gold particles in water-in-oil microemulsions was first reported by Kurihara et al [40] who studied the reaction of $HAuCl_4$ reduction by laser photolysis and pulse radiolysis. Again, higher uniformity of Au particles obtained in micellar solution comparing to those obtained in homogeneous solution was reported.

Studies of particle formation in reverse micelles were initially oriented for the synthesis of highly dispersed catalysts, and in a similar vein was the work of Meyer et al [41] describing synthesis of cadmium sulphide nanoparticles. AOT micelles containing Cd^{2+} were prepared in isooctane and then exposed to H_2S. The possibility to use formed CdS as a photosensitiser was demonstrated. In the study by Lianos and Thomas [42] CdS was obtained by mixing micellar (heptane/AOT/water) solutions of cadmium perchlorate and sodium sulphide. The increase of particle size with water-to-surfactant ratio *w* was reported. This dependence was shown to occur only at $w < 15$, with the particle size being almost constant at higher *w* values [43] (Figure 6). Maximum particle size was reported to be twice that of CdS particles prepared with a two-fold excess of S^{2-}, than for those prepared in a two-fold excess of Cd^{2+}.

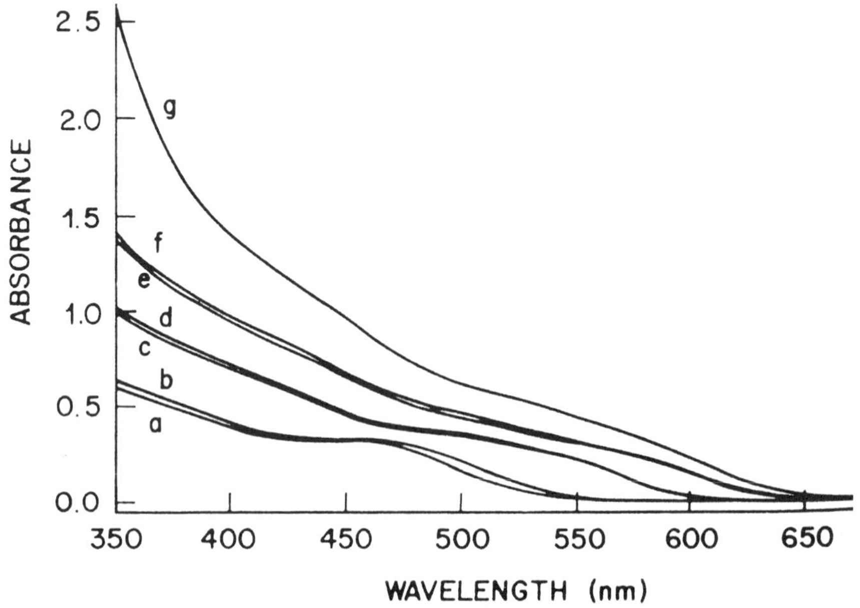

***Figure** 7. Changes in absorbance spectrum during CdSe crystal growth in reverse micelles. Reprinted by permission from J.Am.Chem.Soc. [44] Copyright 1988 ACS Publications.*

In the paper of Steigerwald et al [44] the formation of CdSe in heptane/AOT/water micelles was reported. The synthesis was performed by the fast addition of bis(trimethylsilyl)selenium to the micellar solution of Cd^{2+}. Particle growth was monitored by changes in light absorption spectrum, which exhibited characteristic onset shifting to longer wavelengths on subsequent additions of the selenium derivative (Figure 7). Also, the possibility to change the properties of the particle surface to strongly hydrophobic was demonstrated using phenyl-bis(trimethylsilyl)selenium reacting with excess Cd^{2+} atoms on the surface of CdSe. Although precipitate was formed, it could be subsequently redissolved in non-polar solvents yielding CdSe colloid. Later, the authors demonstrated the use of water-in-oil microemulsions for the synthesis of CdSe/ZnS core-shell structures [45]. Difference in bond length for these compounds is 13 %, and the crystal lattices were found not to match each other. The important result of this study was the synthesis of 35 – 40 Å CdSe particles, covered with 4 Å thick ZnS layer. Deposited ZnS 'fills' deep surface trap states of CdSe, providing strong and stable luminescence of the composite particles.

In the work of Towey et al [46] changes in absorption spectrum during CdS particle growth under a variety of experimental conditions were monitored using stopped-flow technique, and the attempt to analyse growth kinetics quantitatively was presented. The authors concluded that inter-droplet exchange of solubilised reactants was the rate-determining step.

Petit et al [47] proposed to use metal-substituted surfactants, such as cadmium-lauryl sulphate or cadmium-AOT for the synthesis of CdS, as the source of the metal cation. Later, the same approach was employed by the authors for the synthesis of Cu [48] and Ag [49] metal particles. Copper metal particles were obtained by reduction of $Cu(AOT)_2$ in water/isooctane with either hydrazine (added by injection) or borohydride (introduced in the form of water/AOT/isooctane micellar solution). The properties of the particles formed in the reaction were found to depend greatly on the nature of the reducing agent. While in the reaction with hydrazine small (20 – 100 Å) metal particles were formed, whose size, as usual, increased with w value, the reaction with borohydride yielded large (up to 28 nm) particles exhibiting anomalous dependence on w. It was found that the increase of w in the case of borohydride led to a progressive formation of copper oxide instead of Cu metal. Above $w = 8$ pure CuO was reported to appear even in the absence of oxygen.

The synthesis of silver metal particles was demonstrated using the same general approach. Although the average particle size was clearly dependent on w (from 30 Å ($w = 5$) to 70 Å ($w = 15$)) the size distribution was rather broad ($\sigma = 30 – 40$ %).

Chang et al [50] demonstrated synthesis of silica particles by the hydrolysis of tetraethoxysilane in water-in-oil microemulsions in the presence of ammonium hydroxide and hexanol, acting as co-surfactant. SiO_2 had a spherical structure. The size of the formed spheres could be controlled by the reaction conditions in the range 40 – 300 nm with standard deviation of 5%. Also, the possibility to synthesise mixed SiO_2-CdS spheres was demonstrated, with CdS being incorporated in different ways as core, shell, or intermediate sphere, or small (24 Å size) inclusions (volume or surface) (Figure 8), or surface patches.

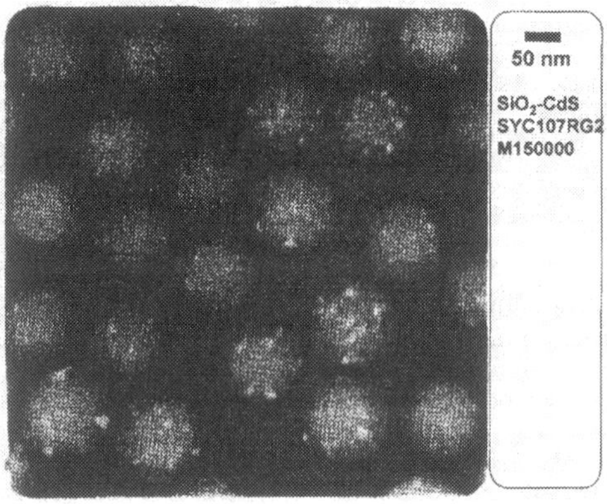

***Figure 8**. Silica spheres containing homogeneously distributed CdS inclusions, formed in water-in-oil microemulsions. Reprinted by permission fromJ.Am.Chem.Soc. [50] Copyright 1994 ACS Publications.*

General strategies described above have been successfully employed for the synthesis of other semiconductors (PbS and CuS [51], TiO_2 [52], Se [53]) as well as magnetic materials (Fe_3O_4 [54], barium ferrite [55], iron ferrite [56], cobalt metal [57]) and other solids (zincophosphate [58], $BaSO_4$ [59]).

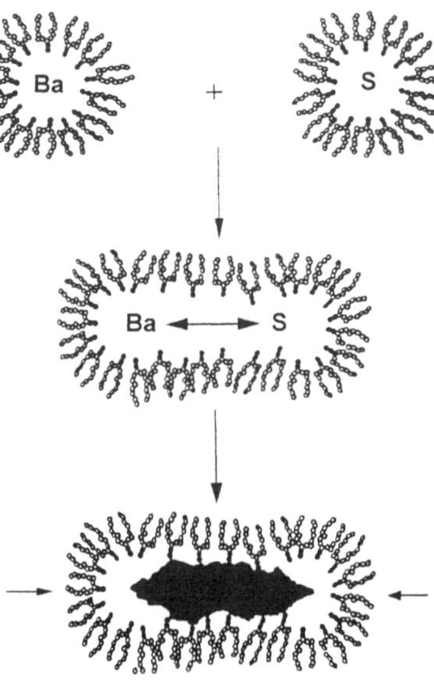

***Figure 9**. Mechanism for the formation of elongated $BaSO_4$ crystal. Reprinted by permission from Chem. Mater. [59] Copyright 1997 ACS Publications.*

The microemulsion method for the inorganic synthesis was shown to allow, in some cases, control of the shape of the inorganic particles. It was demonstrated that the shape of copper metal particles formed in isooctane/AOT/water micelles depends on *w* value [60]. Both at $w < 5.5$ and at $w > 34$ nearly all particles are spherical, however, elongated particles are formed at the intermediate values of *w*. Within this region there is a further dependence of the particle shape on *w*. For example, at $w = 12$ highly elongated cylinders are formed, and at $w = 18$ particles of different shapes and sizes appear. The authors explain these data by changes in micellar solution phase with the variation of *w*.

In the study of Hopwood and Mann [59] on the synthesis of $BaSO_4$ using isooctane/AOT/water reverse micelles the variation of *w* value led to even more drastic changes in the shape of the formed crystals. While at $w = 5$ particles of amorphous $BaSO_4$ were formed, micellar systems with $w > 10$ led to the formation of highly elongated filaments, with lengths up to 100 μm and aspect ratios of 1000. Formation of the filaments was not observed when AOT had been changed to another surfactant. The authors proposed a mechanism for the filaments formation, which involved the anisotropic binding of the surfactant to $BaSO_4$. Since the crystal growth occurs only on the unbound crystal surface, the crystal progressively elongates. (Figure 9).

Recently, the possibility for a self-assembly of nanoparticles formed in reverse micelles was reported [61]. Nanocrystals of barium chromate synthesised in AOT reverse micelles were shown to form periodic arrays due to the interdigitation of surfactant monolayers (Figure 10). By variation of reactant molar ratio it was possible to change the shape of formed $BaCrO_4$ nanoparticles, which, simultaneously, lead to different structures of assembled aggregates.

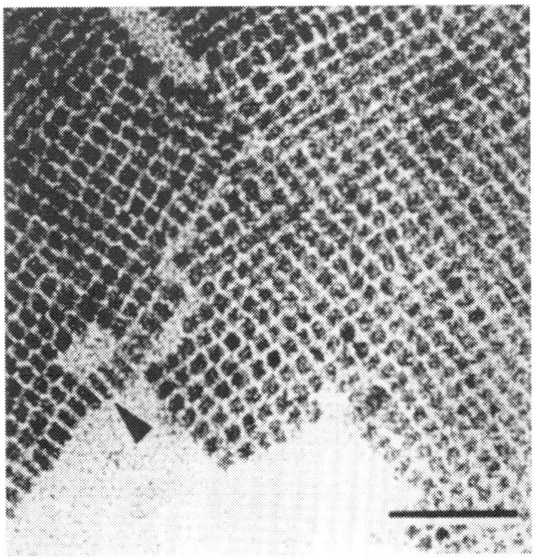

***Figure 10**. Rectangular supperlattice of $BaCrO_4$ nanoparticles prepared in AOT microemulsions ($w = 10$) from equimolar amounts of Ba^{2+} and CrO_4^{2-}. Scale bar = 50 nm. Reprinted by permission from Nature [61]. Copyright 1999 Macmillan Mgazines Ltd.*

3.5.1.2. Oil-in-water micelles

Some surfactants, e. g. sodium dodecyl sulphate (Na(DS)), when dissolved in aqueous solution at concentrations above critical micelle concentration (cmc) form aggregates called normal micelles. In aqueous solutions of functionalised surfactants (with counter-ion replaced by the metal ion participating in the synthesis) the formation of magnetic $CoFe_2O_4$ [62], and Cu metal [63] particles was demonstrated.

Cobalt ferrite was formed by the oxidation of the mixed $Co(DS)_2$ and $Fe(DS)_2$ solution by methylamine. The formed particles were 2 – 5 nm in size, dependent on the initial concentrations of the reactants. Standard deviation in size distribution was 23 – 37%.

Colloidal copper particles were prepared by the reduction of $Cu(DS)_2$ by sodium borohydride. As expected, discrete particles were formed only above the cmc of the surfactant. Below the cmc point, particles formed an interconnected network of either oxide or pure metallic copper aggregates.

Recent paper by Lim et al [64] reports the synthesis of hydroxyapatite nanoparticles in oil-in-water emulsion using petroleum ether, non-ionic surfactant KB6ZA, and aqueous solution of $CaCl_2$. Formation of hydroxyapatite occurred on the addition of ammonium phosphate. The resulting particles were more crystalline than those formed in aqueous solution and in a micellar system containing KB6ZA.

3.5.1.3. Vesicles

Similarly to reverse micelles, vesicles provide a confined environment for particle growth. However, there are a number of distinctions in particle growth. While in the case of reverse micelles one crystal is formed per one assembly, vesicles permit the growth of several crystals bound to either sides of the vesicle bilayer.

First demonstrations [65, 66] showed the occurrence of 100 nm particles of silver oxide formed by precipitation of Ag^+ in phospholipid vesicles at basic pH. Other studies explored the application of single-compartment vesicles as stabilising matrix for CdS-based photocatalysts [67]. Different strategies for CdS synthesis allowed the formation of the crystals on desired size of the phospholipid bilayer [68]. These involved formation of vesicles by sonication in the presence of cadmium salt, and removal of Cd^{2+} ions by passing the sample through cation exchange column. Depending on the preparation procedure, CdS particles ranging from 16 to 26 Å average size, with different size distribution were reported to appear. Unfortunately, the particles were characterised only in an indirect way, on the basis of their absorption spectra.

An attempt to obtain vesicle-bound CdS particles of predictable sizes with narrower size distribution was made by Korgel and Monbouquette [69]. In this study a novel method was applied for the preparation of Cd^{2+}-bound vesicles. Instead of sonication, the authors used detergent dialysis, followed by drying, redissolution in aqueous $CdCl_2$, and subsequent dialysis to remove both the detergent and the excess Cd^{2+}. This preparation led to a more uniform vesicular solution. The average size of formed CdS crystallites could be predicted with the precision up to 2.5 Å, while std. deviation in size distribution was typically less than 10%.

Other inorganic particles synthesised in phospholipid vesicles include selenides [70], mixed semiconductors [71], Al_2O_3 [72] and Al_2O_3 nano-composites with other metal oxides/hydroxides [73], Au and Co metals [74], as well as magnetic iron oxides [75].

3.5.2. Layered surfactant assemblies

3.5.2.1. Surfactant monolayers and Langmuir-Blodgett films

Surfactant molecules, spread on a water surface, align themselves such that the ionic (or polar) part of the molecule interacts with the water while the hydrocarbon tail is oriented away from the surface. Compressing these molecules together at the air-water interface forms an ordered two-dimensional monolayer of the surfactants. These compressed arrays can be transferred, layer-by-layer, onto solid substrates. The resulting films are called Langmuir-Blodgett (LB) films. It must be noted that LB films can be formed not only from surfactants, the same technique may be used to create arrays of fullerenes, polystyrene microspheres and other macromolecules and supramolecular assemblies.

Compressed surfactant monolayers above an aqueous sub-phase, supersaturated with respect to an inorganic solid, have been shown to induce oriented crystal nucleation. It has been demonstrated for a variety of different systems that the adsorption of an amphiphile with the appropriate head group at the air-water interface can induce an oriented nucleation event from supersaturated solutions. Examples of inorganic materials, for which such a molecular templating may be achieved include three different phases of calcium carbonate [76, 77], barium sulphate [78, 79] and calcium sulphate [80]. In these systems there is a high correspondence between the packing of the monolayer and the lattice of the nucleated crystal (geometric factor). Ions adopt unique stereochemical conformations in the crystal lattice that can be mimicked by the monolayer headgroups (stereochemical factor). The headgroup charge results in an accumulation of ions at the interface (electrostatic factor). These three factors are important for oriented nucleation although a degree of mismatch is tolerable.

For example, to synthesise calcium carbonate under surfactant monolayer, monolayer film of amphiphile was spread onto supersaturated calcium bicarbonate solution, and then compressed to an appropriate surface pressure. Crystallisation of $CaCO_3$ proceeded slowly, accompanied by CO_2 evolution. Depending on the experimental conditions (such as the amphiphile used and the concentration of calcium) one or two of three $CaCO_3$ phases (calcite, aragonite or vaterite) was formed. The appearance of a certain $CaCO_3$ phase were first explained by the lattice match between the amphiphile headgroups and inorganic substrate, however, other effects such as the promotional role of water molecules in the nucleation of a specific phase can not be neglected. The concept of direct epitaxial growth of inorganic crystals on organic templates was re-examined in the study of Xu et. al. [81]. The formation of crystalline $CaCO_3$ under amphiphilic porphyrin templates was shown to proceed through an intermediate amorphous phase. The amorphous $CaCO_3$ undergoes phase transformation into the crystalline material with the orientation controlled by the porphyrin template.

Another group of synthetic crystals formed in surfactant layered assemblies is semiconductor and metal nanoparticles. In earlier works on this subject [82, 83] particles of cadmium sulphide of a few nanometer size were formed in surfactant monolayers on liquid-gas interface and then transferred to solid substrates. On the prolonged exposure to hydrogen sulphide disk-shaped particles (20-30 Å thick, 75-100 Å diameter) of CdS were formed. Similar method was employed to synthesise other metal chalcogenides particles in size quantatisation regime [84, 85]. In all cases monoparticulate two-dimensional films were obtained. For several systems, epitaxial matching between growing crystal and the surfactant template was demonstrated. This has been achieved for CdS [86], PbS [87] and PbSe [88] grown under arachidic acid monolayers. Transmission electron microscopy of the PbS - arachidic acid film shows very regular triangular crystals (Figure 11). The authors explain these results by a very close matching between Pb-Pb distance in the PbS {111} plane (4.20 Å) and d{100} spacing in the arachidic acid monolayer (4.16 Å). Doping of the monolayer with another surfactant (octadecylamine) was shown to affect PbS morphology enormously [89].

***Figure 11**. Transmission electron micrographs of PbS crystals grown under arachidic acid monolayers. Scale bar = 50 nm. Reprinted by permission from J. Phys. Chem. [87] Copyright 1992 ACS Publications.*

Surfactant monolayers were shown to organise pre-formed nanoparticles into ordered arrays. Dispersion of colloid suspensions of particles on aqueous sub-phase in a Langmuir film balance produced monoparticulate films of TiO_2 [90], Fe_3O_4 [91] and precious metals [92].

Multilayered, i.e. three-dimensional template of amphiphile molecules may be used for the growth of inorganic nanoparticles in a 3-D array. For example, multilayers of cadmium alkanoates can nucleate cadmium sulphide nanocrystals on the exposure to H_2S [93]. The distance between particles can be adjusted by selecting the alcanoic acid with the appropriate chain length. However, the backscattering experiments conducted on such systems have shown that the semiconductor particles are not really organised into an array, being randomly distributed through the film [94].

3.5.2.2. Self-assembled films

Another approach to the synthesis of organised layered systems is the construction of self-assembled monolayers (SAMs). This synthetic strategy involves the use of strong chemical interactions between adsorbed molecules and clean substrates to obtain well-ordered thin structures on the substrate surfaces.

Keller and co-authors [95] used silicon and gold substrates modified with tetra-n-butylammonium (TBA) to deposit one monolayer of zirconium phosphanate (the possibility to use other compounds such as $Ti_2NbO_7^-$ and $K_2Nb_6O_{17}^{2-}$ was also demonstrated). Then, TBA adsorbed on the outer surface of the inorganic monolayer, was exchanged with added polycation. For this purpose a synthetic polymer (poly(allylamine) hydrochloride) as well as a protein (cytochrome *c*) was used. Then, the next layer of the inorganic substance was deposited. By the alternating deposition of the polymer and the inorganics a sandwich-like heterostructure consisting of up to 16 layers was created (Figure 12).

Similar approach was demonstrated in the work of Kotov et al. [96]. Semiconductor nanoparticles (CdS, PbS, TiO_2) prepared in the presence of a stabiliser were deposited onto solid substrate modified with polycations. The subsequent alternating immersions of the substrate in the solution of the polycation and the colloidal suspension of semiconductor particles produced a multilayered structure with the semiconductor particles ordered within each layer.

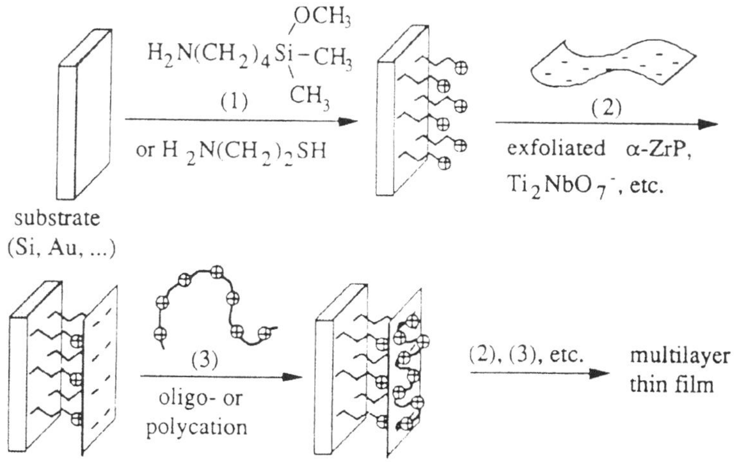

Figure 12. *Method for the preparation of multilayered zirconium phosphanate. Reprinted by permission from J.Am. Chem. Soc. [95] Copyright 1994 ACS Publicatons.*

3.6. SYNTHESIS OF MESOPOROUS MATERIALS

In the early 1990's, ordered surfactant assemblies were shown to be able to direct the assembly of periodic silica structures with mesoporous order. This has resulted in an explosion of mesoporous materials synthesis. Interestingly there have been attempts to direct the hierarchical assemblies into complex forms that approach the complexity often seen in biological mineralization. The pore sizes accessible through this micellar and liquid crystal templating are on the order of 2-10nm but through the use of block co-polymers and colloidal latex beads as templates, pores sizes up to micron dimensions have been realised. In addition, the range of materials that can be utilised in this fabrication approach now includes metal oxides, metal phosphates, and metal sulphides as well as metals [97]. The approach to mesoporous materials formation has been largely empirical although models for the formation of complex morphologies are emerging.

3.6.1. Liquid crystal templating mechanism

A family of mesoporous silicate/aluminosilicate materials designated as M41S were first reported in 1992 [98, 99]. These inorganic materials were made in the presence of cationic surfactants (e.g. hexadecyltrimethylammonium bromide), which were subsequently removed by calcination, and showed a remarkable similarity to the structure of the lyotropic liquid crystal phases of the surfactants themselves. The authors proposed a "liquid crystal templating mechanism" whereby the structure of the organic phase was reproduced as a mineral replica in the templated inorganic phase. The length of the alkyl chain in the surfactant (Figure 13) directly controlled the pore size of these aluminosilicate materials. Two mechanistic pathways were proposed whereby either a) the silicate precursor occupied the regions between the pre-formed cylinders of the lyotropic LC phase and directly coated the micellar rods or b) that the aluminosilicate species mediate the ordering of the surfactants into the hexagonal LC phase. However, since the original work (and much subsequent work) was performed at surfactant concentrations well below the critical micelle concentration (CMC) for formation of the LC phase it seems more likely that the second, cooperative assembly mechanism, is the more prevalent. This work has lead to an explosion of research aimed at developing the paradigm of using the liquid phase behaviour of amphiphilic molecules for the synthesis of novel porous materials which can be synthesised under extremely mild conditions. The generalised synthesis of these surfactant derived mesoporous materials requires a silicate source, which is usually derived from silicic acid ($Si(OH)_4$) or the acid hydrolysis of an organosiloxane (e.g. tetraethyorthosilicate, TEOS) and an amphiphilic surfactant which could be an alkyltrimethylammonium or could include polymers and block copolymers.

The exact nature of the templating mechanism is still debated and includes a number of models. It has been suggested that isolated micellar rods in solution are initially coated with silicate and that these isolated (isotropic) rods assemble into the final hexagonal mesophase. Ageing and heating further promotes the silicate condensation which serves to lock the hexagonal mesostructure [100] in place. The formation of hexagonal mesostructures from solution appears to occur with some heterogeneity as evidenced by formation of silicified micellar rods prior to the precipitation of the mesostructured material. Using low temperature TEM and small angle X-ray scattering, investigators have visualised isolated silicate covered micellar rods. Thus, it seemed that rod-like micelles are formed prior to bulk precipitation and silicate species have been proposed to deposit on individual rods which form the nucleation site for the eventual formation of the bulk mesostructure [101].

Alternatively, it has also been suggested that surfactants assemble directly into a hexagonal LC phase upon addition of silicate species. Silicates are thought to initially organise into layers that pucker and collapse around the rods to form the eventual mesostructure [102]. A mechanistic model which invokes a charge density matching [103, 104] suggests that the hexagonal mesostructures are derived from an initially formed lamellar phase. The layered lamellar phase (seen by XRD) is formed by complementary electrostatic interactions between anionic silicates and surfactant headgroups. Curvature associated with the transformation from layered to hexagonal phase in the mesostructure arises from reduction of charge density in the silica framework, upon condensation of silicates [103, 104].

Under unique conditions where condensation of silicate was prevented (high pH and low temp) a cooperative self-assembly of silicates and surfactants has been shown to occur. Micellar to-hexagonal phase transformation occurred in the presence of silicate anions [105].

Templating of mesoporous materials through hydrogen bonding interactions of alkylamine headgroup and TEOS results in so-called "worm-hole" structures [106, 107] which lack some of the long range ordering of pores present in the M41S family. The silicate framework in these structures is neutral and so the surfactants could be removed by extraction rather than high temperature calcination. This method has also been shown to be effective in forming porous lamellar structures through the use of double-headed alkyldiamines [108]. In addition, a more ordered arrangement of pores could be achieved by templating silicate formation with non-ionic surfactants such as those with polyethylene oxide headgroups attached to alkyl tails. These materials still lack the perfect hexagonal packing but exhibits ordered pores which can be adjusted in size by varying the length of the poly(ethylene oxide) headgroup as well as that of the tail [109]. Recently, mesostructured materials with ultra-large pores have been reported through microemulsion templates using non-ionic surfactants based on poly(ethylene oxide) [110].

Synthetic conditions such as surfactant: silica ratios have been shown to have a profound effect on the form of the mesophase material produced. Thus, at low surfactant: Si ratios the hexagonal phase is favoured whereas at slightly higher ratios the bi-continuous cubic phase is formed while at still higher ratios a lamellar phase is found to form. Stabilisation of intermediate mesoporous phases SBA-8 which transform into the hexagonal MCM-41 upon hydrothermal treatment has been achieved by using bolaform surfactants containing a rigid unit in the hydrophobic chain. The resultant materials are 2-D pore structures for which there is no reported matching lyotropic liquid crystal phase analog.

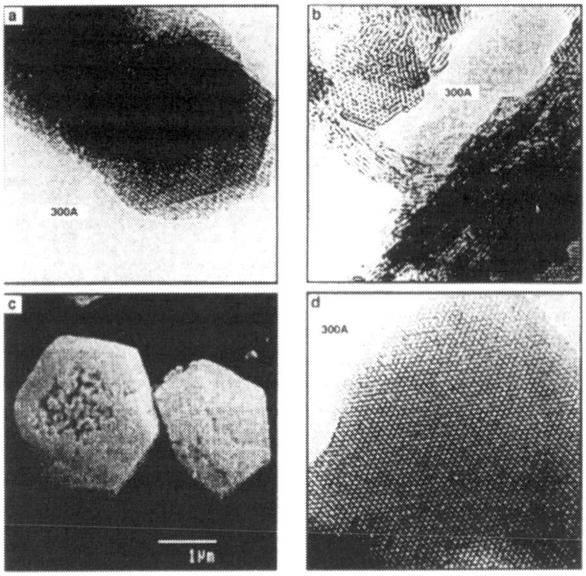

Figure 13. *Transmission electron micrographs (a, b and d) and scanning electron micrograph (c) of MCM-41. Reprinted by permission from J. Am. Chem. Soc. [99] Copyright 1992 ACS Publications.*

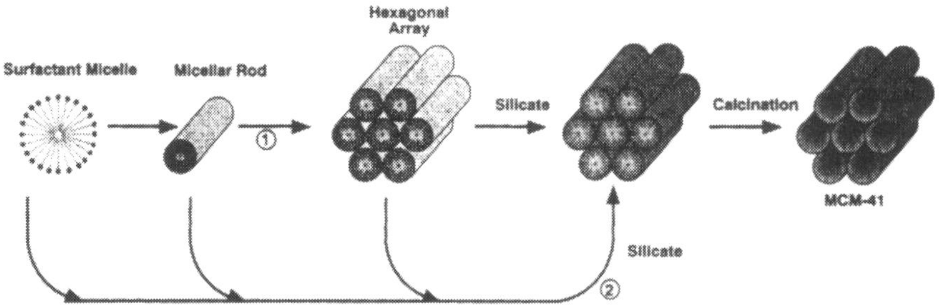

Figure 14. *Mechanistic pathways for the formation of MCM-41: (1) liquid crystal phase initiated and (2) silicate anion initiated. Reprinted by permission from J. Am. Chem. Soc. [99] Copyright 1992 ACS Publications..*

A recent extension of this approach to include complex ternary and quaternary microemulsion mixtures greatly enhances the control over the phase structure and pore size [111] using the direct liquid crystal templating using polyether surfactants and alcohol co-surfactant systems to synthesise monolithic mesoporous structures. Through the use of block co-polymers and colloidal latex beads as templates, mesoporous materials with pore sizes up to micron dimensions have been realised. By mimicking the silicatein-α protein found in the silica spicules of a sponge a synthetic block co-polypeptide of cysteine-lysine show self assembly into structured aggregates which hydrolyse tetraethoxysilane and direct the formation of ordered silica morphologies. Different structures (from hard spheres to columns) can be realised through control of the redox state of the cysteine sulfhydral groups [112].

Mesoporous materials based on transition metal oxide used titanium alkoxides with chelating agents such as acetylacetone in the presence of surfactants to produce stable hexagonal packed pores [113-115]. Recently, ordered large pore mesostructures of TiO_2 ZrO_2 Nb_2O_5 Ta_2O_5 Al_2O_3 SnO_2 SiO_2 WO_3 HfO_2 $ZrTiO_4$ have been synthesised using amphiphilic block copolymers of poly(alkeneoxide) as structure directing agents and inorganic salts as precursors. This is suggested to be a block copolymer self-assembly coupled with alkene oxide complexation of the inorganic metal species [116, 117]. Porous SiO_2 N_2O_5 TiO_2 with three dimensional structures patterned over multiple length scales (10nm to several micrometers) – using polystyrene spheres, amphiphilic triblock copolymers and the cooperative assembly of inorganic sol-gel species [116].

A description of the inorganic-surfactant interactions in the liquid crystal templating was proposed [113] based on the type of electrostatic interaction between surfactant and precursor. The generalised interaction could range from purely electrostatic to covalent interactions which can sometimes be mediated by counterions (Figure 14). The role of the organic phase in these reactions is thus to provide a molecular template for silica (or metal ion) hydrolysis and stabilisation of early intermediates in the hydrolysis reactions. In addition the organic templates provide intermediate and long range ordering which provide a communication between individual nucleation sites to form the extended silicate (or metal oxide) polymer spatially arranged into the observed mesoscopic (or macroscopic) structure and associated characteristics.

3.6.2. Synthesis of biomimetic materials with complex architecture

Biomimetic approach to material synthesis can be also understood as the chemical synthesis of materials with morphologies similar to biominerals. From the point of view of material chemists, the most striking feature of natural biominerals is their complex architecture, ordered over multiple length scale. On the other hand, the method of liquid crystal templating has been proved to be versatile for the creation of materials with complex shape (in the paper of Yang et al. [118] synthesis of various microstructures made of mesoporous silica is described). It is of interest therefore to employ liquid crystal templating for the creation of materials resembling biominerals in their complex structure.

A successful demonstration of such synthesis is the preparation of aluminophosphates resembling microskeletons produced by the single cell marine organisms such as diatoms and radiolarians [119, 120]. In this preparation dodecylammonium hydrogen phosphate template was employed, which formed a smectic liquid crystal. Its dissolution in tetraethyleneglycol resulted in a microemulsion containing water droplets coated with a layer of the phosphate liquid crystal. Lamellar aluminophosphate was formed in the subsequent reaction of $H_2PO_4^-$ with aluminum precursor dissolved in tetraethyleneglycol. The texture of the mesophase was "fossilised" as a lamellar aluminophosphate. The appearing micron-sized surface pattern is very similar to the natural siliceous microskeletons (Figure 15).

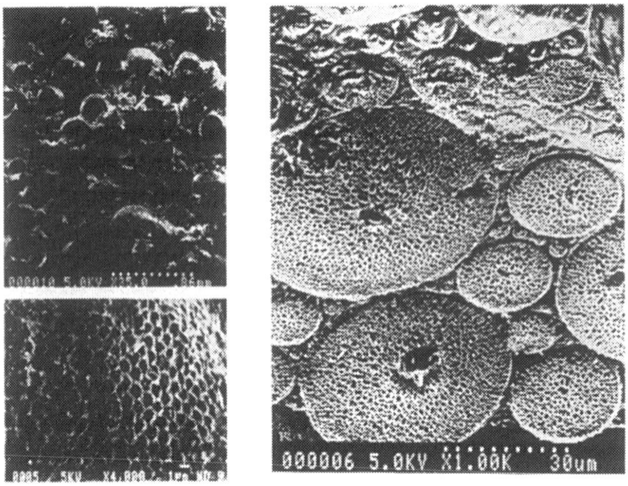

***Figure 15**. Synthetic aluminosilicates resembling biominerals. Reprinted by permission from Adv. Mater. [120] Copyright 1995 Wiley-VCH Verlag.*

3.7. SYNTHESIS OF INORGANIC MATERIALS USING POLYNUCLEOTIDES

3.7.1. Synthesis not involving specific nucleotide-nucleotide interactions

In 1991 Coffer and Chandler [121] demonstrated that polynucleotides might be used to stabilise aqueous colloids of cadmium sulphide nanoparticles. Principle of the CdS synthesis was similar to that used for the synthesis of semiconductor nanoparticles in the presence of conventional stabilising agents such as polymers. To synthesise CdS particles, DNA was first mixed with a solution of a cadmium salt, then an equimolar amount of sodium sulphide was added. Further study [122] showed that the nucleotide content of DNA had a significant effect on the size of formed cadmium sulphide. In the presence of polyadenylic acid or adenine-rich nucleotides, the average size of formed CdS was 38 Å, while the use of other homopolymers resulted in the appearance of larger particles.

Subsequently [123], the authors used a 3455-basepair circular plasmid DNA attached to a solid substrate as a template for CdS synthesis. The rigid structure of the immobilised DNA gave a possibility to obtain an organised assembly of cadmium sulphide nanoparticles. Cadmium sulphide was synthesised by first adding a cadmium salt to the solution of DNA, which then was anchored to a suitably modified glass substrate. Reaction with sulphide resulted in the formation of CdS particles. TEM analysis confirmed that formed nanocrystals were organised in a ring, 1.2 μm circumference, which exactly matched the initial plasmid DNA shape.

In a somewhat different way, the capability of DNA molecules to bind metal cations which could then be used as a basis for the subsequent material synthesis was employed in the work of Braun et al [124]. In this study a linear strand of DNA connecting two gold microelectrodes 12-16 μm apart, was used as a template for the synthesis of a silver nanowire. To provide an anchor for the DNA molecule, a 12-base nucleotide sequence, derivatised with disulfide groups at their 3' end were attached to each gold electrode through S – Au interactions. In this way each of the electrodes was marked by different oligonucleotide sequence. Then, the DNA molecule, modified with oligonucleotides complementary to those attached to the gold electrodes, was introduced to make a bridge. DNA molecule was fluorescently labelled, so the connection could be observed by means of fluorescence microscopy. The synthesis of silver metal involved multiple steps. In the first step silver ions were introduced, to load DNA by means of Na^+/Ag^+ exchange. Adsorbed silver cations were then reduced to yield small silver metal particles by basic hydroquinione solution. To obtain a continuous wire the silver 'image' was 'developed' in the presence of silver salt and acidic solution of hydroquinone under low light conditions (Figure 16).

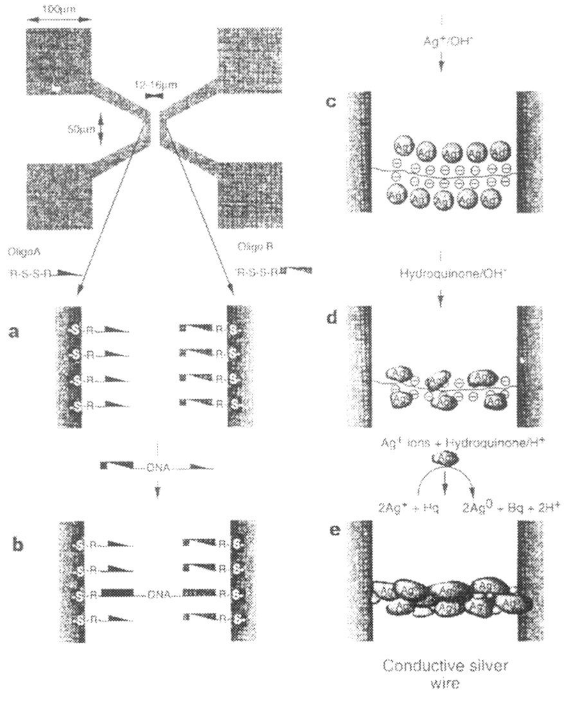

Figure 16. *Method for construction of a silver nanowire that connects two gold electrodes. Reprinted by permission from Nature [124] Copyright 1998 Macmillan Magazines Ltd.*

Current - voltage curves of the resulting wire were quite surprising, and two interesting features were revealed. First, at low voltage zero current, and, consequently, an extremely high resistance was observed. However, at a higher bias the wire showed usual ohmic behaviour. The value of threshold bias was found to depend on the amount of silver metal deposited on DNA. Second, electric current was dependent on the direction of the voltage scan. Possible explanations for these phenomena are based on the concept that the metal wire is not really continuous, but contains a number of silver grains that can require simultaneous charging to provide the wire conductivity. These effects may also appear as a result of chemical processes on grain boundaries.

A different kind of interaction was employed by Cassell et al [125] for the synthesis of DNA – fullerene composites. C_{60} fullerenes modified with N,N-dimethylpyrrolidinium iodide reacted with DNA through the interaction with phosphate groups. The resulting composite was observed by TEM without any additional staining. Fullerene-DNA complexes tended to agglomerate, but this could be prevented by the addition of anionic or zwitterionic surfactants.

It must be noted that although all the above works describe the use of nucleotide chains as templates for the material synthesis, the synthetic procedures does not employ specific nucleotide-nucleotide interactions to form inorganic materials. The synthesis is based on the interactions between specific nucleotide groups and the metal cation (or pyrrole ring). In principle, similar results could be achieved using other polymer molecules containing appropriate active groups. However, DNA has shown to be a suitable template for the crystal growth, and recent progress in alteration of the DNA geometry [126, 127] opens new perspectives for the material synthesis involving the use of nucleotide chains with tuneable and complex architectures.

3.7.2. Synthesis involving nucleotide-nucleotide interactions

The first studies describing the use of specific nucleotide interactions for the creation of ordered structures were devoted to the organisation of gold nanoparticles modified with oligonucleotide ligands through thiol bridges.

Alivisatos et al [128] demonstrated the procedure for the organisation of gold particles into dimers and trimers. In this study, gold particles were first modified with 18-base nucleotide chains. They reacted with a single-stranded DNA molecule that contained two or three sequences complementary to the oligonucleotides attached to Au particles. The reaction with 37-base DNA template yielded particle dimers. Depending on the orientation of the complementary sequences one against the other inside the DNA strain, either "parallel" or "anti-parallel" dimers may be formed. The formation of "parallel" trimers in reaction of the oligonucleotide-modified gold nanoparticles with the 56-base DNA template was also reported.

In later work by Loweth et al [129] this synthetic strategy was developed further to obtain various relative spatial arrangements of gold nanoparticles (5 and 10 nm size) heterodimers and heterotrimers (Figure 17). TEM analysis of these assemblies showed that they are not rigid, even in the case where the DNA does not have single-stranded pieces ("nicks") in it.

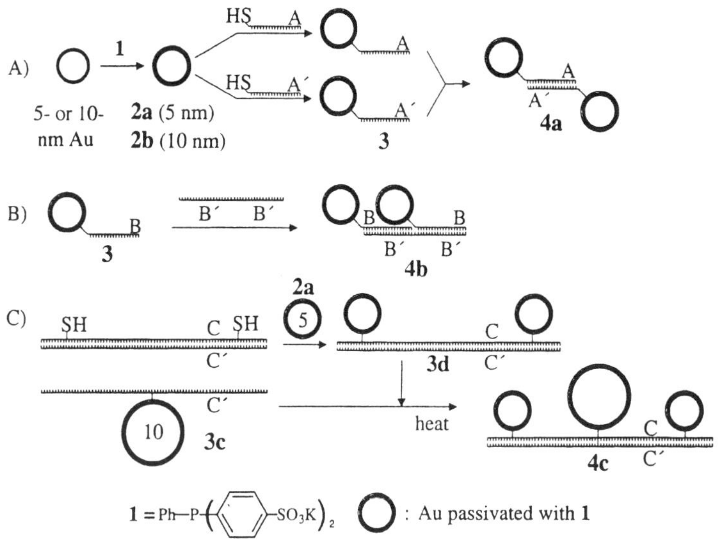

***Figure** 17. Organisation of gold nanoparticles into dimers and trimers using oligonucleotides. Reprinted by permission from Angew. Chem. Int. Ed. [129] Copyright 1999 Viley-VCH Verlag.*

Another approach was employed in the work of Mirkin et al [130]. In this study, 13 nm Au particles were modified with 8-base oligonucleotides of two kinds, which were not complementary. Addition of single-stranded DNA, containing two 8-base sequences, each complementary to either of the oligonucleotides attached to the gold particles, resulted in precipitation of the colloid. TEM analysis revealed the occurrence of a well-ordered network of the gold particles separated uniformly by 60 Å. The Au particles were linked together with DNA duplexes (Figure 18). To confirm this, the aggregate was heated to the temperature of the DNA duplex denaturation. As expected, at the elevated temperature DNA duplex dissociated into single-stranded oligonucleotides yielding the initial reactants.

The particle self-assembly was accompanied by a decrease of the Au surface plasmon band and its shifting to a longer wavelength, changing the initial red colour of the colloid to blue. This feature of the assembly reaction may be used for the colorimetric determination of DNA nucleotide sequence [131]. Using gold nanoparticles with attached appropriate oligonucleotide sequences it is possible to determine whether a DNA strand is exactly complementary to these sequences or not through the colour changes of the gold colloid.

Later, the strategy developed for the nanoparticles assembly was employed for the assembly of structures containing gold particles of two different sizes [132]. Using nucleotide-modified gold particles (8 nm and 31 nm in diameter) in an appropriate ratio, the assembly of structures containing larger Au particle surrounded by many smaller Au particles was demonstrated.

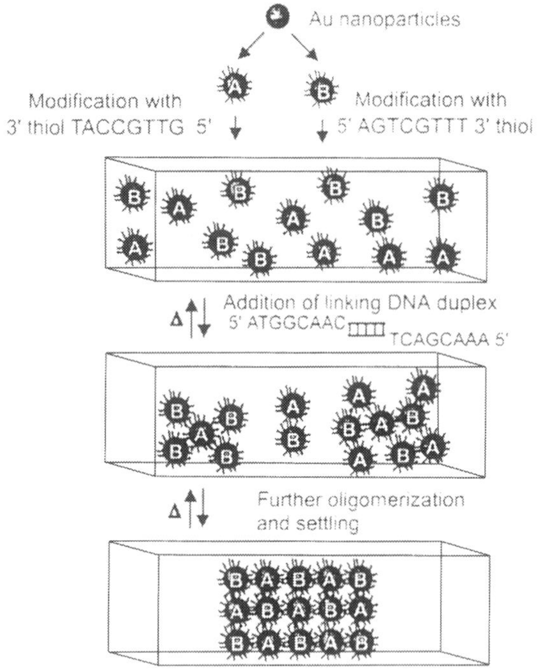

Figure 18. *Strategy to create network of gold particles linked with DNA. Reprinted by permission from J. Cluster Sci [135] Copyright 1997.*

Using nucleotide-nucleotide recognition for nanoparticle assemblies into ordered structures was extended to include semiconductor particles [133]. Specifically, highly luminescent CdSe - core ZnS – shell particles were used. These semiconductor nanoparticles were considered to be of a particular interest since they had been reported to possess a stable fluorescence with a relatively narrow (33 nm at half-maximum) band originating from band-to-band transitions with a quantum yield of 0.5 at room temperature [134]. The semiconductor particles were first synthesised in trioctylphosphine oxide to produce particles with strongly hydrophobic surface. Then, the particle surface was modified with 3-mercaptopropionic acid to provide the solubility in water. This procedure as well as the subsequent attachment of thiol-modified 22-base single strand DNA did not affect the fluorescence of the semiconductor. However, assembly of particles into ordered structure in the presence of complementary DNA template caused some decrease (by 26 %) of the fluorescence intensity. Interestingly, ordered networks of CdSe/ZnS particles linked by DNA duplexes remained in the solution, although they could be separated by centrifugation at relatively low speeds compared to the single semiconductor particles modified with nucleotides and mercaptopropionic acid.

Since both Au and CdSe/ZnS particles may be organised into the DNA-linked assemblies in a similar way, the creation of binary metal-semiconductor networks is possible, and has been demonstrated [133].

For reviews on the use of nucleotide chains in material synthesis see [135, 136].

3.8. BIOLOGICAL SYNTHESIS OF NOVEL MATERIALS

In the pursuit of novel materials the yeasts *Candida glabrata* and *Schizosaccharoyces pombe* have been induced to produce quantum sized crystallites of the semiconductor cadmium sulphide (CdS). The yeasts, cultured in the presence of cadmium salts, produced crystallites (~20Å diameter), coated by short chelating peptides of the general formula $(\gamma\text{-Glu-Cys})_n\text{-Gly}$ which were suggested to control the crystal nucleation and growth as well as stabilising the particles from aggregation [137, 138]. These crystallites were more monodisperse than those produced by conventional chemical means and also were shown to be an unusual polymorph of CdS. Some other microorganisms were found to produce CdS nanoparticles too. Bacteria *Klebsiella pneumonae* were shown to synthesise CdS particles (up to 200 Å size) on the cell surface [139]. The formation of metal sulphide by these bacteria showed cadmium specificity [140]. Biosynthesised CdS nanoparticles could protect the cells against UV radiation [141] and possessed photocatalytic properties similar to CdS nanocrystals obtained by chemical methods [142].

Clearly, this detoxification by the organism has some similarities to the iron sequestration by ferritin but the phytochelatin polypeptide retains significant flexibility to accommodate the chelation of both molecular and nano-materials.

Another interesting example of the synthesis of novel materials by living organisms was demonstrated in the work of Fritz et al. [143]. A highly organised composite material - a "flat pearl" was grown on disks of glass, mica, and MoS_2 inserted between the mantle and shell of *Haliotis rufescens* (red abalone). The biosynthesis of this material proceeded through a developmental sequence closely resembling that occurring at the growth front of the natural shell.

3.9. ORGANIZATION OF NANOPARTICLES INTO ORDERED STRUCTURES

Some examples of the organisation of nanocrystals into the structures ordered at a larger scale have been already mentioned in the previous sections. Langmuir-Blodgett multilayers as well as liquid crystal templates are two examples of techniques used in the synthesis of three-dimensional nanoparticle arrays. Another well-developed procedure for the creation of "nanocrystal superlattices" includes dispersion of highly monodisperse spherical nanoparticles into organic solvent with subsequent solvent evaporation. This method was employed in the work of Murray et al. [144] for the synthesis of assemblies of CdSe nanoparticles of a narrow ($\pm$ 3%) size distribution. The formation of three-dimensional ordered structure was confirmed by TEM, HRSEM and ED. The interparticle distance was shown to depend entirely on the capping substance used in the preparation of CdSe particles. Samples prepared with hexadecyl phosphate, trioctylphosphine oxide and tributylphosphine oxide had interparticle distances of 17, 11, and 7 Å, respectively.

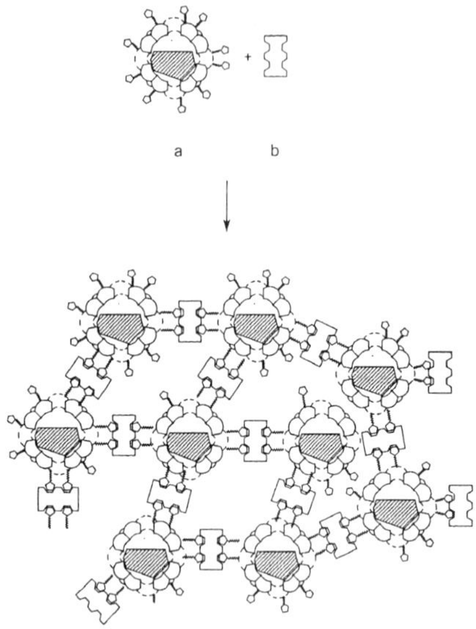

Figure 19. Aggregation of biotinylated ferritin containing iron oxide particles in the presence of streptavidin. Reprinted by permission from Chem. Mater. [147] Copyright 1999 ACS Publications.

Certain organic molecules may be employed to form organised nanoparticle arrays through covalent linking of the particles. This technique was demonstrated, for example, for the synthesis of two-dimensional arrays of 3.7 nm gold particles linked by an aryl di-isonitrile (1,4-di(4-isocyanophenylethynyl)-2-ethylbenzene) [145].

More biological approach to the nanoparticle assembly uses highly specific interactions between certain biomolecules. Besides nucleotide-nucleotide interactions, two other biological recognition systems were used to assemble inorganic nanoparticles. In the study of Shenton et al. [146] 12 nm Au and Ag nanoparticles with attached either IgE or IgG antibodies formed arrays in the presence of bivalent antigens with two appropriate functionalities. Filaments consisting of closely packed metal particles were observed. The other strategy uses streptavidin – biotin interaction. This is of particular interest because of the large association energy of the complex. In the presence of streptavidin the assembly of biotinylated ferritin shells with inorganic cores was demonstrated [147] (Figure 19). Streptavidin was also shown to promote aggregation of gold nanoparticles modified with a disulfide biotin analogue [148].

Most of the above-mentioned methods employ specific organic-organic interactions. A more direct approach to nanoparticle assembly involves specific interactions between the nanoparticle itself and the capping ligands. An important step in this direction has been made in a recent study aimed at the selection of peptides that may bind specifically to a certain semiconductor surface [149]. In this study phage-display libraries, based on a combinatorial library of random peptides, containing 12 amino acids each, were used to probe selective adsorption on specific crystallographic planes of GaAs, InP and Si. This strategy opens the perspectives for nanoparticle assembly with the peptide chains capable to recognise specific solid surfaces.

Biomimetic materials chemistry is at a stage when scientists are able to use the knowledge and tools of a number of disciplines in a pursuit of novel materials. Thus liquid crystal chemistry, polymer chemistry, inorganic materials chemistry, nucleic acid chemistry, protein chemistry and molecular biology all continue to make contributions to this hybrid field.

References

1. Mann, S., Webb, J., Williams, R. J. P. 1989. *Biomineralisation: Chemical and biochemical perspectives*. New York: VCH
2. Mann, S. 1996. *Biomimetic Materials Chemistry*. New York: VCH Publishers
3. Lowenstam, H. A. (1981) Minerals formed by organisms, *Science (Washington, D. C., 1883-) 211*, 1126-1131.
4. Walsh, D., Hopwood, J. D., Mann, S. (1994) Crystal Tectonics: Construction of Reticulated Calcium Phosphate Frameworks in Bicontinuous Reverse Microemulsions, *Science 264*, 1576-1578.
5. Addadi, L., Weiner, S. (1992) Control and design principles in biomineralisation, *Angew. Chem. 104*, 159-176 (See also Angew. Chem., Int. Ed. Engl., 1992, 1931(1992), 1153-1969).
6. Berman, A., Hanson, J., Leiserowitz, L., Koetzle, T. F., Weiner, S., Addadi, L. (1993) Crystal-protein interactions: controlled anisotropic changes in crystal microtexture, *J. Phys. Chem. 97*, 5162-5170.
7. Aizenberg, J., Albeck, S., Falini, G., Levi, Y., Weiner, S., Addadi, L. (1997) Design strategies in mineralised biological materials, *Book of Abstracts, 213th ACS National Meeting, San Francisco, April 13-17* INOR-768.
8. Douglas, T., Young, M. J. (1998) Host -guest encapsulation of materials by assembled virus protein cages, *Nature 393*, 152-155.
9. Harrison, P. M., Andrews, S. C., Artymiuk, P. J., Ford, G. C., Guest, J. R., Hirzmann, J., Lawson, D. M., Livngstone, J. C., Smith, J. M., Treffry, A., Yewdall, S. J. (1991) Probing structure-function relationships in ferritin and bacterioferritin, *Adv. Inorg. Chem. 36*, 449-485.
10. Douglas, T., Ripoll, D. (1998) Electrostatic gradients in the iron storage protein ferritin, *Protein Science 7*, 1083-1091.
11. Yang, X., Arosio, P., Chasteen, N. D. (2000) Molecular diffusion into ferritin: pathways, temperature dependence, incubation time, and concentration effects, *Biophys. J. 78*, 2049-2059.
12. Treffry, A., Zhao, Z., Quail, M. A., Guest, J. R., Harrison, P. M. (1997) Dinuclear center of ferritin: Studies of iron binding and oxidation show differences in the two iron sites, *Biochemistry 36*, 432-441.
13. Hwang, J., Krebs, C., Huynh, B. H., Edmondson, D. E., Theil, E. C., Penner-Hahn, J. E. (2000) A short Fe-Fe distance in peroxodiferric ferritin: control of Fe substrate versus cofactor decay?. [Erratum to document cited in CA132:162578], *Science (Washington, D. C.) 287*, 807.
14. Meldrum, F. C., Heywood, B. R., Mann, S. (1992) Magnetoferritin: In vitro synthesis of a novel magnetic protein, *Science 257*, 522-523.
15. Douglas, T., Bulte, J. W. M., Pankhurst, Q. A., Dickson, D. P. E., Moskowitz, B. M., Frankel, R. B., Mann, S. 1995. In *Hybrid Organic-Inorganic Composites*, ed. J. E. Mark, P. Bianconi, pp. 19-28. Washington, D.C.: American Chemical Society
16. Wong, K. K. W., Douglas, T., Gider, S., Awschalom, D. D., Mann, S. (1998) Biomimetic Synthesis and Characterisation of Magnetic Proteins (Magnetoferritin), *Chem. Mater. 10*, 279-285.

17. Meldrum, F. C., Wade, V. J., Nimmo, D. L., Heywood, B. R., Mann, S. (1991) Synthesis of inorganic nanophase materials in supramolecular protein cages, *Nature 349*, 684-687.
18. Meldrum, F. C., Douglas, T., Levi, S., Arosio, P., Mann, S. (1995) Reconstitution of manganese oxide cores in horse spleen and recombinant ferritins, *J. Inorg. Biochem. 58*, 59-68.
19. Douglas, T., Stark, V. T. (2000) Nanophase Cobalt Oxyhydroxide Mineral Synthesised within the Protein Cage of Ferritin, *Inorg. Chem.* ACS ASAP.
20. Douglas, T., Dickson, D. P. E., Betteridge, S., Charnock, J., Garner, C. D., Mann, S. (1995) Synthesis and structure of an iron(III) sulphide-ferritin bioinorganic nanocomposite, *Science 269*, 54-57.
21. Wong, K. K. W., Mann, S. (1996) Biomimetic synthesis of cadmium sulphide-ferritin nanocomposites, *Adv. Mater. (Weinheim, Ger.) 8*, 928-932.
22. Sleytr, U. B., Messner, P., Pum, D., Sara, M. (1999) Crystalline bacterial cell surface layers (S-layers): From supramolecular cell structures to biomimetics and nanotechnology, *Angew. Chem. Int. Ed. Engl. 38*, 1034-1054.
23. Shenton, W., Pum, D., Sleytr, U. B., Mann, S. (1997) Synthesis of cadmium sulphide superlattices using self-assembled bacterial S-layers, *Nature 389*, 585-587.
24. Dieluweit, S., Pum, D., Sleytr, U. B. (1998) Formation of a gold superlattice on an S-layer with square lattice symmetry, *Supramol. Sci. 5*, 15-19.
25. Douglas, K., Devaud, G., Clark, N. A. (1992) Transfer of biologically derived nanometer-scale patterns to smooth substrates, *Science 257*, 642-644.
26. Shenton, W., Douglas, T., Young, M., Stubbs, G., Mann, S. (1999) Inorganic-Organic Nanotube Composites from Template Mineralization of Tobacco Mosaic Virus, *Adv. Mater. 11*, 253-256.
27. Douglas, T., Young, M. (1999) Virus particles as templates for materials synthesis, *Adv. Mater. (Weinheim, Ger.) 11*, 679-681.
28. Balogh, L., Tomalia, D. A. (1998) Poly(Amidoamine) Dendrimer-Templated Nanocomposites. 1. Synthesis of Zerovalent Copper Nanoclusters, *J. Am. Chem. Soc. 120*, 7355-7356.
29 Zhao, M., Sun, L., Crooks, R. M. (1998) Preparation of Cu Nanoclusters within Dendrimer Templates, *J. Am. Chem. Soc. 120*, 4877-4878.
30. Garcia, M. E., Baker, L. A., Crooks, R. M. (1999) Preparation and characterisation of dendrimer-gold colloid nanocomposites, *Anal. Chem. 71*, 256-258.
31. Zhao, M., Crooks, R. M. (1999) Dendrimer-encapsulated Pt nanoparticles. Synthesis, characterisation, and applications to catalysis, *Adv. Mater. (Weinheim, Ger.) 11*, 217-220.
32. Sooklal, K., Hanus, L. H., Ploehn, H. J., Murphy, C. J. (1998) A blue-emitting CdS/dendrimer nanocomposite, *Adv. Mater. (Weinheim, Ger.) 10*, 1083-1087.
33. Strable, E., Bulte, J. W. M., Vivekanandan, K., Moskowitz, B., Douglas, T. (2000) Synthesis and Characterisation of Soluble Iron-Oxide Dendrimer Composites, *Chem. Mater. submitted*,
34. Henisch, H. K. 1988. *Crystals in Gels and Liesegang Rings*. Cambridge: Cambridge University Press
35. Aizenberg, J., Hanson, J., Ilan, M., Leiserowitz, L., Koetzle, T. F., Addadi, L., Weiner, S. (1995) Morphogenesis of calcitic sponge spicules: a role for specialised proteins interacting with growing crystals, *Faseb J. 9*, 262-268.
36. Aizenberg, J., Lambert, G., Addadi, L., Weiner, S. (1996) Stabilisation of amorphous calcium carbonate by specialised macromolecules in biological and synthetic precipitates, *Adv. Mater. (Weinheim, Ger.) 8*, 222-226.
37. Addadi, L., Aizenberg, J., Albeck, S., Falini, G., Weiner, S. (1995) Structural control over the formation of calcium carbonate mineral phases in biomineralisation, *NATO ASI Ser., Ser. C 473*, 127-139.
38. Weiner, S., Addadi, L. (1991) Acidic macromolecules of mineralised tissues: the controllers of crystal formation, *Trends Biochem. Sci. 16*, 252-256.
39. Boutonnet, M., Kizling, J., Stenius, P., Maire, G. (1982) The preparation of monodisperse colloidal metal particles from microemulsions, *Colloids Surf. 5*, 209-225.
40. Kurihara, K., Kizling, J., Stenius, P., Fendler, J. H. (1983) Laser and pulse radiolytically induced colloidal gold formation in water and in water-in-oil microemulsions, *J. Am. Chem. Soc. 105*, 2574-2579.
41. Meyer, M., Wallberg, C., Kurihara, K., Fendler, J. H. (1984) Photosensitised charge separation and hydrogen production in reversed micelle-entrapped, platinised, colloidal, cadmium sulphide, *J. Chem. Soc., Chem. Commun.* 90-91.

42. Lianos, P., Thomas, J. K. (1986) Cadmium sulphide of small dimensions produced in inverted micelles, *Chem. Phys. Lett. 125*, 299-302.
43. Petit, C., Pileni, M. P. (1988) Synthesis of cadmium sulphide in situ in reverse micelles and in hydrocarbon gels, *J. Phys. Chem. 92*, 2282-2286.
44 Steigerwald, M. L., Alivisatos, A. P., Gibson, J. M., Harris, T. D., Kortan, R., Muller, A. J., Thayer, A. M., Duncan, T. M., Douglass, D. C., Brus, L. E. (1988) Surface derivatisation and isolation of semiconductor cluster molecules, *J. Am. Chem. Soc. 110*, 3046-3050.
45. Kortan, A. R., Hull, R., Opila, R. L., Bawendi, M. G., Steigerwald, M. L., Carroll, P. J., Brus, L. E. (1990) Nucleation and growth of cadmium selendie on zinc sulphide quantum crystallite seeds, and vice versa, in inverse micelle media, *J. Am. Chem. Soc. 112*, 1327-1332.
46. Towey, T. F., Khan-Lodhi, A., Robinson, B. H. (1990) Kinetics and mechanism of formation of quantum-sized cadmium sulphide particles in water-Aerosol OT-oil microemulsions, *J. Chem. Soc., Faraday Trans. 86*, 3757-3762.
47. Petit, C., Lixon, P., Pileni, M. P. (1990) Synthesis of cadmium sulphide in situ in reverse micelles. 2. Influence of the interface on the growth of the particles, *J. Phys. Chem. 94*, 1598-1603.
48. Lisiecki, I., Pileni, M. P. (1993) Synthesis of copper metallic clusters using reverse micelles as microreactors, *J. Am. Chem. Soc. 115*, 3887-3896.
49. Petit, C., Lixon, P., Pileni, M. P. (1993) In situ synthesis of silver nanocluster in AOT reverse micelles, *J. Phys. Chem. 97*, 12974-12983.
50. Chang, S.-Y., Liu, L., Asher, S. A. (1994) Preparation and Properties of Tailored Morphology, Monodisperse Colloidal Silica-Cadmium Sulphide Nanocomposites, *J. Am. Chem. Soc. 116*, 6739-6744.
51. Lianos, P., Thomas, J. K. (1987) Small cadmium sulphide particles in inverted micelles, *J. Colloid Interface Sci. 117*, 505-512.
52. Chhabra, V., Pillai, V., Mishra, B. K., Morrone, A., Shah, D. O. (1995) Synthesis, Characterisation, and Properties of Microemulsion-Mediated Nanophase TiO2 Particles, *Langmuir 11*, 3307-3311.
53. Johnson, J. A., Saboungi, M.-L., Thiyagarajan, P., Csencsits, R., Meisel, D. (1999) Selenium Nanoparticles: A Small-Angle Neutron Scattering Study, *J. Phys. Chem. B 103*, 59-63.
54. Gobe, M., Kon-No, K., Kandori, K., Kitahara, A. (1983) Preparation and characterisation of monodisperse magnetite sols in water/oil microemulsion, *J. Colloid Interface Sci. 93*, 293-295.
55. Pillai, V., Kumar, P., Multani, M. S., Shah, D. O. (1993) Structure and magnetic properties of nanoparticles of barium ferrite synthesised using microemulsion processing, *Colloids Surf., A 80*, 69-75.
56. Feltin, N., Pileni, M. P. (1997) New Technique for Synthesising Iron Ferrite Magnetic Nanosized Particles, *Langmuir 13*, 3927-3933.
57. Chen, J. P., Sorensen, C. M., Klabunde, K. J., Hadjipanayis, G. C. (1994) Magnetic properties of nanophase cobalt particles synthesised in inversed micelles, *J. Appl. Phys. 76*, 6316-6318.
58. Dutta, P. K., Jakupca, M., Reddy, K. S. N., Salvati, L. (1995) Controlled growth of microporous crystals nucleated in reverse micelles, *Nature (London) 374*, 44-46.
59. Hopwood, J. D., Mann, S. (1997) Synthesis of Barium Sulphate Nanoparticles and Nanofilaments in Reverse Micelles and Microemulsions, *Chem. Mater. 9*, 1819-1828.
60. Tanori, J., Pileni, M. P. (1997) Control of the Shape of Copper Metallic Particles by Using a Colloidal System as Template, *Langmuir 13*, 639-646.
61. Li, M., Schnablegger, H., Mann, S. (1999) Coupled synthesis and self-assembly of nanoparticles to give structures with controlled organisation, *Nature (London) 402*, 393-395.
62. Moumen, N., Pileni, M. P. (1996) Control of the Size of Cobalt Ferrite Magnetic Fluid, *J. Phys. Chem. 100*, 1867-1873.
63. Lisiecki, I., Billoudet, F., Pileni, M. P. (1996) Control of the shape and the size of copper metallic particles, *J. Phys. Chem. 100*, 4160-4166.
64. Lim, G. K., Wang, J., Ng, S. C., Gan, L. M. (1999) Formation of Nanocrystalline Hydroxyapatite in Non-ionic Surfactant Emulsions, *Langmuir 15*, 7472-7477.
65. Mann, S., Williams, R. J. P. (1983) Precipitation within unilamellar vesicles. Part 1. Studies of silver(I) oxide formation, *J. Chem. Soc., Dalton Trans.* 311-316, 312 plates.
66. Mann, S., Hannington, J. P., Williams, R. J. P. (1986) Phospholipid vesicles as a model system for biomineralisation, *Nature (London) 324*, 565-567.

67. Tricot, Y. M., Fendler, J. H. (1984) Colloidal catalyst-coated semiconductors in surfactant vesicles: in situ generation of rhodium-coated cadmium sulphide particles in dihexadecyl phosphate vesicles and their utilisation for photosensitised charge separation and hydrogen generation, *J. Am. Chem. Soc. 106*, 7359-7366.
68. Watzke, H. J., Fendler, J. H. (1987) Quantum size effects of in situ generated colloidal cadmium sulphide particles in dioctadecyldimethylammonium chloride surfactant vesicles, *J. Phys. Chem. 91*, 854-861.
69. Korgel, B. A., Monbouquette, H. G. (1996) Synthesis of Size-Monodisperse CdS Nanocrystals Using Phosphatidylcholine Vesicles as True Reaction Compartments, *J. Phys. Chem. 100*, 346-351.
70. Chang, A. C., Pfeiffer, W. F., Guillaume, B., Baral, S., Fendler, J. H. (1990) Preparation and characterisation of selenide semiconductor particles in surfactant vesicles, *J. Phys. Chem. 94*, 4284-4289.
71. Youn, H. C., Baral, S., Fendler, J. H. (1988) Dihexadecyl phosphate, vesicle-stabilised and in situ generated mixed cadmium sulphide and zinc sulphide semiconductor particles: preparation and utilisation for photosensitised charge separation and hydrogen generation, *J. Phys. Chem. 92*, 6320-6327.
72. Bhandarkar, S., Bose, A. (1990) Synthesis of submicrometer crystals of aluminum oxide by aqueous intravesicular precipitation, *J. Colloid Interface Sci. 135*, 531-538.
73. Bhandarkar, S., Bose, A. (1990) Synthesis of nanocomposite particles by intravesicular coprecipitation, *J. Colloid Interface Sci. 139*, 541-550.
74. Markowitz, M. A., Chow, G.-M., Singh, A. (1994) Polymerised phospholipid membrane mediated synthesis of metal nanoparticles, *Langmuir 10*, 4095-4102.
75. Yaacob, I. I., Nunes, A. C., Bose, A., Shah, D. O. (1994) Synthesis and characterisation of magnetic nanoparticles in spontaneously generated vesicles, *J. Colloid Interface Sci. 168*, 289-301.
76. Rajam, S., Heywood, B. R., Walker, J. B. A., Mann, S., Davey, R. J., Birchall, J. D. (1991) Oriented crystallisation of calcium carbonate under compressed monolayers. Part 1. Morphological studies of mature crystals, *J. Chem. Soc., Faraday Trans. 87*, 727-734.
77. Heywood, B. R., Rajam, S., Mann, S. (1991) Oriented crystallisation of calcium carbonate under compressed monolayers. Part 2. Morphology, structure and growth of immature crystals, *J. Chem. Soc., Faraday Trans. 87*, 735-743.
78. Heywood, B. R., Mann , S. (1992) Crystal recognition at inorganic-organic interfaces: Nucleation and growth of oriented BaSO4 under compressed Langmuir Monolayers, *Adv. Mater. 4*, 278-282.
79. Heywood, B. R., Mann, S. (1992) Organic template-directed inorganic crystallisation: oriented nucleation of barium sulphate under compressed Langmuir monolayers, *J. Am. Chem. Soc. 114*, 4681-4686.
80. Douglas, T., Mann, S. (1994) Oriented Nucleation of Gypsum ($CaSO_4 \cdot 2H_2O$) under Compressed Langmuir Monolayers, *Mater. Sci. Eng. C1*, 193-199.
81. Xu, G., Yao, N., Aksay, I. A., Groves, J. T. (1998) Biomimetic Synthesis of Macroscopic-Scale Calcium Carbonate Thin Films. Evidence for a Multistep Assembly Process, *J. Am. Chem. Soc. 120*, 11977-11985.
82. Xu, S., Zhao, X. K., Fendler, J. H. (1990) Ultrasmall semiconductor particles sandwiched between surfactant headgroups in Langmuir-Blodgett films, *Adv. Mater. (Weinheim, Fed. Repub. Ger.) 2*, 183-185.
83. Yi, K. C., Fendler, J. H. (1990) Template-directed semiconductor size quantitation at monolayer-water interfaces and between the headgroups of Langmuir-Blodgett films, *Langmuir 6*, 1519-1521.
84. Scoberg, D. J., Grieser, F., Furlong, D. N. (1991) Control of size and composition of chalcogenide colloids in Langmuir-Blodgett films, *J. Chem. Soc., Chem. Commun.* 515-517.
85. Zhu, R., Min, G., Wei, Y., Schmitt, H. J. (1992) Scanning tunnelling microscopy and UV-visible spectroscopy studies of lead sulphide ultrafine particles synthesised in Langmuir-Blodgett films, *J. Phys. Chem. 96*, 8210-8211.
86. Yang, J., Meldrum, F. C., Fendler, J. H. (1995) Epitaxial Growth of Size-Quantified Cadmium Sulphide Crystals Under Arachidic Acid Monolayers, *J. Phys. Chem. 99*, 5500-5504.
87. Zhao, X. K., Yang, J., McCormick, L. D., Fendler, J. H. (1992) Epitaxial formation of lead sulphide crystals under arachidic acid monolayers, *J. Phys. Chem. 96*, 9933-9939.

88. Yang, J., Fendler, J. H., Jao, T. C., Laurion, T. (1994) Electron and atomic force microscopic investigations of lead selenide crystals grown under monolayers, *Microsc. Res. Tech. 27*, 402-411.
89. Yang, J., Fendler, J. H. (1995) Morphology Control of PbS Nanocrystallites, Epitaxially Grown under Mixed Monolayers, *J. Phys. Chem. 99*, 5505-5511.
90. Kotov, N. A., Meldrum, F. C., Fendler, J. H. (1994) Monoparticulate Layers of Titanium Dioxide Nanocrystallites with Controllable Interparticle Distances, *J. Phys. Chem. 98*, 8827-8830.
91. Meldrum, F. C., Kotov, N. A., Fendler, J. H. (1994) Preparation of Particulate Mono- and Multilayers from Surfactant-Stabilised, Nanosized Magnetite Crystallites, *J. Phys. Chem. 98*, 4506-4510.
92. Meldrum, F. C., Kotov, N. A., Fendler, J. H. (1995) Formation of Thin Films of Platinum, Palladium, and Mixed Platinum: Palladium Nanocrystallites by the Langmuir Monolayer Technique, *Chem. Mater. 7*, 1112-1116.
93. Guo, S., Konopny, L., Popovitz-Biro, R., Cohen, H., Porteanu, H., Lifshitz, E., Lahav, M. (1999) Thioalkanoates as Site-Directing Nucleating Centers for the Preparation of Patterns of CdS Nanoparticles within 3-D Crystals and LB Films of Cd Alkanoates, *J. Am. Chem. Soc. 121*, 9589-9598.
94. Yamaki, T., Asai, K., Ishigure, K. (1997) RBS analysis of Langmuir-Blodgett films bearing quantum-sized CdS particles, *Chem. Phys. Lett. 273*, 376-380.
95. Keller, S. W., Kim, H.-N., Mallouk, T. E. (1994) Layer-by-Layer Assembly of Intercalation Compounds and Heterostructures on Surfaces: Toward Molecular "Beaker" Epitaxy, *J. Am. Chem. Soc. 116*, 8817-8818.
96. Kotov, N. A., Dekany, I., Fendler, J. H. (1995) Layer-by-Layer Self-Assembly of Polyelectrolyte-Semiconductor Nanoparticle Composite Films, *J. Phys. Chem. 99*, 13065-13069.
97. Ying, J. Y., Mehnert, C. P., Wong, M. S. (1999) Synthesis and applications of supramolecular-templated mesoporous materials, *Angew. Chem. Int. Ed. 38*, 56-77.
98. Kresge, C. T., Leonowicz, M. E., Roth, W. J., Vartuli, J. C., Beck, J. S. (1992) Ordered mesoporous molecular sieves synthesised by a liquid-crystal template mechanism, *Nature, 359*, 710-712.
99. Beck, J. S., Vartuli, J. C., Roth, W. J., Leonowicz, M. E., Kresge, C. T., Schmitt, K. D., Chu, C. T.-W., Olson, D. H., Sheppard, E. W., McCullen, S. B., Higgins, J. B., Schlenker, J. L. (1992) A New Family of Mesoporous Molecular Sieves Prepared with Liquid Crystal Templates, *J. Am. Chem. Soc. 114*, 10834-10843.
100. Chen, C. Y., Burkett, S. L., Li, H. X., Davis, M. E. (1993) Studies on mesoporous materials. II. Synthesis mechanism of MCM-41, *Microporous Mater. 2*, 27-34.
101. Regev, O. (1996) Nucleation Events during the Synthesis of Mesoporous Materials Using Liquid Crystalline Templating, *Langmuir 12*, 4940-4944.
102. Steel, A., Carr, S. W., Anderson, M. W. (1994) 14N NMR study of surfactant mesophases in the synthesis of mesoporous silicates, *J. Chem. Soc., Chem. Commun.* 1571-1572.
103. Monnier, A., Schüth, F., Huo, Q., Kumar, D., Margolese, D., Maxwell, R. S., Stucky, G. D., Krishnamurty, M., Petroff, P., Firouzi, A., Janicke, M., Chmelka, B. F. (1993) Cooperative Formation of Inorganic-Organic Interfaces in the Synthesis of Silicate Mesostructures, *Science 261*, 1299-1303.
104. Stucky, G. D., Monnier, A., Schueth, F., Huo, Q., Margolese, D., Kumar, D., Krishnamurty, M., Petroff, P., Firouzi, A., et al. (1994) Molecular and atomic arrays in nano- and mesoporous materials synthesis, *Mol. Cryst. Liq. Cryst. Sci. Technol., Sect. A 240*, 187-200.
105. Attard, G. S., Glyde, J. C., Goltner, C. G. (1995) Liquid-crystalline phases as templates for the synthesis of mesoporous silica, *Nature (London) 378*, 366-368.
106. Tanev, P. T., Pinnavaia, T. J. (1995) A neutral templating route to mesoporous molecular sieves, *Science (Washington, D. C.) 267*, 865-867.
107. Tanev, P. T., Pinnavaia, T. J. (1996) Mesoporous Silica Molecular Sieves Prepared by Ionic and Neutral Surfactant Templating: A Comparison of Physical Properties, *Chem. Mater. 8*, 2068-2079.
108. Tanev, P. T., Pinnavaia, T. J. (1996) Biomimetic templating of porous lamellar silicas by vesicular surfactant assemblies, *Science (Washington, D. C.) 271*, 1267-1269.
109. Bagshaw, S. A., Prouzet, E., Pinnavaia, T. J. (1995) Templating of mesoporous molecular sieves by non-ionic polyethylene oxide surfactants, *Science (Washington, D. C.) 269*, 1242-1244.
110. Schmidt-Winkel, P., Glinka, C. J., Stucky, G. D. (2000) Microemulsion Templates for Mesoporous Silica, *Langmuir 16*, 356-361.

111. Feng, P., Bu, X., Stucky, G. D., Pine, D. J. (2000) Monolithic mesoporous silica templated by microemulsion liquid crystals, *J. Am. Chem. Soc. 122*, 994-995.
112 Cha, J. N., Stucky, G. D., Morse, D. E., Deming, T. J. (2000) Biomimetic synthesis of ordered silica structures mediated by block co-polypeptides, *Nature (London) 403*, 289-292.
113. Huo, Q., Margolese, D. I., Ciesla, U., Feng, P., Gier, T. E., Sieger, P., Leon, R., Petroff, P. M., Schueth, F., Stucky, G. D. (1994) Generalised synthesis of periodic surfactant/inorganic composite materials, *Nature (London) 368*, 317-321.
114 Ciesla., U., Demuth, D., Leon, R., Petroff, P., Stucky, G., Unger, K., Schueth, F. (1994) Surfactant controlled preparation of mesostructured transition-metal oxide compounds, *J. Chem. Soc., Chem. Commun.* 1387-1388.
115. Antonelli, D. M., Ying, J. Y. (1995) Synthesis of hexagonally packed mesoporous TiO2 by a modified sol-gel method, *Angew. Chem., Int. Ed. Engl. 34*, 2014-2017.
116. Yang, P., Deng, T., Zhao, D., Feng, P., Pine, D., Chmelka, B. F., Whitesides, G. M., Stucky, G. D. (1998) Hierarchically ordered oxides, *Science (Washington, D. C.) 282*, 2244-2247.
117. Yang, P., Zhao, D., Margolese, D. I., Chmelka, B. F., Stucky, G. D. (1999) Block Copolymer Templating Syntheses of Mesoporous Metal Oxides with Large Ordering Lengths and Semicrystalline Framework, *Chem. Mater. 11*, 2813-2826.
118. Yang, H., Ozin, G. A., Kresge, C. T. (1998) The role of defects in the formation of mesoporous silica fibres, films and curved shapes, *Adv. Mater. (Weinheim, Ger.) 10*, 883-887.
119. Oliver, S., Kuperman, A., Coombs, N., Lough, A., Ozin, G. A. (1995) Lamellar aluminophosphates with surface patterns that mimic diatom and radiolarian microskeletons, *Nature (London) 378*, 47-50.
120. Ozin, G. A., Oliver, S. (1995) Skeletons in the beaker. Synthetic hierarchical inorganic materials, *Adv. Mater. (Weinheim, Ger.) 7*, 943-947.
121. Coffer, J. L., Chandler, R. R. (1991) Nucleotides as structural templates for the self-assembly of quantum-confined cadmium sulphide crystallites, *Mater. Res. Soc. Symp. Proc. 206*, 527-531.
122. Bigham, S. R., Coffer, J. L. (1995) The influence of adenine content on the properties of Q-CdS clusters stabilised by polynucleotides, *Colloids Surf., A 95*, 211-219.
123. Coffer, J. L., Bigham, S. R., Li, X., Pinizzotto, R. F., Rho, Y. G., Pirtle, R. M., Pirtle, I. L. (1996) Dictation of the shape of mesoscale semiconductor nanoparticle assemblies by plasmid DNA, *Appl. Phys. Lett. 69*, 3851-3853.
124. Braun, E., Eichen, Y., Sivan, U., Ben-Yoseph, G. (1998) DNA-templated assembly and electrode attachment of a conducting silver wire, *Nature (London) 391*, 775-778.
125. Cassell, A. M., Scrivens, W. A., Tour, J. M. (1998) Assembly of DNA/fullerene hybrid materials, *Angew. Chem., Int. Ed. 37*, 1528-1531.
126. Seeman, N. C. (1998) DNA nanotechnology: novel DNA constructions, *Annu. Rev. Biophys. Biomol. Struct. 27*, 225-248.
127. Shchepinov, M. S., Mir, K. U., Elder, J. K., Frank-Kamenetskii, M. D., Southern, E. M. (1999) Oligonucleotide dendrimers: stable nano-structures, *Nucleic Acids Res. 27*, 3035-3041.
128. Alivisatos, A. P., Johnsson, K. P., Peng, X., Wilson, T. E., Loweth, C. J., Bruchez, M. P., Jr., Schultz, P. G. (1996) Organisation of 'nanocrystal molecules' using DNA, *Nature (London) 382*, 609-611.
129. Loweth, C. J., Caldwell, W. B., Peng, X., Alivisatos, A. P., Schultz, P. G. (1999) DNA-based assembly of gold nanocrystals, *Angew. Chem., Int. Ed. 38*, 1808-1812.
130. Mirkin, C. A., Letsinger, R. L., Mucic, R. C., Storhoff, J. J. (1996) A DNA-based method for rationally assembling nanoparticles into macroscopic materials, *Nature (London) 382*, 607-609.
131. Storhoff, J. J., Elghanian, R., Mucic, R. C., Mirkin, C. A., Letsinger, R. L. (1998) One-Pot Colorimetric Differentiation of Polynucleotides with Single Base Imperfections Using Gold Nanoparticle Probes, *J. Am. Chem. Soc. 120*, 1959-1964.
132. Mucic, R. C., Storhoff, J. J., Mirkin, C. A., Letsinger, R. L. (1998) DNA-Directed Synthesis of Binary Nanoparticle Network Materials, *J. Am. Chem. Soc. 120*, 12674-12675.
133. Mitchell, G. P., Mirkin, C. A., Letsinger, R. L. (1999) Programmed Assembly of DNA Functionalised Quantum Dots, *J. Am. Chem. Soc. 121*, 8122-8123.
134. Hines, M. A., Guyot-Sionnest, P. (1996) Synthesis and Characterisation of Strongly Luminescent ZnS-Capped CdSe Nanocrystals, *J. Phys. Chem. 100*, 468-471.
135. Storhoff, J. J., Mucic, R. C., Mirkin, C. A. (1997) Strategies for organising nanoparticles into aggregate structures and functional materials, *J. Cluster Sci. 8*, 179-216.

136. Storhoff, J. J., Mirkin, C. A. (1999) Programmed Materials Synthesis with DNA, *Chem. Rev. (Washington, D. C.) 99*, 1849-1862.
137. Dameron, C. T., Smith, B. R., Winge, D. R. (1989) Glutathione-coated cadmium sulphide crystallites in Candida glabrata, *J. Biol. Chem. 264*, 17355-17360.
138. Dameron, C. T., Reese, R. N., Mehra, R. K., Kortan, A. R., Carroll, P. J., Steigerwald, M. L., Brus, L. E., Winge, D. R. (1989) Biosynthesis of cadmium sulphide quantum semiconductor crystallites, *Nature (London) 338*, 596-597.
139. Holmes, J., Smith, P. R., Evans-Gowing, R., Richardson, D. J., Russel, D. A., Sodeau, J. R. (1995) Energy-dispersive X-ray analysis of the extracellular cadmium sulphide crystallites of *Klebsiella aerogenes*, *Arch. Microbiol. 163*, 143-147.
140. Holmes, J. D., Richardson, D. J., Saed, S., Evans-Gowing, R., Russell, D. A., Sodeau, J. R. (1997) Cadmium-specific formation of metal sulphide "Q-particles" by Klebsiella pneumoniae, *Microbiology (Reading, U. K.) 143*, 2521-2530.
141. Holmes, J. D., Smith, P. R., Evans-Gowing, R., Richardson, D. J., Russell, D. A., Sodeau, J. R. (1995) Bacterial photoprotection through extracellular cadmium sulphide crystallites, *Photochem. Photobiol. 62*, 1022-1026.
142. Smith, P. R., Holmes, J. D., Richardson, D. J., Russell, D. A., Sodeau, J. R. (1998) Photophysical and photochemical characterisation of bacterial semiconductor cadmium sulphide particles, *J. Chem. Soc., Faraday Trans. 94*, 1235-1241.
143. Fritz, M., Belcher, A. M., Radmacher, M., Walters, D. A., Hansma, P. K., Stucky, G. D., Morse, D. E., Mann, S. (1994) Flat pearls from biofabrication of organised composites on inorganic substrates, *Nature (London) 371*, 49-51.
144. Murray, C. B., Kagan, C. R., Bawendi, M. G. (1995) Self-organisation of CdSe nanocrystallites into three-dimensional quantum dot superlattices, *Science (Washington, D. C.) 270*, 1335-1338.
145. Andres, R. P., Bielefeld, J. D., Henderson, J. I., Janes, D. B., Kolagunta, V. R., Kubiak, C. P., Mahoney, W. J., Osifchin, R. G. (1996) Self-assembly of a two-dimensional superlattice of molecularly linked metal clusters, *Science (Washington, D. C.) 273*, 1690-1693.
146. Shenton, W., Davis, S. A., Mann, S. (1999) Directed self-assembly of nanoparticles into macroscopic materials using antibody-antigen recognition, *Adv. Mater. (Weinheim, Ger.) 11*, 449-452.
147. Li, M., Wong, K. K. W., Mann, S. (1999) Organisation of Inorganic Nanoparticles Using Biotin-Streptavidin Connectors, *Chem. Mater. 11*, 23-26.
148. Connolly, S., Fitzmaurice, D. (1999) Programmed assembly of gold nanocrystals in aqueous solution, *Adv. Mater. (Weinheim, Ger.) 11*, 1202-1205.
149. Whaley, S. R., English, D. S., Hu, E. L., Barbara, P. F., Belcher, A. M. (2000) Selection of peptides with semiconductor binding specificity for directed nanocrystal assembly, *Nature (London) 405*, 665-668.
150. Fendler, J. H. (1987) Atomic and molecular clusters in membrane mimetic chemistry, *Chem. Rev. 87*, 877-899.
151. Pileni, M. P. (1997) Nanosized Particles Made in Colloidal Assemblies, *Langmuir 13*, 3266-3276.

DENDRIMERS:
Chemical principles and biotechnology applications

L. HENRY BRYANT, JR. AND JEFF W.M. BULTE

Laboratory of Diagnostic Radiology Research (CC), National Institutes of Health, 10 Center Drive, Bethesda, MD 20892-1074.

Summary

Dendrimers have received an enormous amount of attention in the last ten years and several recent review articles have appeared in the literature that address their potential applications [1-3]. Stoddart et al [1] have stated that: "We are now approaching a time when the study of dendrimers becomes inextricably linked with many other fields, leaving the comprehensive reviewer of the subject a near-impossible task to fulfil". On that note, this review provides a brief introduction to the chemical principles of dendrimers by highlighting main synthetic strategies and methods for characterisation. Dendrimers containing heteroatoms will not be reviewed per se since these have recently been reviewed [4]. The major thrust of this review is the potential applications of dendrimers in such areas as boron neutron capture therapy, as contrast agents in magnetic resonance imaging, as vaccines, as cellular transfection agents and as bioconjugate dendrimers, i.e., in-vitro immunoassays for antigens. The outline used in this review proved to be effective in classifying most published papers about dendrimers, but it must be kept in mind that some articles not only transcended two different classifications, such as synthesis and characterisation, but several classifications such as synthesis, characterisation and at least one potential application covered in this review.

1. Synthesis

Dendrimers, a distinct class of macromolecules, are highly branched polyfunctional polymers. Tomalia et al [5] consider dendrimers to be one major class of macromolecular architecture. Dendrimers have been synthesised using the divergent or convergent methods or combinations thereof. In addition to being called dendrimers they have other trivial names such as arborols, bow-tied, ball-shaped, bolaform,

M. De Cuyper and J.W.M. Bulte (eds.), Physics and Chemistry Basis of Biotechnology, 47–69.

cascade (macro)molecules, cascade (multibranched; dendritic) polymers, cascadol, cauliflower polymers, crowned arborols, molecular fractals, polycules, silvanols and starburst dendrimers. An attempt has been made to devise a systematic nomenclature [6]. Synthetic strategies have been developed to vary the terminal groups, the internal blocks and the core.

1.1. DIVERGENT

The divergent method (cascade-type reaction; "inside out") has been extensively utilised by Tomalia et al to achieve geometric growth of dendrimers [7]. The synthesis involves the reaction of a central initiator core molecule with a second organic molecule which is used as the chemical building block. The chemical building blocks are then reacted with a third organic molecule which could either be identical to the central core molecule or different. The third organic molecule has reactive groups on both ends. One functional group reacts with the building block while the functional group on the other end provides the terminal or surface group of the dendrimer. The product from the reaction has reactive groups for the reaction of more of the initiator core molecule to be on the periphery. The isolated product from the first reaction sequence is termed generation 0 (G0). The process can be iterated where the number of coupling reactions increases exponentially with each subsequent generation.

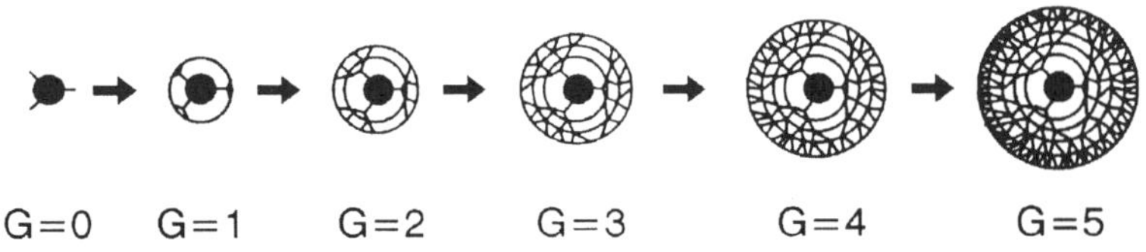

Figure 1. *Generalised divergent reaction sequence for the propagation of dendrimer generations (G).*

Some of the first dendrimers prepared by the divergent method involved the reaction of ammonia and methyl acrylate followed by extensive amidation of the resulting esters with large excesses of ethylenediamine. The starburst PAMAM dendrimers have been prepared up to G10 and are commercially available from Dendritech, Inc (Midland, MI). Meijer at al have synthesised poly(propyleneimine) (PPI) dendrimers having a diaminobutane (DAB)-core on a large scale using the divergent method [8, 9]. They are available from DSM Research (Netherlands) up to generation 5 under the trade name Astramol™. In addition to these, other poly(R) dendrimer families have been synthesised [10] (R = ethers, siloxanes, thioethers, arylethers, amidoalcohol, amines, phosphonium, alkanes, nucleic acids, organometallics, and fluorinated carbosilanes [11]).

1.2. CONVERGENT

The convergent method ("outside in") has been pioneered by Fréchet et al [12, 13] who heard the work of the divergent method by Tomalia at a presentation in 1987 (Fréchet, personal communication). As the name of the synthetic methodology implies, which was inspired from the classical organic disconnection approach, growth begins at what will become the surface or terminal groups of the dendrimer termed dendrons. These pre-formed dendritic fragments are then attached to a central polyfunctional core molecule. Unlike the divergent method, the number of coupling reactions is constant per generation. An accelerated convergent scheme, based on double exponential growth, has emerged which results in the same degree of polymerisation to occur at the third generation as opposed to the seventh generation based on the traditional convergent method [14, 15]. The first dendrimers to be synthesised using the convergent method involved phenoxide-based, benzylic bromide displacements [12]. Great care is needed to generate the bromide dendrons in high yield and an alternative route utilising mesylates has recently been reported [16].

There is more control over the focal point and the surface or terminal functional groups with the convergent approach compared to the divergent approach. The surface groups do not have to be of the same functionality using the convergent method which allows for the synthesis of dendritic block copolymers based on the non-uniform functionalised surfaces [17]. These block copolymers are described as surface-, segment-, and layer-block. In addition, the need for large excesses of reagents required for the divergent method is avoided which simplifies purification.

Above a particular dendrimer generation the structure is expected to be more or less spherical [18]. An attempt has been made to synthesise dendrimers which have a cylindrical shape in solution using dendrons of the Fréchet-type (convergent method); however, the cylindrical shape in solution still remains to be verified [19].

1.3. HETEROATOM

The heteroatom dendrimers (other than N and O which comprise the bulk of the literature) include P (second most reported type of heteroatom dendrimer), Si and Ge. In addition, a variety of metal ions such as Ru, Rh, Os and Ir have comprised the core (endo-), as connectors, at the branching unit or at the surface (exo-) of the dendrimer. These metallodendrimers have been reviewed [20]. The metallodendrimers can also have the metal ion at more than one of these positions [21] and represent metallosupramolecular chemistry [22] which results in the synthesis of supramolecular nanostructures with specific functions [23]. The co-ordination geometry and co-ordination number of the metal ions are matched with the ligand assemblage (i.e., matching preferred metal ions to preferred ligands) in advance. These metal-centred dendrimers (endo-) can be chiral based on the optical rotation of the co-ordinated ligand. A carbon-centred dendrimer is achiral when it has four similar branching units. The chirality and thus stereochemistry of these metallodendrimers presents an inherent problem for their characterisation [24]. Although high generation carbon dendrimers are readily characterised, the organometallic dendrimers are not as readily characterised because of stereochemistry, high charge and low solubility. One approach is to use cyclic voltammetry for characterisation. One such dendrimer is based on oligopyridines which provide the desired co-ordination chemistry and donor atoms to a wide variety of metal ions. These metal-centred dendrimers are of interest as magneto-, electro-, or photo-physical materials for harvesting solar energy. Their synthesis is based on the 'complexes as metals, complexes as ligands' approach and involve both divergent and convergent approaches which allow a precise control of the synthesis and the ability to introduce pre-determined building blocks at each step [25]. The complex metals are mono- or poly-nuclear metal complexes which have labile ligands, thus providing metal ions with unsaturable co-ordination sites. The complex ligands are mono- or poly-nuclear complexes with free chelating sites on the ligand(s). The approach has allowed the design of polynuclear (Ru/Os; Ru/Rh; Os/Rh; Ru/Ir) dendrimers [26]. An Fe-S G4 core dendrimer, based on acetylene-type linkages, which is redox-active has been reported [27]. Porphyrins and poly-pyridine metal complexes, as well as ferrocene, have served as the central core of dendrimers [21].

Phosphorus-containing dendrimers (both exo- and endo-) are important because of the potential to graft a phosphate nucleotide onto the dendrimer [28]. Phosphorus-containing dendrimers up to G12 with a MW greater than 3 million and a diameter of about 20 nm with 12288 chlorine atoms at the periphery have been synthesised using the divergent approach [29]. The solubility and reactivity very much depends on the particular substituent at the periphery. In a series of grafted tri- and tetra-co-ordinated phosphorus-containing dendrimers from G1 to G5 with azinephosphate, azinephosphinite and ylide linkages on the periphery, there was observed a dramatic decrease in solubility in organic solvents, whereas long-chain hydrocarbons increased solubility [28]. Phosphorus-containing dendrimers allow for the ability to exo-co-ordinate Au, Pt, Pd, Ru, Rh, Fe and W [30] as well as the ability to graft at the periphery a large number of sets of 2, 3 or 4 functional groups (multiplurifunctionalisation). Exo-grafting of polyaza macrocycles with aldehydes or PCl terminal groups has been possible because these groups were found to be as reactive as their monomer counterparts [31]. A dendrimer having an Os(II) core and Ru(II) in the branches has been reported [32].

The Si-containing dendrimers are comprised of [33]:

- Silicones –(O-SiMe$_x$O$_y$),
- Carbosilanes –(Si-alkyl) and
- Polysilane –(Si-alkyl)$_y$

A series of organometallic silicon dendritic macromolecules containing a controlled number of redox active centres have been synthesised and characterised [34]. The kinetic and thermodynamic C-S bond found in carbosilanes makes them one of the most important classes of Si-based dendrimers. A G5 polysilane dendrimer having both endo- and exo- Si atoms has been prepared using repetitive alkenylation-hydrosilylation cycles [35]. The synthesis of the first water-soluble carbosilane dendrimer has been reported [36]. The synthesis of carbosilane dendrimers are achieved by either the divergent or convergent approach [37] which also allowed the synthesis of the G2 organogermanium dendrimer [38]. Dendrimers based on transition-metal complexes have been reviewed [39].

1.4. SOLID PHASE

The solid-phase synthesis of polyamidoamine dendrimers up to G4 has recently been achieved [40]. The solid phase reaction allows the use of a large excess of reagents followed by facile purification by washing the resin. The solid-phase synthesis allowed for peptides and small molecules to be "grown" directly onto the periphery of the dendrimer while it was still attached to the resin bead. Incomplete reactions only required a repeated treatment with reagents which could be recycled. The solid state synthesis allows for the construction of a small combinatorial library using the dendrimers.

Another interesting approach, termed DCC: dendrimer-supported combinatorial chemistry, has been introduced [41]. Instead of the dendrimer attached to a resin, it is allowed to float freely in solution. The large sizes of the dendrimers allow for easy size-exclusion purification of the dendrimer intermediates. The combinatorial approach to the solution phase synthesis of diverse dendrimers was achieved by reaction of various isocyanate and amine monomers which were simultaneously added in solution to form dendrimers. One such dendrimer had a 50:50 mixture of amine and benzyl ether groups on the periphery which allowed solubility in both water and chloroform resulting in a potential “universal micelle” [42].

1.5. OTHER

The first radiosynthesis of dendrimers involved the C-14 methyl acrylate addition to the G2 and G5 starburst PAMAM dendrimers in a divergent reaction sequence [43]. The products from the reaction are the G2.5 and G5.5 starburst PAMAM dendrimers. The addition of the radioactively labelled methyl acrylate in the final step to the existing purified dendrimers results in less radioactive C-14 methyl acrylate to be used.

The orthogonal coupling scheme (OCS) has been used for the synthesis of dendrimers [44]. It involves the covalent attachment of two different building blocks (monomers) in two orthogonal coupling reactions. The AB2 monomers minimise the number of steps for the actual dendrimer synthesis, but the monomer synthesis can be laborious.

Chiral carbon core molecules have been investigated for the synthesis of chiral dendrimers [45]. The ‘Tris(hydroxymethyl)methane’ derivatives are enantiomerically pure.

Without carrying out the actual covalent bond formation reaction, hydrophilic, positively charged amines on the periphery of the dendrimer were found to electrostatically interact with the carboxy-terminated dodecanoic acid to produce a hydrophobic dendrimer which allows the encapsulation of methyl orange dye and its transfer to toluene from an aqueous solution [46].

2. Characterisation

Concurrent with developments in the synthesis of dendrimers are novel analytical techniques for their characterisation [47]. The various analytical techniques that have been used to characterise dendrimers have been reviewed [48]. The techniques include multinuclear NMR spectroscopy, various mass spectrometry methods, chromatographic methods, electrophoresis, X-ray scattering, neutron scattering, small-angle neutron scattering (SANS), viscosimetry, electron microscopy and molecular dynamics with two different end groups [49]. These analytical techniques have been used for monitoring a perfect dendritic structure from defective structures. It has recently been noted that a hyperbranched polymer and a dendrimer are not synonymous and the different properties of the two have been reported [50]. The hyperbranched dendrimers could be considered trivially as a mixture of dendrimers. Dendrimers contain no linear segments whereas hyperbranched polymers are intermediate between dendrimers and linear polymers. Dendrimers are more soluble than hyperbranched polymers, which in turn, are more soluble than the linear polymers in solvents such as acetone, which were able to solvate the peripheral surface groups [51]. New terminology has recently been introduced to distinguish the structural composition of dendrimer preparations. A mononuclear dendrimer preparation means that all of the dendrimers in that batch have the same structural composition, i.e., and no defects. A monodisperse dendrimer preparation means that the dendrimers in the same batch are structurally related but not of the exact structural composition. However, as pointed out by Newkome et al these various analytical techniques cannot guarantee the exact structural composition for the entire sample thus leading to "monodisperse" dendrimers which are a closely related mixture though not perfectly defined as the "mononuclear" dendrimers [52]. The monodisperse or mononuclear nature of a dendrimer preparation was performed using a qualitative procedure based on the fluorescent property of anthracene with a detection limit of less than 1 %. Unfortunately, a quantitative analysis was not possible. NMR spectroscopy cannot distinguish perfect dendrimers from those having small defects. Conventional mass spectrometry only works for low MW (generation) dendrimers. Gel permeation chromatography (GPC) can result in errors as large as 30 %, although it is useful in monitoring the reaction. Ionisation MS cannot easily be used for analytical purposes. MALDI-TOF-MS for a series of aromatic polyester dendrimers gave the molecular mass to within one amu of the calculated molecular mass with peak half-widths between 4 and 8 amu with excellent S/N ratios [53]. The shape and size of the monodendritic building blocks of dendrimers were analysed by X-ray analysis of the liquid crystalline (LC) lattice which supported the change from a cylindrical to spherical shape going from G2 to G3 dendrimer [54]. The terminal groups are at the periphery of the dendrimer as determined by neutron scattering based upon the radius of the G7 dendrimer (39.3 +-1.0 Å) compared to the radius of gyration (34.4 +-0.2 Å) for the whole dendrimer [55]. C-13 NMR spectroscopy has been used to measure the spin-lattice relaxation times of the internal and terminal carbon atoms of a series of PAMAM dendrimers. The relaxation times of the terminal carbons decrease as the generation increases indicating that potential steric crowding as the generation increases does not affect the dynamics. The T1 relaxation time of internal carbon atoms was essentially independent of the generation above G2, indicating the slowing of the internal carbon atoms as the generation increases [56]. Dendrimer conformation in solution has been probed by measurement of the T1 relaxation times of the dendrimer protons after incorporation of a paramagnetic core [57]. The terminal protons were found to be less mobile than the inner-core protons suggesting more congestion near the surface.

The first observation of a single dendrimer has been reported [58]. The incorporation of a fluorescent molecule, dihydropyrrolopyrroledione, as the core by the divergent method allowed the detection of the single dendrimer molecules with a modified fluorometer. The use of fluorescein showed that for the positively charged full generation ethylenediamine or ammonia core dendrimers, both had the same structure going from an open to a closed structure between G2 to G3 dendrimer[59]. A fluorescent probe, pyrene, which does not fluoresce in water, when taken up by a G2 PAMAM dendrimer in aqueous solution, still fluoresced indicating that water hardly penetrates the cavity of the dendrimer [60].

The characterisation of a systematic PS dendrimer series in the solid state has been carried out using XPS (X-ray photoelectron spectroscopy) [61]. For determining Ru and Os in metallodendrimers, electrothermal atomic absorption spectrometric (ETAAS) methods have been developed [62]. Chiral-core dendrimers have been analysed by scanning tunnelling microscopy (STM) to investigate the size and symmetry of dendrimers by dissolving them in CH_2Cl_2 and film casting onto a Pt(100) surface where it was found that the bulky phenyl rings comprising the dendrimers formed from a chiral dendrimer had different STM's [63].

3. Biotechnology applications

3.1. BIOMOLECULES

Dendrimers have been coupled to various biomolecules including monoclonal antibodies. The attachment is referred to as "exo-receptor-targeted molecules". That is, molecules attached to the dendrimer surface or at the termini of the dendrimer branches as opposed to "endo-receptor-targeted molecules" which make up the dendrimer core. The coupling of the dendrimer to the an antibody did not result in the loss of protein immunoreactivity [64]. The methodology may allow for the production of cell-specific targeting devices for biological systems.

Attachment of a tripeptide growth factor (GHK) to a dendrimer periphery resulted in an enhancement on the GHK ability to facilitate the growth of hepatoma cells in solid-support systems which may be useful in bioartificial liver support systems for liver transplantation [65]. The biodistribution of a G2 dendrimer coupled to indium or yttrium chelates has been evaluated in mice [66]. The biodistribution in mice was first in the kidneys followed by uptake in the liver, spleen and bone at most time points. Conjugation of the antibody did not change the biodistribution patterns. However, it has been noted that a greater understanding of whole body and cellular pharmacokinetics of naked and modified dendrimers is needed for optimal design of organ- and cell-specific structures [67].

The attachment of quaternary dimethyldodecylamine to the surface of G3 dendrimers yields dimethyl dodecylammonium chloride derivatised dendrimers which inhibited the growth of gram negative E. coli and gram positive S. aureus as detected by a bioluminescence method. The reduced luminescence quantitatively verified their use as dendrimer-based antibacterial agents [68]. The most widely used functionalities for the attachment of drug molecules include amide, carbonate, carbamate and ester bonds which are hydrolytically labile. A PEG-dendrimer allows for the water-soluble conjugation of cholesterol and two amino acids [69].

The synthesis of dendrimers which contain reactive terminal groups by the convergent method for attachment of a biomolecule has been reported [70]. However, there were difficulties in preparing large quantities of high-generation dendrimers with reactive terminal groups, so terminally reactive dendrimers prepared from the divergent method were employed.

Generalised methods for the functionalisation of methyl carboxylate and primary amino groups on the surface of dendrimers for coupling of biomolecules have been reported [71]. The surface groups were converted to either electrophiles such as the iodoacteamido, epoxy or N-hydroxysuccinimidyl groups or the nucleophiles such as the sulfhydryl group. These reactive groups allowed the covalent attachment of alkaline phosphatase. These protein-conjugated dendrimers could then be further conjugated with other antibodies such as the Fab' fragment of anti-CKMB antibody resulting in dendrimer-based multifunctional activities for use as immunoassay reagents. The appropriate selection of activating conditions for a dendrimer allows for the conjugation of two similar or dissimilar proteins. The presence of one protein does not affect the biological activity of the other protein on the same dendrimer [72]. Biotin, a molecule with high binding specificity for streptavidin, has been conjugated to dendrimers in efforts to pretarget radionuclides for cancer therapy [73]. A new biotinylation reagent was coupled to dendrimers, radioactively labelled with streptavidin and the in-vivo biodistribution and pharmacokinetics evaluated in mice. The dendrimers were rapidly cleared from the blood via both renal and hepatobiliary excretion. Kidney concentration increased with increasing dendrimer generation up to G3, being almost 50 % ID/g.

Potentially, localisation can be achieved by binding with an antibody conjugate previously localised on tumour cells. Several specific antibodies have been coupled to PAMAM dendrimers without losing their stability and immunological binding, both in solution and when immobilised onto a solid support [74] which holds promise in the production of immunodiagnostic products. The solubility was not altered significantly when antibodies were attached thus allowing improved sensitivity and shorter assay times of solution-phase binding of antibody to analyte compared to commercial solid-phase or microparticulate reagents used in many automated immunoassay systems. (The dendrimer-based system showed analytical sensitivity and speed that was equivalent to or greater than commercial-based systems.)

3.2. GLYCOBIOLOGY

Glycobiology is the study of carbohydrate-protein interactions [75] which are prevalent at cellular levels. Most cells are coated with carbohydrates to which lectins or carbohydrate-binding proteins can attach. Interest is due to the ubiquitous and essential roles which sugar moieties of complex carbohydrates play in living systems and high hopes for glycodendrimers in the prevention of pathogenic infections and other related diseases. The attachment allows the adherence of various pathogens which can then take over the host tissues. Carbohydrates have multiple functional groups on each monomer unit which are capable of forming a myriad of structures, each one capable of a different specific biological message. The glycoside cluster effect is the binding of many sugar residues by a lectin which has clustered sugar-binding sites. By using convergent methods glycodendrimers have been synthesised by build-up of a gallic acid trivalent core to which carbohydrate residues were attached [76]. The convergent synthesis allows for the incorporation of other carbohydrates. The attachment of alpha-thiosialosides to a dendrimer allows the cluster of the active groups which significantly increases the inhibitory capacities compared to the monosialoside. Boronic acid groups, which facilitate the complexation of a variety of saccharides, have been exo-attached to dendrimers. The binding can be monitored by the fluorescence intensity of the host. Various symmetrically tethered sialodendrimers have been synthesised to generate families of multivalent glycoconjugates which may be used as inhibitors of haemaglutination of human erythrocytes by influenza viruses [77]. The influenza virus has two envelope glycoproteins, HA (hemagglutinin) and NA (neuraminidase). HA binds to sialic acid receptors (sialosides; cell surface carbohydrates) on cells and thus the virus gets internalised by receptor-mediated endocytosis. Competition with sialosides to HA using sialic acid coupled to dendrimer (i.e., vaccine) may allow the potential interaction with many HA moieties on a single virus. Multiple sialic acid (SA) residues coupled to dendrimers inhibit influenza-induced agglutination of red blood cells. The linear polyacrylamide backbones are cytotoxic. The ability of the SA coupled dendrimers to inhibit virus haemaglutination (HA) and to block viral infection of mammalian cells in-vitro has been reported [78]. Dendrimer polymers were not cytotoxic to mammalian cells at therapeutic levels. Variation in inhibitory effects were observed with different viruses.

A series of alpha-D-mannopyranose-containing dendrimers were synthesised using the convergent reaction scheme and were investigated with respect to the efficiency of these dendrimers in inhibiting the binding of a lectin to a purified yeast mannan fraction which is found in the serum of patients with Crohn's disease and may cause allergic responses from overexposure by bakers and brewers [79]. The preparation of dendroclefts (dendritic cleft-type receptors) up to G2 allow the chiral molecular complexation of mannosaccharides by H-bonding. The H-bonding observed mimics bacterial and protein binding to sugars (80]. The study of the interaction of sugar-binding proteins such as human IgG fraction to lactose has been undertaken by attaching *p*-aminophenyl-β-D-lactoside onto the dendrimer surface. The attachment of the glycodendrimers to solid-phase supports such as microtiter plate wells provided binding studies to be carried out with the proteins. There was a potential for marked selectivity of certain glycodendrimers towards distinct classes of sugar receptors [81]. The binding properties of PAMAM dendrimers ending with mannopyranoside residues were determined. They form insoluble carbohydrate-lectin complexes and can selectively precipitate a carbohydrate-binding protein from a lectin mixture; thus, they constitute new biochromatography materials [82].

In microtiter wells, dendrimer-based oligosaccharides which are recognised by cholera toxin and the enterotoxin of E. coli inhibited their binding to the wells which were coated with just the oligosaccharides in a competitive inhibition study. The toxins were labelled with I-125 and the activity measured in the wells. These toxins cause traveller's diarrhoea and if left untreated will result in death. These in-vitro experiments imply that administering the dendrimer-based oligosaccharide in-vivo may remove these toxins [83-85]. The cluster effect is more evident in small glycodendrimers possibly because of steric interaction as the generation of the glycodendrimer increases. Spacer arms have been placed between the dendrimer and the saccharide which allows for not only flexibility (relief from steric strain) but also the potential to control hydrophobic/hydrophilic interactions [86]. The mannose-binding protein which is an acute phase protein of immune response may be competitively challenged with mannoside-based dendrimers although no data was presented [87]. The scaffolding of poly-L-lysine allowed the covalent attachment of alpha-D-mannopyranoside glycodendrimers. Inhibition of the binding of yeast mannan to concanvalin A was up to 2000-fold higher than for the mannopyranoside alone [88]. Rather than being attached to the periphery of the dendrimer, the chemoenzymatic syntheses of N-acetyllactosamine has been reported based on the scaffolding of L-lysine. These disaccharides cores are associated with tumours, thyroid disorders and a sexually transmitted disease of the *H. ducreyi* pathogen and their binding properties studied showing the variability in the binding interactions suggesting additional research [89].

3.3. PEPTIDE DENDRIMERS

Antibodies are potentially of great value in targeted drug therapy because of the inherent specificity of the antibody-antigen interaction. The conventional approach to preparing antibodies is to conjugate a peptide to a known protein or synthetic polymer, in order to mimic the macromolecular structure of the native protein. However, this method generates macromolecular carriers that are ambiguous in structure and composition. To improve on this approach, multiple antigenic peptide (MAP) systems were developed as efficient and chemically defined systems to produce immunogens in the absence of protein carriers. The MAP system consists of an oligomeric branching lysine core of seven lysine units and eight arms of peptides that contain antigenic epitopes. The overall structure of the MAP system is a polymer with a high density of surface peptide antigens and a molecular weight greater than 10 K [90]. The almost mutually exclusive desire to attach a large number of drug molecules to an antibody while still retaining maximal antibody immunoreactivity may be overcome by dendrimers which act as intermediate linkers between the drugs (capable of covalent attachment to the dendrimer) and the antibody (which binds to the dendrimer by only one modified site on the antibody).

A MAP dendrimer with a lipophilic surface has been synthesised for possible drug delivery. These lipoamino acids while lipophilic also retain the solvation properties of amino acids and peptides [91]. MAP dendrimers are peptide dendrimers which amplify peptide immunogenicity. Unlike most vaccines, MAPS can be stored or shipped as powders. There is an excellent review on MAPS [92]. They have been used in-vitro as immunogens, vaccines, immunodiagnostics, serodiagnostics, ligands, inhibitors, artificial proteins, epitope mapping, affinity purification, presentation of T-cell epitopes and intracellular delivery. The synthesis of MAP dendrimers has been improved by the use of a N-acetylation capping step which allows the direct (stepwise) synthesis and purification of MAPS. The automated peptide synthesiser was programmed to provide a N-acetylation capping reaction following each amino acid coupling reaction thus serving as a protecting group for further reaction with unwanted amino acids in the crude mixture [93].

3.4. BORON NEUTRON CAPTURE THERAPY

Boron neutron capture therapy is based on the nuclear reaction that occurs when a stable B-10 isotope is irradiated with low-energy neutrons to yield high LET radiation consisting of alpha particles and recoiling Li-7 which are energetic and cytotoxic [94]. To deliver the approximately 10^9 number of B-10 atoms needed to effectively eradicate a tumour cell, dendrimers have been conjugated with a polyhedral borane and subsequently attached to a monoclonal antibody. The number of boron atoms range from 250 to 1000 per dendrimer molecule. Unfortunately, in-vivo studies with mice revealed hepatic and splenic uptake over tumour localisation. Instead of attaching the polyhedral borane to the periphery or surface of the dendrimer, the borane cluster has been incorporated into the interior of the dendrimer [95]. Decaborane was reacted with the alkyne functionality located in the interior of the dendrimer to give 0-carboranes. The incorporation into the interior of the dendrimer increased their aqueous solubility.

An interesting application of boronated dendrimers is in electron spectroscopic imaging-based immunocytochemistry [96]. The dendrimers contain a boron cluster on one side of the dendrimer and an antibody fragment on the other side. These antibody-dendrimer-boranes were used to allow the visual detection of BSA in epithelial cells of ileum which had been internalised by endocytotic vesicles of ileal enterocytes in new-born piglets after administration of BSA. As determined by electron spectroscopic imaging of boron, a G4 starburst dendrimer bearing an epidermal growth factor was bound to the cell membrane and endocytosed in-vitro of the human malignant glioma U-343MG cell line expressing EGF receptors [97].

3.5. MR IMAGING AGENTS

Dendrimer gadolinium poly-chelates are a new class of MR imaging agents with large proton relaxation enhancements and high molecular relaxivity (relaxation rates per mM metal ion). Wiener at al [98] first introduced that the covalent attachment of gadolinium chelates to dendrimers have the potential to be blood-pool MR T1 imaging agents for use in MR angiography. The synthesis involved the covalent attachment of the acyclic GdDTPA chelate to a G2 and G6 dendrimer utilising a stable thiourea linkage between the chelate and the dendrimer. These dendrimer-based MR imaging agents had a molar relaxivity that was up to six times higher than for clinically used gadolinium chelates because of the high molecular weight of the dendrimer. The authors demonstrated the potential usefulness of these agents for vascular imaging by being able to delineate the vascular system of a rat for at least up to one hour. Not necessarily confined to complexation of Gd for T1-weighted MRI, a dysprosium chelate has been attached to a dendrimer in a similar fashion and opens up the opportunity for T2 MR imaging agents to utilise the macromolecular characteristics that the dendrimer provides [99]. The incorporation of Dy provides an unique T2 relaxation agent which may be important for tissue perfusion studies using MRI.

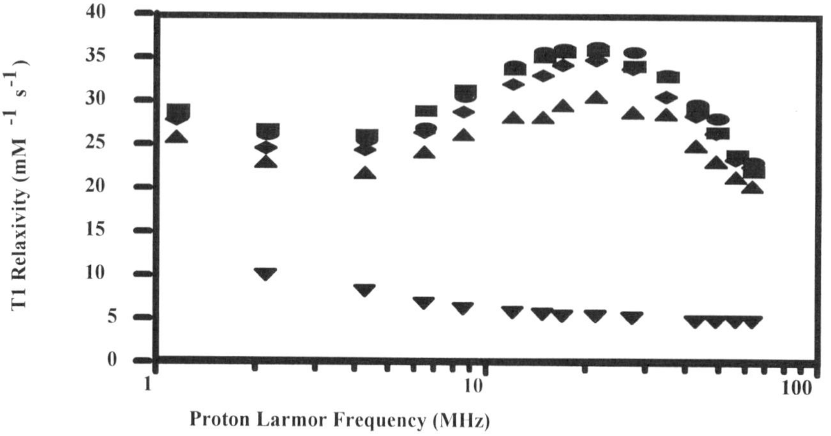

Figure 2. *T1 MRD relaxivity profiles for the G5 (upper triangles), G7 (diamonds), G9 (circles), and G10 (blocks) PAMAM dendrimers with attached Gd(p-SCN-Bz-DOTA) chelates. For comparison, the MRD profile of Gd(p-NO_2-Bz-DOTA) is also shown (lower triangles).*

The three main parameters that dictate achieving maximum T1 relaxivity for dendrimer-based gadolinium chelates are the amount of time that the water molecule interacts with the gadolinium, how fast the gadolinium tumbles in solution and how fast the paramagnetic electron spin density relaxes back to the ground state [100]. A limitation in achieving the full expected relaxivity for dendrimer-based MR contrast agents was observed and verified by O-17 NMR studies of the G3, G4, and G5 dendrimer-based MR imaging agents [101]. It was concluded that modification of the chelating ligand may result in faster water exchange and therefore higher relaxation rates. A plateau in the relaxivity was observed as the generation of the dendrimer increases from G5 to G10, providing evidence of even more severe limitation of achieving full relaxivity as the generation of the dendrimer increases [102]. Since contrast depends on the co-ordinated water molecule interacting with the bulk water, a long residence time at the gadolinium limits the relaxivity and therefore would limit the observed contrast.

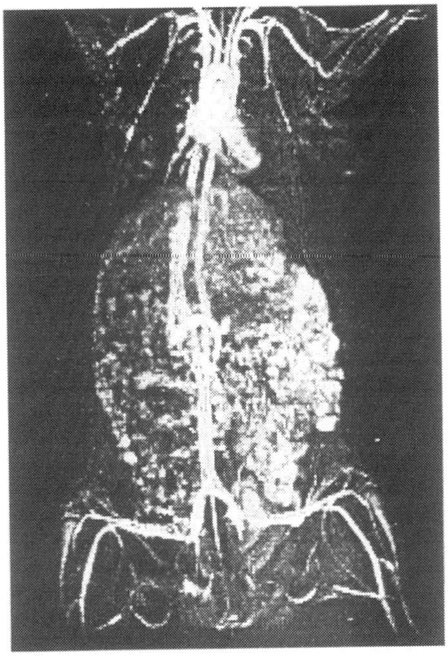

Figure 3. *3D-TOF MR angiogram of a rat 50 minutes following injection of 0.05 mmol/kg of Gd(-p-SCN-Bz-DOTA) attached to the surface of a G9 dendrimer. The nanomolecular size of the dendrimer construct permits an extended vascular visualisation compared to the clinically used MR imaging agents.*

The potential of these dendrimer-based gadolinium chelates as blood pool imaging agents has been explored in pigs [103]. The dendrimer-based MR imaging agent was found to have the same blood pool properties as Gd-DTPA-polylysine. No statistical differences in relative signal intensities were observed in various pig organs between the two imaging agents. The MR angiographic properties of these dendrimer-based gadolinium chelates has been explored. In rats they are able to provide strong tumour rim enhancement and detailed angiographic definition of peritumoural vessels [104]. In the MRI of canine breast tumours a delayed tumour clearance was observed compared to the clinically used gadopentate dimeglumine [105]. The minimum effective dose of

0.02 mmol/kg of dendrimer-based gadolinium chelate was observed for visualisation of the mediastinum, abdomen and lower limbs of rabbits on 3D time of flight magnetic resonance angiography of the body [106].

For modified PAMAM dendrimers coupled to Gd chelates, the pharmacokinetic and biodistribution are found to depend on dendrimer size as well as the type of terminal groups [107]. Hepatic localisation decreased and blood half-life increase by the covalent attachment of polyethylene glycol (PEG).

The ability for these dendrimer-based MR imaging agents to be site-specific has been shown [108]. The attachment of folate to dendrimer-based chelates targets these particles to folate binding proteins which exist in the serum as well as on the surface of many cancer cells.

A kinetic theory for describing the dynamic properties of an intermediate-sized MR imaging agent, cascade-Gd-DTPA-24 polymer, (Schering AG, Berlin, Germany, MW<30 kDA) has been developed [109]. The dendrimer-based MR imaging agent was considered intermediate in size relative to Gd-DTPA and albumin-Gd-DTPA-30. The method has clinical applications based on its potential for pixel-by-pixel mapping. The first covalent attachment of tetraaza macrocycles to the terminal phosphorous group of a G1 and G3 P-S containing dendrimer has been achieved [110]. The co-ordination of Gd to the grafted macrocycle still needs to be explored, but if possible, it would open up the possibility of having two nuclei, Gd-157 and P-31, which are detectable by MRI and may be useful in multinuclear MRS.

3.6. METAL ENCAPSULATION

The complexation of divalent cations such as Cu, Zn, Ni and Au on the periphery of dendrimers has been reported [111, 112]. However, recently cooper nanoclusters have been trapped within dendrimers [113]. Ten weeks later Tomalia published his article on copper-dendrimer nanocomposites [114]. The trapping involves zerovalent copper nanoclusters within dendrimers as well as other elemental metals or metal sulphides such as Ag(I), Pt(II), Pd(II), Ru(III) and Ni(II) [115]. These nanocomposites could conceptually incorporate any metal ion which could be chemically reduced once inside the dendrimer shell thus trapping the metal ions as clusters giving rise to intra-dendrimer metal nanoparticles. Instead of discrete metal atoms within dendrimers, 2-3 nm gold colloids have been stabilised by multiple amine-terminated dendrimers, thus giving rise to inter-dendrimer nanoparticles [116]. The potential applications of these intra- or inter-dendrimer nanoparticles could be in catalysis and as use as nanodevices, although one could also envision the delivery of metal ions for therapeutic or diagnostic applications, including MR imaging [117].

3.7. Transfection agents

Dendrimers have been shown to form physiologically stable complexes with DNA and to mediate transfection of the DNA into a wide variety of cells in culture [118]. It was found that the transfection efficiency was both a function of the particular dendrimer generation and the type of cell. By using twenty different types of polyamidoamine dendrimers with a broad range of cell lines it was found that the size, shape and number of surface groups on the dendrimer affects the degree of transfection. The use of heat-activated dendrimers has increased the efficiency of cellular transfection of DNA by more than 50-fold [119]. Heating in a solvolytic solvent such as water for various times degraded the dendrimers. The highest molecular weight component of the degraded products, termed a 'fractured' dendrimer, mediated transfection. The degradation supposedly occurs by cleavage of the amide bonds within the dendrimer. The results suggest that other physical properties such as dendrimer flexibility is important in transfection ability. These heat-activated dendrimers are commercially available as SuperFect™ (Qiagen). SuperFect-DNA complexes possess a net positive charge which allows them to bind to negatively charged receptors, such as sialylated glycoproteins on the surface of eukaryotic cells. It has been proposed that once inside the cell, SuperFect™ buffers the lysosome after it has fused with the endosome, leading to pH inhibition of lysosomal nucleases. This ensures stability of SuperFect-DNA complexes and the transport of intact DNA to the nucleus. In general, significantly higher transfection efficiencies were found than for widely used liposomal reagents. SuperFect™ is suitable for both adherent and suspension cells. These include primary cells such as pig endothelial, human smooth muscle, and HUVEC or sensitive cell lines such as HaCaT and HT-1080.

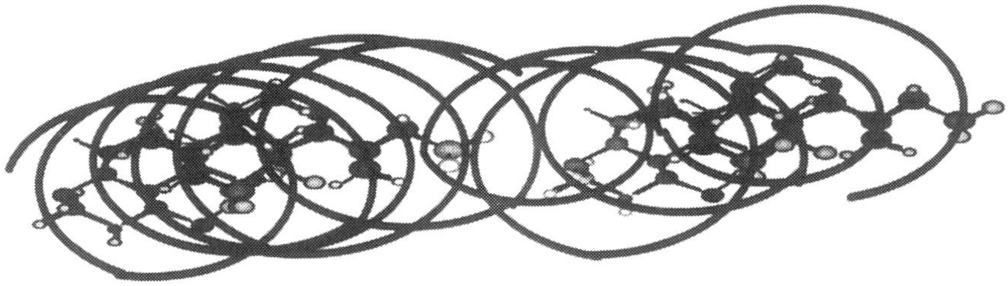

Figure 4. *Illustration of dendrimer-DNA complex. DNA wraps around the dendrimers which condense the DNA and allows cellular incorporation (transfection).*

Dendrimers have also been found to transfect cells with antisense oligonucleotides and plasmid expression vectors coding antisense mRNA (antisense nucleic acids) through energy-dependent endocytosis which allowed higher transfection efficiency compared to DNA alone or lipid-mediated transfection [120]. The cells were assayed for transfection based on the inhibition of luciferase expression. The unmodified oligonucleotides were found to form stable complexes with dendrimers and to function as native oligonucleotides, thus requiring no chemical modification of the oligonucleotide to prevent degradation or intracellular destruction. The transfection of plasmid DNA by dendrimers in-vivo has been demonstrated [121]. The G5 ethylenediamine-core dendrimer enhanced the transfer of plasmid DNA which encodes for viral interleukin-10, a gene which regulates immune responses, into transplanted mouse cardiac isografts in-vivo. Compared to the DNA alone, the transfection of higher amounts of the DNA for longer periods of time by the dendrimer resulted in longer graft survival times. By varying the charge ratio of DNA to dendrimer the survival time was extended to 39 days compared to 14 days for the control. Interestingly, the use of the G9 dendrimer showed a decrease in survival time from 39 days to 27 days suggesting that many factors including the dendrimer generation must also be taken into account when used in-vivo. The possibility of using dendrimers to overcome corneal allograft rejection or to treat disorders of the corneal endothelium appears promising [122]. The immune response to transplanted corneal allografts produces high levels of tumour necrosis factor (TNF). Blocking the action of TNF may prolong some graft survival times. A dendrimer was used to deliver the gene encoding for tumour necrosis factor receptor immunoglobulin (TNFR-Ig), a TNF blocker, to rabbit corneal endothelium ex-vivo. The gene was expressed for up to 9 days in the transfected corneas.

Dendrimers allow for the transfection of mammalian cells with various genes. The interactions have been explored by EPR and the idea is to be able to extend the work to the mechanism of gene transfer [123]. Using nitroxide-labelled dendrimers in the presence of various oligonucleotides, the G2 dendrimer at a pH of 5.5 showed significant interaction with polynucleotides which decreased with an increase in dendrimer concentration presumably because of the aggregation of the dendrimers. The interaction of the G6 dendrimer with the polynucleotides increased with dendrimer concentration until the interacting sites were saturated. It has been proposed that coulombic interaction between the negatively charged phosphate groups of the nucleic acid and positively charged amino groups of the dendrimer (at physiological pH) allow ionic complex formation and that the current polymeric polyelectrolyte theory does not allow a reliable theoretical calculation of the compositions [124]. It has been noted that an excess of the cationic dendrimer increases transfection efficiency. However, further studies need to be conducted in vivo to assess the cellular localisation and fate of transfected nucleotide and dendrimer [125].

3.8. DENDRITIC BOX

Meijer stumbled on the idea of the dendritic box in his quest for a chiral dendrimer encapsulating amino acids by reaction of the terminal amines with BOC-protected amino acids. [126]. The dendritic box is made up of PPI dendrimers which have diameters of 5 nm as revealed by EPR and NMR spectroscopy [127]. The synthesis of a rigid, dense outer shell around the dendrimer in the presence of guest molecules results in a dendritic box which encapsulates the guest molecules. Hydrolysis (i.e., enzymatic) of the outer shell allows the release of the trapped guest molecule. A second class of dendritic host involves an alkyl chain-propagating dendrimer which behaves as an inverted micelle which transfers the guest from the aqueous to the organic phase using pH as the control parameter. The use of low MW surfactants solubilises the dendrimer-guest in water. The lowering of the pH organises the globular inverted dendritic micelles into cylindrical amphoteric vesicles which can then assemble into unimolecular micelles with the subsequent release of the guest—thus providing a novel drug-delivery system, especially the encapsulation and Solubilisation of hydrophobic drugs. One industrial application is the enhancement of the dyeability of polyolefins and to improve the dispersion of silica in rubber formulations [128].

4. Concluding remarks

Although there are almost unlimited potential applications of dendrimers, it is interesting that the only commercial application of dendrimers to date are as transfection agents. With research in the 1990's directed toward applications, it is envisioned that other viable commercial applications will be announced in the new millennium. The interest in dendrimers has been so profound that a web-site has been set up at www.dendrimers.com.

References

1. Matthews, O.A., Shipway, A.N., and Stoddart, J.F. (1998) Dendrimers-branching out from curiosities into new technologies, *Prog Polym Sci 23*, 1-56.
2. Kim, Y. and Zimmerman, S.C. (1998) Applications of dendrimers in bio-organic chemistry, *Curr Opin Chem Biol 2*, 733-742.
3. Fischer, M. and Vögtle, F. (1999) Dendrimers: from design to application-a progress report, *Angew Chem Int Edit 38*, 885-905.
4. Majoral, J.-P. and Caminade, A.-M. (1999) Dendrimers containing heteroatoms (Si, P, B, Ge, or Bi), *Chem Rev 99*, 845-880.
5. Tomalia, D.A. and Esfand, R. (1997) Dendrons, dendrimers and dendrigrafts, *Chem Ind-London 11*, 416-420.
6. Newkome, G.R., Baker, G.R., Young, J.K., and Traynham, J.G. (1993) A systematic nomenclature for cascade polymers, *J Polym Sci Pol Chem 31*, 641-651.
7. Tomalia, D.A., Baker, H., Dewald, H., Hall, M., Kallos, G., Roeck, S.M.J., Ryder, J., and Smith, P. (1985) A new class of polymers: starburst-dendritic macromolecules, *Polym J 17*, 117-132.
8. Brabander-van den Berg, E.M.M., Nijenhuis, A., Mure, M., Keulen, J., Reintjens, R., Vandenbooren, F., Bosman, B., de Raat, R., Frijns, T., v.d. Wal, S., Castelijns, M., Put, J., and Meijer, E.W. (1994) Large-scale production of polypropylenimine dendrimers, *Macromol Symp 77*, 51-62.
9. de Brabander, E.M.M., Brackman, J., Mure-Mak, M., de Man, H., Hogeweg, M., Keulen, J., Scherrenberg, R., Coussens, B., Mengerink, Y., and van der Wal, S. (1996) Polypropylenimine dendrimers: improved synthesis and characterisation, *Macromol Symp 102*, 9-17.
10. Tomalia, D.A. (1993) Starburst ™/cascade dendrimers. Fundamental building blocks for a new nanoscopic chemistry set, *Aldrichimica Acta 26*, 91-107.
11. Omotowa, B.A., Keefer, K.D., Kirchmeier, R.L., and Shreeve, J.M. (1999) Preparation and characterisation of nonpolar fluorinated carbosilane dendrimers by APcI mass spectrometry and mall-angle x-ray scattering, *J Am Chem Soc 121*, 11130-11138.
12. Hawker, C.J. and Fréchet, J.M.J. (1990) Preparation of polymers with controlled molecular architecture-a new convergent approach to dendritic macromolecules, *J Am Chem Soc 112*, 7638-7647.
13. Hawker, C.J. and Fréchet, J.M.J. (1990) Control of surface functionality in the synthesis of dendritic macromolecules using the convergent-growth approach, *Macromolecules 23*, 4726-4729.
14. Wooley, K.L., Hawker, C.J., and Fréchet, J.M.J. (1991) Hyperbranched macromolecules via a novel double-stage convergent growth approach, *J Am Chem Soc 113*, 4252-4261.
15. Kawaguchi, T., Walker, K.L., Wilkins, C.L., and Moore, J.S. (1995) Double exponential dendrimer growth, *J Am Chem Soc 117*, 2159-2165.

16. Forier, B. and Dehaen, W. (1999) Alternative convergent and accelerated double-stage convergent approaches towards functionalised dendritic polyethers, *Tetrahedron 55*, 9829-9846.
17. Hawker, C.J. and Fréchet, J.M.J. (1992) Unusual macromolecular architectures: the convergent growth approach to dendritic polyesters and novel block copolymers, *J Am Chem Soc 114*, 8405-8413.
18. Naylor, A.M., Goddard III, W.A., Kiefer, G.E., and Tomalia, D.A. (1989) Starburst dendrimers. 5. Molecular shape control., *J Am Chem Soc 111*, 2339-2341.
19. Karakaya, B., Claussen, W., Gessler, K., Saenger, W., and Schulter, A.-D. (1997) Toward dendrimers with cylindrical shape in solution, *J Am Chem Soc 119*, 3296-3301.
20. Stoddart, F.J. and Welton, T. (1999) Metal-containing dendritic polymers, *Polyhedron 18*, 3575-3591.
21. Venturi, M., Serroni, S., Juris, A., Campagna, S., and Balzani, V. (1998) Electrochemical and photochemical properties of metal-containing dendrimers, *Top Curr Chem 197*, 193-228.
22. Constable, E.C. (1997) Metallodendrimers: metal ions as supramolecular glue, *Chem Commun*, 1073-1080.
23. Serroni, S., Denti, G., Campagna, S., Juris, A., Ciano, M., and Balzani, V. (1992) Arborols based on luminescent and redox-active transition metal complexes, *Angew Chem Int Edit 31*, 1493-1495.
24. Achar, S., Immoos, C.E., Hill, M.G., and Catalano, V.J. (1997) Synthesis, characterisation, and electrochemistry of heterometallic dendrimers, *Inorg Chem 36*, 2314-2320.
25. Balzani, V., Campagna, S., Denti, G., Juris, A., Serroni, S., and Venturi, M. (1995) Harvesting sunlight by artificial supramolecular antennae, *Sol Energ Mat Sol C 38*, 159-173.
26. Balzani, V., Campagna, S., Denti, G., Juris, A., Serroni, S., and Venturi, M. (1994) Bottom-up strategy to obtain luminescent and redox-active metal complexes of nanometric dimensions, *Coordin Chem Rev 132*, 1-13.
27. Gorman, C.B., Smith, J.C., Hager, M.W., Parhurst, B.L., Sierzputowska-Gracz, H., and Haney, C.A. (1999) Molecular structure-property relationships for electron-transfer rate attenuation in redox-active core dendrimers, *J Am Chem Soc 121*, 9958-9966.
28. Prévôté, D., Caminade, A.-M., and Majoral, J.-P. (1997) Phosphate-, phosphite-, ylide-, and phosphonate-terminated dendrimers, *J Org Chem 62*, 4834-4841.
29. Majoral, J.-P. and Caminade, A.-M. (1998) Divergent approaches to phosphorus-containing dendrimers and their functionalisation, *Topp Curr Chem 197*, 79-124.
30. Caminade, A.-M., Slany, M., Launay, N., Lartigue, M.-L., and Majoral, J.-P. (1996) Phosphorus dendrimers: a new class of macromolecules, *Phosphorus Sulfur 110*, 517-520.
31. Launay, N., Slany, M., Caminade, A.-M., and Majoral, J.-P. (1996) Phosphorus-containing dendrimers. Easy access to new multi-difunctionalised macromolecules, *J Org Chem 61*, 3799-3805.
32. Serroni, S., Juris, A., Venturi, M., Campagna, S., Resino, I.R., Denti, G., Credi, A., and Balzani, V. (1997) Polynuclear metal complexes of nanometer size. A versatile synthetic strategy leading to luminescent and redox-active dendrimers made of an osmium(II)-based core and ruthenium(II)-based units in the branches, *J Mater Chem 7*, 1227-1236.
33. Veith, M., Elsässer, R., and Krüger, R.-P. (1999) Synthesis of dendrimers with a N-Si-C framework, *Organometallics 18*, 656-661.
34. Alonso, B., Cuadrado, I., Morán, M., and Losada, J. (1994) Organometallic silicon dendrimers, *J Chem Soc Chem Comm*, 2575-2576.
35. van der Made, A.W. and van Leeuwen, P.W.N.M. (1992) Silane dendrimers, *J Chem Soc Chem Comm*, 1400-1401.
36. Krska, S.W. and Seyferth, D. (1998) Synthesis of water-soluble carbosilane dendrimers, *J Am Chem Soc 120*, 3604-3612.
37. Gossage, R.A., Muñoz-Martínez, E., Frey, H., Burgath, A., Lutz, M., Spek, A.L., and van Koten, G. (1999) A novel phenol for use in convergent and divergent dendrimer synthesis: access to core fuctionalisable trifurcate carbosilane dendrimers-the x-ray crystal structure of [1,3,5-tris{4-(triallylsilyl)phenylester]benzene], *Chem-Eur J 5*, 2191-2197.
38. Huc, V., Boussaguet, P., and Mazerolles, P. (1996) Organogermanium dendrimers, *J Organomet Chem 521*, 253-260.

39. Balzani, V., Campagna, S., Denti, G., Juris, A., Serroni, S., and Venturi, M. (1998) Designing dendrimers based on transition-metal complexes. Light-harvesting properties and predetermined redox patterns, *Accounts Chem Res 31*, 26-34.
40. Wells, N.J., Basso, A., and Bradley, M. (1998) Solid-phase dendrimer synthesis, *Biopolymers 47*, 381-396.
41. Kim, R.M., Manna, M., Hutchins, S.M., Griffin, P.R., Yates, N.A., Bernick, A.M., and Chapman, K.T. (1996) Dendrimer-supported combinatorial chemistry, *P Natl Acad Sci USA 93*, 10012-10017.
42. Newkome, G.R., Childs, B.J., Rourk, M.J., Baker, G.R., and Moorefield, C.N. (1999) Dendrimer construction and macromolecular property modification via combinatorial methods, *Biotechnol Bioeng 6*, 243-253.
43. Maxwell, B.D., Fujiwara, H., Habibi-Goudarzi, S., Ortiz, J.P., and Logusch, S.J. (1998) Preparation of [^{14}C]G2.5 and [[^{14}C]G5.5 Starburst® PAMAM dendrimers: the first example of dendrimer radiosynthesis, *Comp Rad 41*, 935-939.
44. Zeng, F. and Zimmerman, S.C. (1996) Rapid synthesis of dendrimers by an orthogonal coupling strategy, *J Am Chem Soc 118*, 5326-5327.
45. Lapierre, J.-M., Skobridis, K., and Seebach, D. (1993) 173. Preparation of chiral building blocks for starburst dendrimer synthesis, *Helv Chim Acta 76*, 2419-2432.
46. Chechik, V., Zhao, M., and Crooks, R.M. (1999) Self-assembled inverted micelles prepared from a dendrimer template: phase transfer of encapsulated guests, *J Am Chem Soc 121*, 4910-4911.
47. Frey, H., Lach, C., and Lorenz, K. (1998) Heteroatom-based dendrimers, *Adv Mater 10*, 279-293.
48. Roovers, J. and Comanita, B. (1999) Dendrimers and dendrimer-polymer hybrids, *Adv Polym Sci 142*, 179-228.
49. Scherrenberg, R., Coussens, B., van Vliet, P., Edouard, G., Brackman, J., de Brabander, E., and Mortensen, K. (1998) The molecular characteristics of poly(propyleneimine) dendrimers as studied with small-angle neutron scattering, viscosimetry, and molecular dynamics, *Macromolecules 31*, 456-461.
50. Fréchet, J.M.J., Hawker, C.J., Gitsov, I., and Leon, J.W. (1996) Dendrimers and hyperbranched polymers: two families of three-dimensional macromolecules with similar but clearly distinct properties, *Pure Appl Chem A33*, 1399-1425.
51. Wooley, K.L., Fréchet, J.M.J., and Hawker, C.J. (1994) Influence of shape on the reactivity and properties of dendritic, hyperbranched and linear aromatic polyesters, *Polymer 35*, 4489-4495.
52. Newkome, G.R., Weis, C.D., Moorefield, C.N., and Weis, I. (1997) Detection and functionalisation of dendrimers possessing free carboxylic acid moieties, *Macromolecules 30*, 2300-2304.
53. Sahota, H.S., Lloyd, P.M., Yeates, S.G., Derrick, P.J., Taylor, P.C., and Haddleton, D.M. (1994) Characterisation of aromatic polyester dendrimers by matrix-assisted laser desorption ionisation mass spectrometry, *J Chem Soc Chem Comm*, 2445-2446.
54. Percec, V., Cho, W.-D., Mosier, P.E., Ungar, G., and Yeardley, D.J.P. (1998) Structural analysis of cylindrical and spherical supramolecular dendrimers quantifies the concept of monodendron shape control by generation number, *J Am Chem Soc 120*, 11061-11070.
55. Topp, A., Bauer, B.J., Klimash, J.W., Spindler, R., Tomalia, D.A., and Amis, E.J. (1999) Probing the location of the terminal groups of dendrimers in dilute solution, *Macromolecules 32*, 7226-7231.
56. Meltzer, A.D., Tirrell, D.A., Jones, A.A., Inglefield, P.T., Hedstrand, D.M., and Tomalia, D.A. (1992) Chain dynamics in poly(amido amine) dendrimers. A study of C13 NMR relaxation parameters, *Macromolecules 25*, 4541-4548.
57. Gorman, C.B., Hager, M.W., Parkhurst, B.L., and Smith, J.C. (1998) Use of a paramagnetic core to affect longitudinal nuclear relaxation in dendrimers-a tool for probing dendrimer conformation, *Macromolecules 31*, 815-822.
58. Hofkens, J.H., Verheijen, W., Shukla, R., Dehaen, W., and De Schryver, F.C. (1998) Detection of a single dendrimer macromolecule with a fluorescent dihydropyrrolopyrroledione (DPP) core embedded in a thin polystyrene polymer film, *Macromolecules 31*, 4493-4497.
59. Jockusch, S., Ramirez, J., Sanghvi, K., Nociti, R., Turro, N.J., and Tomalia, D.A. (1999) Comparison of nitrogen core and ethylenediamine core starburst dendrimers through photochemical and spectroscopic probes, *Macromolecules 32*, 4419-4423.
60. Pistolis, G., Malliaris, A., Paleos, C.M., and Tsiourvas, D. (1997) Study of poly(amidoamine) starburst dendrimers by fluorescence probing, *Langmuir 13*, 5870-5875.

61. Demathieu, C., Chehimi, M.M., Lipskier, J.-F., Caminade, A.-M., and Majoral, J.-P. (1999) Characterisation of dendrimers by X-ray photoelectron spectroscopy, *Appl Spectrosc 53*, 1277-1281.
62. Taddia, M., Lucano, C., and Juris, A. (1998) Analytical characterisation of supramolecular species-determination of ruthenium and osmium in dendrimers by electrothermal atomic absorption spectrometry, *Anal Chim Acta 375*, 285-292.
63. Hermann, B.A., Hubler, U., Jess, P., Lang, H.P., Guntherodt, H.-J., Greiveldinger, G., Rheiner, P.B., Murer, P., Sifferlen, T., and Seebach, D. (1999) Chiral dendrimers on a Pt(100) surface investigated by scanning tunnelling microscopy, *Surf Interface Anal 27*, 507-511.
64. Wu, C., Brechbiel, M.W., Kozak, R.W., and Gansow, O.A. (1994) Metal-chelate-dendrimer-antibody constructs for use in radioimmunotherapy and imaging, *Bioorg Med Chem Lett 4*, 449-454.
65. Kawase, M., Kurikawa, N., Higashiyama, S., Miura, N., Shiomi, T., Ozawa, C., Mizoguchi, T., and Yagi, K. (1999) Effectiveness of polyamidoamine dendrimers modified with tripeptide growth factor, glycly-L-histidyl-L-lysine, for enhancement of function of hepatoma cells, *J Biosci Bioeng 88*, 433-437.
66. Kobayashi, H., Wu, C., Kim, M., Paik, C.H., Carrasquillo, J.A., and Brechbiel, M.W. (1999) Evaluation of the in vivo biodistribution of indium-111 and yttrium-88 labelled dendrimer-1B4M-DTPA and its conjugation with anti-tac monoclonal antibody, *Bioconjugate Chem 10*, 103-111.
67. Malik, N., Evagorou, E.G., and Duncan, R. (1999) Dendrimer-platinate: a novel approach to cancer chemotherapy, *Anti-Cancer Drug 10*, 767-776.
68. Chen, C.Z., Tan, N.C.B., and Cooper, S.L. (1999) Incorporation of dimethyldodecylammonium chloride functionalities onto poly(propylene imine) dendrimers significantly enhances their antibacterial properties, *Chem Commun*, 1585-1586.
69. Liu, M., Kono, K., and Fréchet, J.M.J. (1999) Water-soluble dendrimer-poly(ethylene glycol) starlike conjugates as potential drug carriers, *J Polym Sci Pol Chem 37*, 3492-3503.
70. Leon, J.W., Kawa, M., and Fréchet, J.M.J. (1996) Isophthalate ester-terminated dendrimers: versatile nanoscopic building blocks with readily modifiable surface functionalities, *J Am Chem Soc 118*, 8847-8859.
71. Singh, P. (1998) Terminal groups in starburst dendrimers: activation and reactions with proteins, *Bioconjugate Chem 9*, 54-63.
72. Singh, P., Moll III, F., Lin, S.H., and Ferzli, C. (1996) Starburst dendrimers: a novel matrix for multifunctional reagents in immunoassays, *Clin Chem 42*, 1567-1569.
73. Wilbur, D.S., Pathare, P.M., Hamlin, D.K., Buhler, K.R., and Vessella, R.L. (1998) Biotin reagents for antibody pretargeting. 3. Synthesis, radioiodination, and evaluation of biotinylated starburst dendrimers, *Bioconjugate Chem 9*, 813-825.
74. Singh, P., Moll III, F., Lin, S.H., Ferzli, C., Yu, K.S., Koski, K., Saul, R.G., and Cronin, P. (1994) Starburst dendrimers: enhanced performance and flexibility for immunoassays, *Clin Chem 40*, 1845-1849.
75. Lee, Y.C. and Lee, R.T. (1995) Carbohydrate-protein interactions: basis of glycobiology, *Accounts Chem Res 28*, 321-327.
76. Meunier, S.J., Wu, Q., Wang, S.-N., and Roy, R. (1997) Synthesis of hyperbranched glycodendrimers incorporating alpha-thiosialosides based on a gallic acid core, *Can J Chem 75*, 1472-1482.
77. Zanini, D. and Roy, R. (1997) Synthesis of new alpha-thiosialodendrimers and their binding properties to the sialic acid specific lectin from *Limax flavus*, *J Am Chem Soc 119*, 2088-2095.
78. Reuter, J.D., Myc, A., Hayes, M.M., Gan, Z., Roy, R., Qin, D., Yin, R., Piehler, L.T., Esfand, R., Tomalia, D.A., and Baker, J.R., Jr. (1999) Inhibition of viral adhesion and infection by sialic-acid-conjugated dendritic polymers, *Bioconjugate Chem 10*, 271-278.
79. Ashton, P.R., Hounsell, E.F., Jayaraman, N., Nilsen, T.M., Spencer, N., Stoddart, J.F., and Young, M. (1998) Synthesis and biological evaluation of alpha-D-mannopyranoside-containing dendrimers, *J Org Chem 63*, 3429-3437.
80. Smith, D.K., Zingg, A., and Diederich, F. (1999) Dendroclefts: optically active dendritic receptors for the selective recognition and chiroptical sensing of monosaccharide guests, *Helv Chim Acta 82*, 1225-1241.

81. Andre, S., Ortega, P.J.C., Perez, M.A., Roy, R., and Gabius, H.-J. (1999) Lactose-containing starburst dendrimers: influence of dendrimer generation and binding-site orientation of receptors (plant/animal lectins and immunoglobulins) on binding properties, *Glycobiology 9*, 1253-1261.
82. Pagé, D. and Roy, R. (1997) Synthesis and biological properties of mannosylated starburst poly(amidoamine) dendrimers, *Bioconjugate Chem 8*, 714-723.
83. Thompson, J.P. and Schengrund, C.-L. (1997) Oligosaccharide-derivatised dendrimers: defined multivalent inhibitors of the adherence of the cholera toxin B subunit and the heat labile enterotoxin of E. coli to GM1, *Glycoconjugate J 14*, 837-845.
84. Thompson, J.P. and Schengrund, C.-L. (1998) Inhibition of the adherence of cholera toxin and the heat-labile enterotoxin of E. coli to cell-surface GM1 by oligosaccharide-derivatised dendrimers, *Biochem Pharmacol 56*, 591-597.
85. Zanini, D. and Roy, R. (1998) Practical synthesis of starburst PAMAM alpha-thiosialodendrimers for probing multivalent carbohydrate-lectin binding properties, *J Org Chem 63*, 3486-3491.
86. Peerlings, H.W.I., Nepogodiev, S.A., Stoddart, J.F., and Meijer, E.W. (1998) Synthesis of spacer-armed glycodendrimers based on the modification of poly(propylene imine) dendrimers, *Eur J Org Chem 9*, 1879-1886.
87. Langer, P., Ince, S.J., and Ley, S.V. (1998) Assembly of dendritic glycoclusters from monomeric mannose building blocks, *J Chem Soc Perkin T 1 23*, 3913-3915.
88. Pagé, D., Zanini, D., and Roy, R. (1996) Macromolecular recognition: effect of multivalency in the inhibition of yeast mannan to concanvalin A and pea lectins by mannosylated dendrimers, *Bioorgan Med Chem 4*, 1949-1961.
89. Zanini, D. and Roy, R. (1997) Chemoenzymatic synthesis and lectin binding properties of dendritic N-acetyllactosamine, *Bioconjugate Chem 8*, 187-192.
90. Zhang, L.S., Torgerson, T.R., Liu, X.Y., Timmons, S., Colosia, A.D., Hawiger, J., and Tam, J.P. (1998) Preparation of functionally active cell-permeable peptides by single-step ligation of two peptide modules, *P Natl Acad Sci USA 95*, 9184-9189.
91. Sakthivel, T., Toth, I., and Florence, A.T. (1998) Synthesis and physiochemical properties of lipophilic polyamide dendrimers, *Pharmaceut Res 15*, 776-782.
92. Tam, J.P. (1996) Recent advances in multiple antigen peptides, *J Immunol Methods 196*, 17-32.
93. Keah, H.H., Kecorius, E., and Hearn, M.T.W. (1998) Direct synthesis and characterisation of multi-dendritic peptides for use as immunogens, *J Pept Res 51*, 2-8.
94. Barth, R.F., Adams, D.M., Soloway, A.H., Alam, F., and Darby, M.V. (1994) Boronated starburst dendrimer-monoclonal antibody immunoconjugates: evaluation as a potential delivery system for neutron capture therapy, *Bioconjugate Chem 5*, 58-66.
95. Newkome, G.R., Moorefield, C.N., Keith, J.M., Baker, G.R., and Escamilla, G.H. (1994) Chemistry within a unimolecular micelle precursor: boron superclusters by site- and depth-specific transformations of dendrimers, *Angew Chem Int Edit 33*, 666-668.
96. Qualmann, B., Kessels, M.M., Klobasa, F., Jungblut, P.W., and Sierralta, W.D. (1996) Electron spectroscopic imaging of antigens by reaction with boronated antibodies, *J Microsc-Oxford 183*, 69-77.
97. Capala, J., Barth, R.F., Bendayan, M., Lauzon, M., Adams, D.M., Soloway, A.H., Fenstermaker, R.A., and Carlsson, J. (1996) Boronated epidermal growth factor as a potential targeting agent for boron neutron capture therapy of brain tumours, *Bioconjugate Chem 7*, 7-15.
98. Wiener, E.C., Brechbiel, M.W., Brothers, H., Magin, R.L., Gansow, O.A., Tomalia, D.A., and Lauterbur, P.C. (1994) Dendrimer-based metal chelates: a new class of magnetic resonance imaging contrast agents, *Magn. Reson. Med. 31*, 1-8.
99. Bulte, J.W.M., Wu, C., Brechbiel, M.W., Brooks, R.A., Vymazal, J., Holla, M., and Frank, J.A. (1998) Dysprosium-DOTA-PAMAM dendrimers as macromolecular T2 contrast agents: preparation and relaxometry, *Invest Radiol 33*, 841-845.
100. Lauffer, R.B. (1987) Paramagnetic metal complexes as water proton relaxation agents for NMR imaging: theory and design., *Chem Rev 87*, 901-927.
101. Toth, E., Pubanz, D., Vauthey, S., Helm, L., and Merbach, A.E. (1996) The role of water exchange in attaining maximum relaxivities for dendrimeric MRI contrast agents, *Chem-Eur J 2*, 1607-1615.

102. Bryant, L.H., Jr., Brechbiel, M.W., Wu, C., Bulte, J.W.M., Herynek, V., and Frank, J.A. (1999) Synthesis and relaxometry of high-generation (G=5, 7, 9, and 10) PAMAM dendrimer-DOTA-gadolinium chelates, *J Magn Reson Imaging 9*, 348-352.
103. Adam, G., Neuerburg, J., Spuntrup, E., Muhler, A., Scherer, K., and Gunther, R.W. (1994) Gd-DTPA-Cascade-Polymer: potential blood pool contrast agent for MR imaging, *J Magn Reson Imaging 4*, 462-466.
104. Schwickert, H.C., Roberts, T.P.L., Muhler, A., Stiskal, M., Demsar, F., and Brasch, R.C. (1995) Angiographic properties of Gd-DTPA-24-cascade-polymer--a new macromolecular MR contrast agent, *Eur J Radiol 20*, 144-150.
105. Adam, G., Muhler, A., Spuntrup, E., Neuerburg, J.M., Kilbinger, M., Bauer, H., Fucezi, L., Kupper, W., and Gunther, R.W. (1996) Differentiation of spontaneous canine breast tumours using dynamic magnetic resonance imaging with 24-gadolinium-DTPA-cascade-polymer, a new blood-pool agent-preliminary experience, *Invest Radiol 31*, 267-274.
106. Bourne, M.W., Margerun, L., Hylton, N., Campion, B., Lai, J.-J., Derugin, N., and Higgins, C.B. (1996) Evaluation of the effects of intravascular MR contrast media (gadolinium dendrimer) on 3D time of flight magnetic resonance angiography of the body, *J Magn Reson Imaging 6*, 305-310.
107. Margerum, L.D., Campion, B.K., Koo, M., Shargill, N., Lai, J.-J., Marumoto, A., and Sontum, P.C. (1997) Gadolinium(III) DO3A macrocycles and polyethylene glycol coupled to dendrimers-effect of molecular weight on physical and biological properties of macromolecular magnetic resonance imaging contrast agents, *J Alloy Compd 249*, 185-190.
108. Wiener, E.C., Konda, S., Shadron, A., Brechbiel, M., and Gansow, O. (1997) Targeting dendrimer-chelates to tumours and tumour cells expressing the high-affinity folate receptor, *Invest Radiol 32*, 748-754.
109. Demsar, F., Shames, D.M., Roberts, T.P.L., Stiskal, M., Roberts, H.C., and Brasch, R.C. (1998) Kinetics of MRI contrast agents with a size ranging between Gd-DTPA and albumin-Gd-DTPA: use of cascade -Gd-DTPA-24 polymer, *Electro Magnetobio 17*, 283-297.
110. Prévôté, D., Donnadieu, B., Moreno-Manas, M., Caminade, A.-M., and Majoral, J.-P. (1999) Grafting of tetraazamacrocycles on the surface of phosphorous-containing dendrimers, *Eur J Org Chem*, 1701-1708.
111. Bosman, A.W., Schenning, A.P.H.J., Janssen, R.A.J., and Meijer, E.W. (1997) Well-defined metallodendrimers by site-specific complexation, *Chem Ber-Rec 130*, 725-728.
112. Lange, P., Schier, A., and Schmidbaur, H. (1996) Dendrimer-based multinuclear gold(I) complexes, *Inorg Chem 35*, 637-642.
113. Zhao, M., Sun, L., and Crooks, R.M. (1998) Preparation of Cu nanoclusters within dendrimer templates, *J Am Chem Soc 120*, 4877-4878.
114. Balogh, L. and Tomalia, D.A. (1998) Poly(amidoamine) dendrimer-templated nanocomposites. 1. Synthesis of zerovalent copper nanoclusters, *J Am Chem Soc 120*, 7355-7356.
115. Zhao, M. and Crooks, R.M. (1999) Dendrimer-encapsulated Pt nanoparticles: synthesis, characterisation, and applications to catalysis, *Adv Mater 11*, 217-220.
116. Garcia, M.E., Baker, L.A., and Crooks, R.M. (1999) Preparation and characterisation of dendrimer-gold colloid nanocomposites, *Anal Chem 71*, 256-258.
117. Bulte, J.W.M., Douglas, T., Strable, E., Moskowitz, B.M., and Frank, J.A. (2000) Magnetodendrimers as a new class of cellular contrast agents, *Proc Intl Soc Mag Reson Med 8*, 2062.
118. Kukowska-Latallo, J.F., Bielinska, A.U., Johnson, J., Spindler, R., Tomalia, D.A., and Baker, J.R., Jr. (1996) Efficient transfer of genetic material into mammalian cells using Starburst polyamidoamine dendrimers, *P Natl Acad Sci USA 93*, 4897-4902.
119. Tang, M.X., Redemann, C.T., and Szoka, F.C., Jr. (1996) In vitro gene delivery by degraded polyamidoamine dendrimers, *Bioconj Chem 7*, 703-714.
120. Bielinska, A., Kukowska-Latallo, J.F., Johnson, J., Tomalia, D.A., and Baker, J.R., Jr. (1996) Regulation of in vitro gene expression using antisense oligonucleotides or antisense expression plasmids transfected using starburst PAMAM dendrimers, *Nucleic Acids Res 24*, 2176-2182.
121. Qin, L., Pahud, D.R., Ding, Y., Bielinska, A.U., Kukowska-Latallo, J.F., Baker, J.R., Jr, and Bromberg, J.S. (1998) Efficient transfer of genes into murine cardiac grafts by starburst polyamidoamine dendrimers, *Human Gene Therapy 9*, 553-560.

122. Hudde, T., Rayner, S.A., Comer, R.M., Weber, M., Isaacs, J.D., Waldmann, H., Larkin, D.F.P., and George, A.J.T. (1999) Activated polyamidoamine dendrimers, a non-viral vector for gene transfer to the corneal endothelium, *Gene Therapy 6*, 939-943.
123. Ottaviani, M.F., Sacchi, B., Turro, N.J., Chen, W., Jockusch, S., and Tomalia, D.A. (1999) An EPR study of the interactions between starburst dendrimers and polynucleotides, *Macromolecules 32*, 2275-2282.
124. Bielinska, A.U., Chen, C., Johnson, J., and Baker, J.R., Jr. (1999) DNA complexing with polyamidoamine dendrimers: implications for transfection, *Bioconjugate Chem 10*, 843-850.
125. Bielinska, A.U., Kukowska-Latallo, J.F., and Baker, J.R., Jr. (1997) The interaction of plasmid DNA with polyamidoamine dendrimers: mechanism of complex formation and analysis of alterations induced in nuclease sensitivity and transcriptional activity of the complexed DNA, *Biochim Biophys Acta 1353*, 180-190.
126. Dagani, R. (1996) Chemists explore potential of dendritic macromolecules as functional materials, *C&EN June 23*, 30-38.
127. Jansen, J.F.G.A., de Brabander-van den Berg, E.M.M., and Meijer, E.W. (1994) Encapsulation of guest molecules into a dendritic box, *Science 266*, 1226-1229.
128. Stanssens, D. (1999) Applications of apolar-modified Astramol poly(propyleneimine) dendrimers, *Workshop on Properties and Applications of Dendritic Polymers*, NIST.

RATIONAL DESIGN OF P450 ENZYMES FOR BIOTECHNOLOGY

SHEILA J. SADEGHI, GEORGIA E. TSOTSOU, MICHAEL FAIRHEAD, YERGALEM T. MEHARENNA AND GIANFRANCO GILARDI*
Department of Biochemistry, Imperial College of Science, Technology and Medicine, London SW7 2AY, UK

Abstract

Nanobiotechnology is a novel field where bio-molecules are assembled on devices for exploitation in bio-analytical applications. The increased understanding of the structure-function relationship of redox proteins and enzymes combined with the progress made in protein engineering, molecular spectroscopy and structural biology allows today the possibility of creating genetically engineered proteins/enzymes to be used in arrays for high-through-put screening.

This paper reports on the use of small and well characterised electron transfer proteins/enzymes, such as flavodoxin, cytochrome c_{553} and cytochrome P450 as modules to design and construct covalently linked, artificial electron transfer chains. Functional characterisation of these molecular wires will increase our understanding on the structure-function relationships in electron transfer systems. This approach has been named "molecular lego", and its application to cytochromes P450, an important class of enzymes responsible for the metabolism of a large number of drugs and xenobiotics, is particularly relevant to biotechnology. An efficient, artificial electron transfer chain was obtained by fusing the flavodoxin from *D. vulgaris* and the soluble haem domain of cytochrome P450 from *B. megaterium*. Moving to a higher level of complexity, the scaffold of this soluble enzyme was also used to insert the key structural and functional elements of the human cytochrome P450 2E1. The chimeric protein containing the fused bacterial and human domains was successfully engineered. Finally, a method designed to identify active P450 mutants to be used for the assembly of arrays with different activity/specificity is presented.

M. De Cuyper and J.W.M. Bulte (eds.), Physics and Chemistry Basis of Biotechnology, 71–104.

1. Introduction

Electron transfer (ET) is essential for life: not only it provides means for harnessing solar energy through photosynthesis but it is also of crucial importance for the generation of metabolic energy in living cells [1-3]. This process is extremely interesting from a technological point of view as many reactions of biotechnological interest such as degradation of pollutants and drug and food processing are based on redox systems. The biotechnological exploitation of this process together with the rapidly growing field of biosensors and bioelectronics leads to fascinating possibilities especially in the creation of amperometric biosensors.

However, the key factor for the successful exploitation of new and interesting redox enzymes in amperometric devices remains the efficient interaction with electrode surfaces. Much success has been achieved in the modification of electrode surfaces with various strategies aiming at rendering the electrode surface more compatible with the biological matrix, in its delicate balance of the folded and active state [4-7]. In some cases the rate of success is limited by the fact that the relevant redox centre is deeply buried in the biological matrix, that in this way can control the properties and reactivity of such centre, but with the detriment of the bio-electrochemical signal. A notable example of such case is the well-known and highly relevant family of cytochromes P450 [8-9].

Recent progress made in the area of protein engineering, molecular spectroscopy and structural biology allows today the possibility of rationally designing site-directed mutants of proteins and enzymes with properties tailored, for example, for the oriented immobilisation on electrode surfaces [10-12].

Despite the great amount of reactions carried out simultaneously by many enzymes on the numerous substrates in the cell, the living systems exhibit highly efficient redox chains, where electrons are tunnelled in specific directions to sustain life. This success is the result of the slow process of evolution that can today be speeded up and mimicked by the protein engineer towards targets for the benefit of bio-electrochemistry. In particular, novel protein engineering methods based on random mutagenesis and *in vivo* selection and/or *in vitro* screening allows the creation of combinatorial libraries of proteins and enzymes that can be evolved for specific biotechnological targets [13-14]. This approach to engineering proteins is complementary to the rational, site-directed approach (Figure 1), and it can extend the capabilities in the generation of new proteins, beyond their physiological functions.

This paper reports on the progress made on selected aspects of protein engineering applied to the amperometric biosensor area, with particular reference to P450 enzymes and its implications in the pharmaceutical and bioremediation areas of application.

1.1. INTERPROTEIN ELECTRON TRANSFER

The fundamental importance of protein ET in biology and the potential biotechnological implications that it underpins have attracted much research effort over the last decade. This has mostly been directed towards elucidating the nature of the interactions between hypothetical physiological ET partners demonstrating how inter-protein ET is the result of a number of events that occur in several stages [15-16]. The initial event leading to the reaction is the diffusional encounter of the protein pairs stirred by electrostatic and steric forces communicated through the solvent electrolyte medium; among these stirring parameters, long-range electrostatics are often the overruling forces [15]. The net charges on the proteins not only enhance the rate of collision by diffusion, but also cause their dipole moments [17-19] to line up in a specific orientation. The second event leading to inter-protein ET is the formation of a complex between the redox pairs, where a specific interaction is required to produce a configuration that actually results in rapid ET. The complex formation is expected to depend upon the nature of the protein surface topology and the distribution of the charged residues. A good fit between the surfaces of the two proteins allows their redox centres to be brought into close proximity under the control of short range forces such as hydrogen bonds, van der Waals forces, electrostatic and hydrophobic interactions [15]. The balance between these forces can vary markedly in different ET complexes. Once an ET competent complex has been formed, the ET process occurs, leading to the third stage of the reaction. This in turn depends on a variety of factors including the geometric disposition of donor and acceptor sites in the complex, the difference in free energy of the oxidised and reduced states, the activation energy, reorganisation energy and the electronic coupling between the two redox centres. These parameters and their influence over the kinetics and thermodynamics of this process have been described by Marcus and Sutin [20].

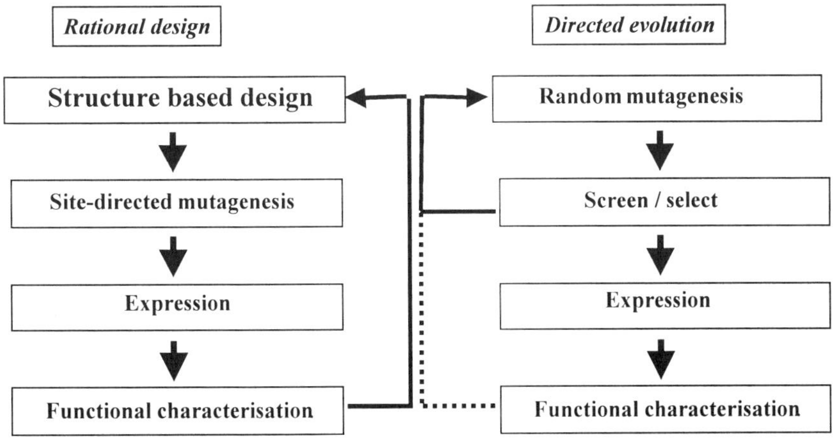

Figure 1. *Flow chart depicting the several experimental stages involved in rational design and directed evolution.*

When the rates of ET are very/extremely fast, other processes may become rate limiting. For instance, when the pre-complexes have to undergo dynamic reorientation to give the best ET complex, the reaction becomes gated. This kind of ET has been reported by several groups [21-24]. In other cases, the ET reaction could be coupled to a process which precedes it such as substrate induced conformational changes or redox potential modulations and examples of this type are also reported in literature [24-26].

In general, an important step in the ET between electron carriers is believed to be the formation of a specific complex. Progress in the structural analysis of protein ET complexes can provide valuable information about the complexes themselves plus the ET process which ensues. In the first instance, the interaction geometry of the two proteins within the complex can be ascertained, indicating the most relevant features of the proteins. The latter will have a bearing on the electrostatic coupling between the two centres which may depend simply on their separation distance or on structural features such as the presence of one or more networks of covalent, non-covalent or van der Waals interactions linking the two redox centres.

In light of this, a significant effort has been put on the study and characterisation of ET complexes such as photosynthetic reaction centre of *Rhodobacter sphaeroides*-cytochrome c_2 [27], the cytochrome *c* peroxidase-cytochrome *c* [28], methylamine dehydrogenase-amicyanin-cytochrome *c*-551$_i$ [29] and plastocyanin-cytochrome *f* [30]. It should be emphasised that in the case of the co-crystals, these may overlap with the highest affinity binding geometry, but may not necessarily reflect the complex competent in terms of ET. Indeed, kinetics and binding measurements in some cases suggest that the highest affinity binding complexes are not the fastest in terms of ET [31-33]. Another point to be addressed is that in the case of co-crystal structures, these represent static views of the interaction geometries between ET partners and give very little information on the role of dynamics in the ET processes when generally it is believed that these proteins form dynamic binary complexes.

However, in the absence of structures for ET protein complexes, a great potential is offered by the generation of hypothetical models in which protein dynamics can be included. Hypothetical structures for the complexes of *c*-type cytochromes-flavodoxin [34-35], cytochrome b_5-methaemoglobin [36], cytochrome P450-cytochrome b_5 [37], putidaredoxin-putidaredoxin reductase [38], putidaredoxin-cytochrome P450cam [39], flavocytochrome b_2-cytochrome *c* [40] and flavodoxin-cytochrome c_{553} [41] have been proposed. As a general rule, electrostatic interactions have been used in these models to facilitate the correct orientation of the redox centres, in a step prior to protein complex formation.

This wealth of knowledge now available on physiological systems has encouraged new studies on inter-protein ET between non-physiological partners, carried out in this laboratory and reviewed in the following sections. Understanding inter-protein ET and the parameters that regulate this phenomenon also provide an insight into the regulation process adopted by the living cell, in making sure that electrons are efficiently channelled in the correct redox chain. Understanding how this process is achieved bears implications in the construction of artificial systems for biotechnological applications.

1.2. STRUCTURE-FUNCTION OF CYTOCHROME P450 ENZYMES

1.2.1. P450 redox chains

Cytochromes P450 is a superfamily of enzymes containing a cysteinyl-co-ordinated protoporphyrin IX and acting as the terminal oxidases in the P450-dependent mono-oxygenase systems [42]. These systems are commonly found in all aerobic organisms varying from bacteria, fungi, to plants, insects and mammals. Apart from the P450 enzymatic component, the mono-oxygenase systems contain a NADPH-dependent P450-reductase, consisting of one or two protein components, which shuttles reducing equivalents from the external source (most commonly NAD(P)H) to the P450 component. Thus in eukaryotes the mitochondrial membrane-bound P450s require a two-protein electron-transfer system, comprising a NAD(P)H-dependent flavin-containing reductase and a second Fe/S-containing ferredoxin which shuttles the reducing equivalents from the flavoprotein to the P450. A very similar mono-oxygenase system, where the P450 moiety is soluble, is found in bacteria. Bacterial and mitochondrial P450s are classified as class I P450s. In contrast, the eukaryotic P450s anchored to the membranes of the endoplasmatic reticulum (class II) are dependent on a single reductase, which contains both FAD and FMN. Cytochromes P450 have generally broad and usually overlapping specificities. They generally show high enantio-selectivity not only during substrate recognition, but also in the position of attack and product formation. This feature of P450s has high commercial importance in terms both of chiral discrimination and asymmetric product formation. In the presence of oxygen and NAD(P)H, P450s typically catalyse the mono-oxygenation of their substrates, following the general reaction:

$$RH + O_2 + 2e^- + 2H^+ \rightarrow R\text{-}OH + H_2O$$

where RH is the substrate.

Apart from catalysing carbon hydroxylation or dehydrogenation reactions, P450s are known to attack heteroatoms. Nitrogen, sulphur and oxygen atoms are attacked by P450s with the formation of hydroxylation, oxidation and Cα-dealkylation products [43-44]. They are also known to catalyse olefin epoxidation and reductive dehalogenation. Less common reactions reported to be catalysed by P450s include dehydration, 1,2-aryl migration and dimerisation [44].

Based on the available 3D structures of P450s [8, 45-50] cytochromes P450 appear to have a conserved 3D fold. This is in spite of their low sequence identity and high diversity in terms of substrate specificity and catalysed reactions. The conservation of the 3D fold is very high in particular regions of the P450 structures, including the region involved in haem binding, while some variable regions also exist. The latter are attributed to play a role in binding of redox partners and substrate binding, explaining in this way the variability in the substrate specificity of P450s [50-51]. High similarity in the haem-binding region suggests a well-conserved mechanism of catalysis.

1.2.2. P450 catalysis

A common mechanism of catalysis known as the "P450 cycle" has been proposed long before the structure of any P450 was defined [52]. A detailed, recent description of the P450 cycle has been given by Halkier [42]; a simplified version of this cycle is given in Figure 2. Whereas for the initial stages of the cycle (stages A-D) there is structural information available, there are additional events in the transition from stage D to A which include transient radicals, and are not fully understood. When the enzyme is in its resting state the ferric iron is hexa-co-ordinated, with a water molecule as the distal ligand (A). Upon substrate binding the water molecule is displaced, with the iron becoming penta-co-ordinated (B). A concomitant shift in the spin state of the iron occurs from low to high, which is depicted in the absorbance spectrum of the cytochrome as a shift of the Soret peak from the 420 nm to the 390 nm. Electron reduction occurs first (C), followed by molecular oxygen binding to form the ferrous-oxy complex (D). The reduction of the ferric haem to the ferrous is facilitated by substrate binding [53-54]. Carbon monoxide is a common inhibitor of P450 activity. Its inhibitory action is due to competition with molecular oxygen to ligate to the reduced iron in P450s. Upon carbon monoxide binding to the reduced P450 (C') a diagnostic peak at 450 nm appears, to which the cytochrome P450 family owe their name. Second electron reduction leads to the active ferryl intermediate, through a sequence of events, including cleavage of molecular oxygen and reduction of an oxygen atom into water. The ferryl species is considered to attack the substrate, when insertion of the other oxygen atom into the substrate occurs. Dissociation of the product from the complex with the enzyme follows, leaving a water molecule as sixth ligand to the ferric iron and regenerating the resting state of the enzyme (A).

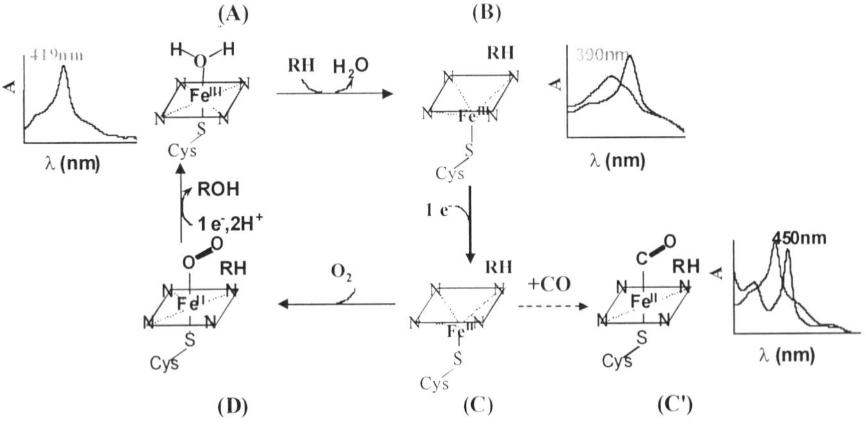

Figure 2. A simplified version of the catalytic cycle of cytochromes P450.

1.2.3. Bacterial P450s in biotechnology

From both a biochemical and a commercial point of view, cytochromes P450 are considered a class of enzymes of major importance due to their predominant role in

- the clearance of drugs and pollutants [55];
- chemical carcinogenesis [56];
- metabolism of arachidonic acid to metabolites with a role in cation/anion fluxes, mitogenesis, regulation of membrane-bound transporters and activation of intracellular signalling pathways [57]
- biotransformation including racemic mixture separation and chemical synthesis [55, 58-59].

The first aspect, the significance of the P450 superfamily in the degradation of drugs and environmental contaminants, is going to be addressed in more detail in this review.

CYP 102 (P450 BM3), a soluble P450 mono-oxygenase isolated from *Bacillus megaterium*, [60-61] has attracted our interest due to the fact that this P450 is not dependent on a redox partner for the transfer of reducing equivalents from NADPH, a feature shared only with P450foxy from *Fusarium oxysporum* [62]. P450 BM3 owes its catalytic self-sufficiency to the fact that it includes on a single polypeptide chain both an FMN/FAD-containing reductase and a P450 moiety. P450 BM3 catalyses, with high rates and a particularly low uncoupling, the hydroxylation of amphiphilic molecules with a long hydrocarbon chain. However, its specificity seems to be quite broad involving hydrophobic molecules, with a structure significantly different from the "natural" amphiphilic substrates [63-65]. Its catalytic sufficiency, the mainly hydrophobic pocket, together with its proven broad substrate specificity and the availability of an efficient recombinant expression system in *E. coli* [66] have also directed our choice towards this cytochrome as a model system for the generation of novel catalytically self-sufficient proteins for biosensing purposes.

Moreover, P450 BM3 shares a number of common features with the mammalian microsomal counterparts that justify its use as a surrogate for human P450s:

- the haem domain of P450 BM3 shows higher sequence similarity to the microsomal P450s compared to the bacterial ones,
- the corresponding functional domains of P450 BM3 and the microsomal mono-oxygenases have comparable size in terms of amino acid sequence length
- they contain the same cofactors. Based on these similarities, P450 BM3 has been widely used as a template in modelling studies of microsomal P450s [67-68]. When the first mammalian microsomal P450 structure of an engineered CYP 2C5 was resolved, there appeared to be a quite structurally conserved binding site and overall fold between 2C5 and BM3 [50], giving proven support to the use of P450 BM3 as a prototype for mammalian cytochromes. A significant degree of divergence was observed in the N-terminal regions, which however is expected since the N-terminal region is the one involved in the anchoring of the microsomal P450s to the membrane of the endoplasmatic reticulum.

Due to the increased concern of the threat to the human and the ecosystem deriving from the accumulation of various lipophilic substances, from industrial and agricultural practices, the potential of cytochromes P450 in the area of bioremediation and biosensing has recently started to be investigated. Thus the usage of cultured, human, rat and quail hepatic cells has been proposed for the *in vitro* screening and evaluation of the short-term toxicity of environmental contaminants on humans and other members of the ecosystem, with focus on pesticide toxicity [69]. A number of P450 mono-oxygenases in fungi were shown to be involved in the degradation of polycyclic aromatic hydrocarbons (PAHs) [70]. The mutant F87G of P450 BM3 was found to oxidise the PAHs pyrene and benzo[α]pyrene to phenolic derivatives [71]. A number of other successful projects to engineer cytochromes P450 for bioremediation purposes have been reviewed by Kellner and co-workers [9]. These included the introduction by different groups of specific mutations to P450cam (CYP 101) from *Pseudomonas putida,* using a rational approach to yield cytochromes with novel specificities against pollutants. Furthermore, the bacterial cytochromes P450cam and P450 BM3 were shown to be responsible for the attack of polyhaloethanes [64, 72], whereas other P450 enzymes were reported to catalyse the oxidation of haloaromatics [73]. The successful engineering of P450cam to a polychlorinated benzene-oxidising enzyme was also reported, with a further prospect of introducing the P450cam into *Pseudomonas* bacteria able to proceed to the degradation of the metabolites of the P450cam attack on polychlorinated benzenes [74]. The proven potential of cytochromes P450 in attacking environmental pollutants can be used for the development of microbial systems able to process the complete mineralisation of pollutants. Wackett [75] has engineered such system by introducing the toluene dioxygenase system to a *P. putida* strain expressing P450cam. The system was able to metabolise pentachloroethane to non-chlorinated end products after initial attack by the P450cam. Finally applications of P450s in engineering herbicide tolerance in plants and in using recombinant soil bacteria for safeguarding the environment from pesticide pollution, have also been demonstrated within the last years [76].

1.2.4. P450s in drug metabolism

Mammalian P450s are involved in many physiological processes such as biosynthesis of steroid hormones and bile acids, metabolism of both endogenous compounds such as fatty acids and hormones, and xenobiotics. The latter are hydrophobic compounds, whose elimination from the foreign organism is difficult due to their low solubility in water. Such exogenous compound include an array of natural products such as terpenes, alkaloids and plant toxins, as well as synthetic chemicals such as drugs and environmental pollutants. Mammalian P450s are expressed in many tissues but are relatively much more abundant in the liver, where xenobiotic metabolism occurs. The process of xenobiotic metabolism is initiated by the introduction of a polar functional group on the foreign compound, increasing its hydrophilicity. Cytochromes P450 are involved in such process. Completion of the solubilisation process follows by the conjugation of moieties such as glutathione, glucose or cysteine to the metabolite of the initial stage. However, in some cases, during the process of xenobiotic elimination, reactive intermediates are formed which have toxic or carcinogenic properties [77].

Cytochromes P450 have been identified as the main family of enzymes responsible for the metabolism of drugs in humans [78]. During the last years it has become clear that the prediction of clearance of drugs in humans by experimental results obtained using non-human laboratory animals has limitations due to species differences in P450 content [77]. In order to circumvent this problem, *in vitro* investigation of the metabolic fate of potential drugs has been introduced and is currently being applied by the pharmaceutical industry at an early stage of trials of potential therapeutic agents. This is because the metabolic pattern of a drug influences its toxicological and therapeutic action. Thus decreased clearance of a candidate drug, due to being poorly metabolised by the host organism, will result to increased bioavailability and possibly toxicity. Investigation of the toxicity of the reactive metabolites formed is also very important. Finally, studying drug-drug interactions is highly necessary to predict the effect of P450 inhibition due to co-administration of an additional drug in multiple drug therapies [77]. Early investigation of the metabolic pattern of a candidate drug is mainly dictated by the need to reject potential drugs not fulfilling the pharmaceutical standards at the very initial stages of the trials, prior to clinical studies, for obvious financial and time-saving reasons. Different experimental systems for *in vitro* investigation of drug metabolism have been widely used, ranging from enzyme expression systems, to human hepatocytes and liver slices, as well as cell-free systems, such as liver homogenates and microsomes [79-80]. Each of these systems presents certain disadvantages and advantages, however, enzyme expression systems seem to be more widely accepted by the scientific community, as mimicking better the *in vivo* drug metabolism pathways and drug-drug interactions.

Only a limited number of the P450s, present in the human liver, are involved in drug metabolism namely CYP 1A2, CYP 2C9, CYP 2C19, CYP 2D6 and CYP 3A4 [81-82]. However few others, including CYP 2E1, are reported to play a less significant role [83]. Due to the availability of the clone for the heterologous expression of CYP 2E1 in our laboratory, we focussed our studies on cytochrome CYP 2E1 [84]. CYP 2E1 is present in the liver and other tissues in many mammalian species. To date it has been shown to oxidise over 80 compounds including ethanol as well as carcinogens, environmental pollutants and drugs [85]. CYP 2E1 has been implicated to carcinogenesis, through the activation of procarcinogens, including potent tobacco-specific procarcinogens [86]. Finally the induction and stabilisation of CYP 2E1 by ethanol consumption and the enzyme's role in activating carcinogens is suspected to account for the increased carcinogenecity of certain compounds in alcoholics and this link has been shown in model systems for several compounds including bromobenzene, isoniazid and phenylbutazone [87].

1.3. CHIMERAS OF P450 ENZYMES

The production of catalytically self-sufficient P450 enzymes in bacterial hosts has a wide range of possible exciting applications. For instance cytochromes P450 metabolise a range of drugs and chemicals so bacteria containing these enzymes may help to further our understanding of human drug metabolism, the consequences of drug-drug interactions and the activation of carcinogens. The fusion protein hosts may also allow the mass production of compounds of interest by incorporating P450 biosynthetic pathways into one or a variety of hosts. Another application maybe the breakdown of environmental toxins, many of which are P450 substrates. The proteins may also further our understanding of the effect of P450 polymorphisms on human health by incorporating various allelic variants into fusion protein systems and studying their different activities.

1.3.1. Bacterial P450-P450-reductase fusion protein systems

So far only two naturally occurring catalytically self-sufficient bacterial P450s are known of the P450-P450-reductase fusion protein P450 BM3 from *Bacillus megaterium* and the fungal P450foxy from *Fusarium oxysporum.* De Montellano's group [88] have also produced an artificial P450 from P450cam and its two reductase components putidaredoxin (Pd, a Fe-S protein) and putidaredoxin reductase (PdR, a FAD protein). P450cam was the first bacterial P450 to be crystallised and subsequently is the most studied and well understood of all P450s. Four fusion proteins were made by the same investigators, consisting of various combinations of the P450 and the reductase components. They found the PdR-Pd-P450cam construct to have the highest activity and also to be well coupled to oxygen consumption. It also had comparable activity to the reconstituted system at low protein concentrations (<0.3 μM).

1.3.2. Plant P450-P450-reductase fusion proteins

The main focus of plant P450 fusion protein system research has been to engineer herbicide tolerance into plants [76] and to harness their incredible biosynthetic activities for human purposes.

Fusion proteins derived from plants have been used to study the effect of glyphosate on the CYP 71B1 from *Thlaspi arvensae* fused to the NADPH-cytochrome P450-reductase (CPR) domain of *Catharanthus roseus* [90]. They showed that glyphosate inhibited the turnover of the polycyclic aromatic hydrocarbon benzo(α)pyrene (another herbicide) by the enzyme. They argue that this observation has important implications for the development of CYP inhibitors which act as fungicides (CYP 51 inhibitors), herbicides (Plant CYP 51 inhibitors) and plant growth regulators (gibberelin biosynthesis).

The expression of plant cytochromes P450 in bacteria has also been done for purposes of trying to mass-produce medically important compounds that plants produce naturally but at very low levels. One step along this route is the isolation and expression of Tabersonine 16-hydroxylase (CYP T16H) [91]. The latter is the first enzyme in a pathway in *Catharanthus roseus* that produces the medically important bisindoles vinblastine and vincristine, both of which have causes in the treatment of leukaemia. The construction of an artificial pathway is prompted by the fact that very little of these bisindoles are naturally produced in the plant (0.0005%) and their complex structure renders chemical synthesis costly and difficult.

1.3.3. Plant/mammalian P450-P450-reductase fusion proteins

These fusion proteins offer the interesting possibility of engineering animal P450s into plants. This idea has been used for purposes of herbicide resistance by engineering Rat CYP 1A1:Yeast CPR fusion protein into tobacco plants [92]. The fusion protein was shown to provide the plants with resistance to the herbicide chlortoluron. Another group of investigators [93] have produced a protein consisting of Rat CYP 1A1 fused to maize ferredoxin (Fd, a Fe-S protein) and pea ferredoxin $NADP^+$ reductase (FNR, a FAD protein). These investigators, like the former ones, fused the P450 and reductase components in various orders but found their most efficient construct to be CYP 1A1-FNR-Fd. They found this fusion protein (expressed in yeast) to be active towards 7-ethoxycoumarin and the herbicide chlortoluron. The group also found that the activity of the protein was limited by the availability of free FAD in the reaction mixture, which suggests their protein is not fully coupled. This limiting factor may actually have the exciting possibility of linking the fusion protein activity to photosynthetic system, if expressed in chloroplasts.

The active form of Rat CYP 1A1 has already been expressed in the chloroplasts of tobacco plants [93]. The chloroplasts exhibit a P450 dependent mono-oxygenation activity when exposed to light. This suggests CYP 1A1 is coupled to some of the photosynthetic proteins. This maybe possible because Fd and FNR have potentials to form an ET chain similar to that found in the mitochondrial mono-oxygenase system which consists of Fd and NADPH-ferredoxin-oxidoreductase.

1.3.4. Mammalian fusion proteins

Several of these fusion proteins have been produced with the main aim of studying substrate metabolising properties of P450 enzymes [94]. One area where these fusion proteins have proved particularly useful is in the study of the effect of site-directed mutagenesis on P450 activity. Harlow and associates [95] substituted the Asp290 of CYP 2B11 from dog liver (fused to Rat CPR) with a variety of other amino acids to study the role of this amino acid in catalysis. They found that substitution with glutamate reduced androstenedione hydroxylation activity to 55% of the wild type enzyme, while substitution with a positive charge (Arg) gave an enzyme with virtually no measurable activity. The regio-selectivity of the enzyme was unaffected, suggesting the involvement of other key residues, but stereo-selectivity was altered.

The most important human drug metabolising enzyme CYP 3A4 has also been turned into a fusion protein [96]. The latter investigators fused the human CYP 3A4 to the CPR of Rat and used this protein to study the formation of metabolite-inhibitor complex during the metabolism of triacetyloleandomycin (TAO). This complex formation with TAO and other compounds such as the macrolide antibiotic erythromycin has important implications for drug-drug interactions during multiple drug therapies. The CYP 3A4 fusion protein therefore shows the value of fusion systems in predicting such negative interactions. Many other studies have also been carried out studying the effect of various compounds on fusion protein systems [97-98].

1.4. BIOSENSING

Electrical contacting of redox proteins and electrode surfaces is the fundamental pre-requisite for applications of redox active biomaterials in bioelectronic devices such as biosensors. A biosensor is an analytical device that uses the specificity of a biological element in sensing target molecules. It is composed of :

- a selector, which singles out the target molecule from the sample solution,

- a transducer which converts the signal generated by the selector into a signal fit for amplification,
- a detector, an electronic component designed to give a quantitative signal as depicted in Figure 3.

The coupling step between selector and transducer is usually the bottleneck in the engineering of a biosensor; in electrochemical sensors this step is represented by the electrical contact between the redox protein and the electrode surface.

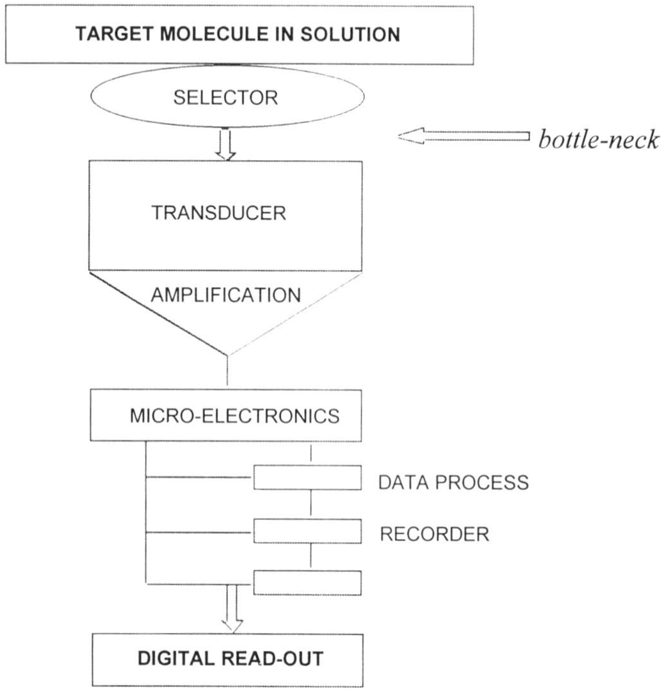

Figure 3. *Schematic representation of a biosensor.*

More than half of all sensors appearing in the literature can be classified as electrochemical sensors. These can be subdivided into either amperometric, potentiometric or conductometric sensors. In the amperometric category, an enzyme is typically coupled to an amperometric electrode and as the enzyme reacts with the substrate, a current is produced that is correlated to the analyte concentration. In this case, the bottleneck is the coupling between the enzyme and the electrode so that efficient conduction of the electrons is achieved. The second classification, potentiometric sensors, can be sub-divided into one group that utilises an immobilised enzyme on the surface of a glass pH electrode and a second type based on the production or consumption of protons by the enzyme as measured by the electrode.

However, the dynamic response of this type of sensors is usually slower than with the amperometric ones.

The tendency of proteins to adsorb strongly and often denature at surfaces is a well-known problem in the development of electrochemical sensors. Denaturation occurs because of a change in the balance of forces that normally favour the native folded conformation of proteins. Distortion may occur as a result of the large electric field that is generated across the double layer. In addition, part of the protein surface will be in contact with the electrode and the normal ionic and hydration shell may be broken. Since proteins differ greatly in their structure, there can be no general formula for minimising denaturation. However, ET proteins normally have relatively robust structures and therefore denaturation may not be a major factor in biosensor development. Nevertheless, biosensors using adsorbed enzymes or proteins are insensitive and, except for a few cases, this procedure alone is rarely used in biosensor construction.

One way of minimising the denaturation of proteins on electrodes has been by modification of the electrode surfaces. The realisation that suitably modified or functionalised electrode surfaces could interact in a specific and non-degradative manner with proteins came about in the late 70's. Eddowes and Hill [100] modified gold electrodes with 4,4'-bipyridyl to provide a suitable surface for interaction with cytochrome *c*.

A critical step in the development of biosensors is effective enzyme immobilisation, while maintaining free diffusion of substrates and products into and out of the enzyme layer. Various methods have been described for protein immobilisation including entrapment in a gel, cross-linking by a multifunctional reagent (for example glutaraldehyde) and covalent linkage [101]. New methods continue to be developed to deal with the ever-increasing demands of sophisticated bioanalytical chemistry. Combining protein engineering with the development of new coupling techniques capable of immobilising specific sites on the protein to the support surface may lead to new and more effective biosensors.

Over the years, many different solutions to the problem of electronic coupling of proteins and electrodes have been proposed and tested leading to three generations of biosensors, although not always based on direct ET. In the first generation, the transduction was based on the detection of suitable secondary metabolites. One such biosensor was developed for glucose by Clark and Lyons [102]. In the second generation of biosensors, contact between the electrode and the enzyme was made possible through the use of mediators. Diffusional electron mediators such as ferrocenes [103-105], ferricyanide [106-107], N,N'-bipyridinium salts [108] and quinone derivatives [109-110] were used as charge transporters that connect the active redox centre and the electrode interface. Commercial biosensors are based on either the first or second-generation biosensors. A very successful example of which is the commercial hand-held ferrocene-based enzyme electrode for glucose [103]. More recently, attempts have been made to couple the enzyme directly to the electrode leading to the third generation of biosensors, in which direct ET occurs in the absence of any mediators. These biosensors have superior selectivity and are less prone to interfering reactions. The most important feature of these systems is the possibility of modulating the desired properties of the biosensor using protein engineering on the one hand, together with novel interfacial technologies on the other.

2. Engineering artificial redox chains

For potential applications of cytochromes P450 in biosensors it is more desirable to replace the biological electron delivery and transport system by artificial ones like electrochemical [99] or photochemical systems [111-112]. Both methods have been applied to cytochrome P450 since the early years of P450 research. Several laboratories have used various methods to reduce cytochromes P450 electrochemically [113-117]. Although some electrochemical aspects of P450s were reported more than 20 years ago [118-119], the direct, non-promoted electrochemistry of P450 is rather difficult to obtain with unmodified electrodes. The enzyme does not interact with the electrode and is denatured.

The first direct electrochemistry in solution at the edge-plane graphite electrode was reported by Hill's group [113]. Rustling's group has found that P450cam incorporated in lipid or polyelectrolyte film displayed the well-defined redox behaviour from its haem Fe(II/III) [114]. More recently, Hill's group [115] demonstrated cyclic voltammograms on an edge-plane graphite electrode for various P450cam mutants.

The P450 enzyme of interest in the work carried out in this laboratory is P450 BM3 whose characteristics already covered in the introduction make it very interesting for biotechnological applications. However, P450 BM3 does not react with electrodes mainly due to its buried haem. The strategy adopted to tackle this problem makes use of an engineered, artificial redox chain, where electrons are conveyed to the catalytic unit *via* a protein known to interact with the electrode surfaces. This strategy plans to exploit the knowledge of biological ET for biotechnological purposes. In this strategy, a redox protein with well-characterised electrochemistry, flavodoxin, is used as a module to transfer electrons to the P450 unit (Figure 4).

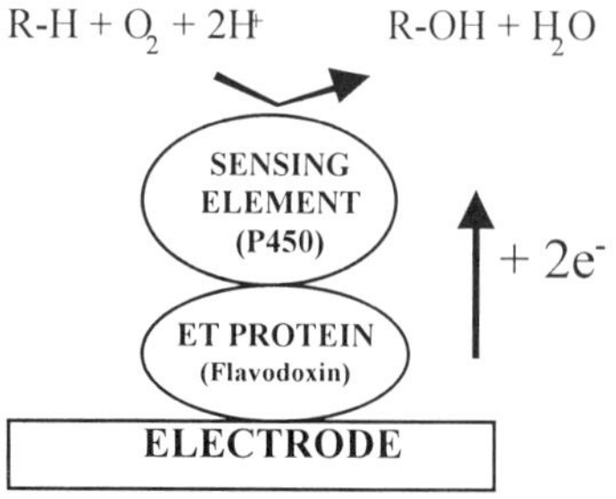

***Figure 4**. Schematic representation of the proposed artificial redox chain assembly.*

In order to establish the functionality of the chosen building blocks to be used for the covalent assembly of the artificial redox chains, the ET between the separate proteins was studied by stopped-flow spectrophotometry [120]. In the first instance the ability of flavodoxin from *Desulfovibrio vulgaris* (fld) to transfer electrons to simple cytochromes, cytochrome *c* from horse heart (hh*c*) and cytochrome c_{553} from *Desulfovibrio vulgaris* (c_{553}), was investigated. The knowledge gained from these systems was then applied to the more complex enzymatic system namely that of the haem domain of P450 BM3 from *Bacillus megaterium* (BMP) [121].

Flavodoxin (fld_q) was reduced anaerobically to its semiquinone form (fld_{sq}) in one syringe of the stopped-flow apparatus by the semiquinone radical of deazariboflavin ($dRfH^{\bullet}$) produced by photo-irradiation in the presence of EDTA (Figure 5).

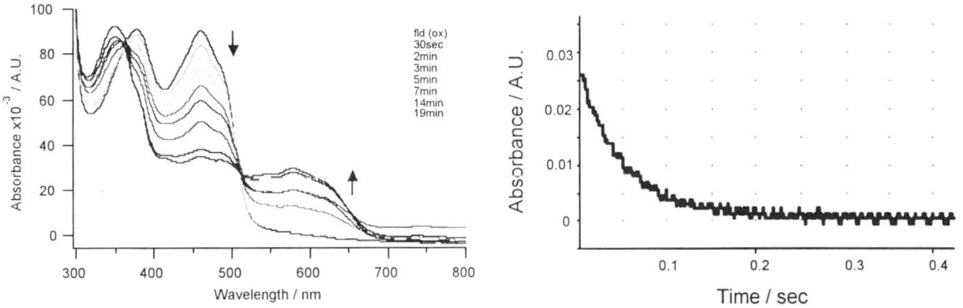

Figure 5. *Absorption spectra following the photoreduction of oxidised fld under anaerobic conditions to its semiquinione form (left) and reoxidation of the latter by oxidised cytochrome followed at 580 nm by stopped flow spectrophotometry (right).*

The reaction scheme studied is summarised in the following equations, where equation [3] applies to hh*c* and c_{553}, and equation [3]' applies to BMP:

$$dRf \xrightarrow[EDTA]{h\nu} dRfH^{\bullet} \qquad [1]$$

$$dRfH^{\bullet} + fld_q \rightarrow dRf + fld_{sq} \qquad [2]$$

$$fld_{sq} + (cyt\ c)_{ox} \rightleftharpoons [fld_{sq}\bullet(cytc)_{ox}] \rightarrow fld_q + (cytc)_{red} \qquad [3]$$

$$fld_{sq} + (BMP\text{-}S)_{ox} \rightleftharpoons [fld_{sq}\bullet(BMP\text{-}S)_{ox}] + CO \rightarrow fld_q + (BMP\text{-}S\text{-}CO)_{red} \qquad [3']$$

Under pseudo-first order and saturating conditions, the ET process of the fld/hh*c* redox pair showed two components with k_{lim} of 41.45±4.75 s^{-1} and 12.15±2.14 s^{-1} and K_{app} of 32±10 μM and 44±18 μM for the fast and slow processes, respectively. For the fld/c_{553}

and fld/BMP redox pairs a single component was found with a k_{lim} of 0.48±0.05 s^{-1} and 43.77±2.18 s^{-1} and a K_{app} of 21±6 μM and 1.23±0.32 μM, respectively [89, 120].

An important factor for achieving efficient ET is the formation of an ET competent complex between the redox pairs. The effect of the electrostatic forces in producing the complexes was studied by changing the ionic strength of the protein solutions. The second-order rate constants for the reaction of fld/hh*c* were two orders of magnitude higher (10^6 $M^{-1}s^{-1}$) than the same rates measured for fld/c_{553} (10^4 $M^{-1}s^{-1}$). Furthermore, these rates decreased monotonically with increasing ionic strength for the fld/hh*c* as expected for a reaction occurring between two molecules with opposite charges. The fld/c_{553} and fld/BMP redox couples showed a bell-shaped behaviour due to hydrophobic as well as electrostatic interactions. The effect of solution ionic strength on the electrostatic surface potentials of all three redox couples was also calculated as shown in Figure 6. As expected, the proteins with most charged surface residues, fld and hh*c*, are more affected by the changes in ionic strength than c_{553} and BMP.

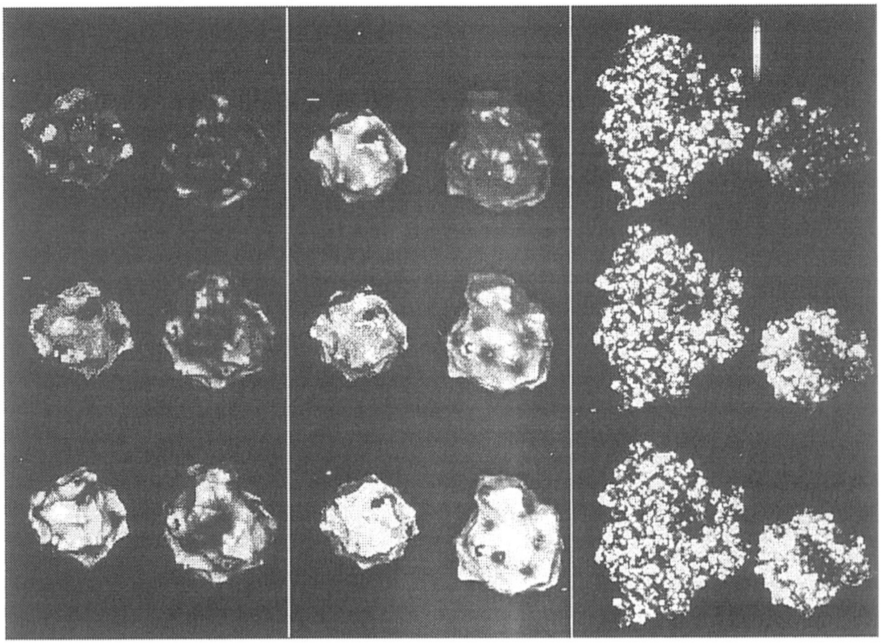

Figure 6. *Surface potentials calculated for the three ET pairs; fld/hhc (left), fld/c553 (centre) and fld/BMP (right) at three different solution ionic strengths, 1mM (top), 60 mM (centre) and 500 mM (right). These calculations were carried out using the program DelPhi with a probe radius of 10Å for hhc and c553 with 1.4Å for BMP.*

The ET data were analysed further using the parallel plate model developed by Tollin and co-workers [122]. This model takes into account the asymmetric distribution of charges on the surface of the protein and emphasises local electrostatic interactions between the charged moieties at the site of ET. The parameters obtained from fitting the kinetic data to this model are reported in Table 1.

Table 1*. Parameters calculated from the Parallel Plate Model.*

	fld/hhc		fld/c_{553}		fld/BMP
Parameter	Monopole-monopole	Monopole-dipole	Monopole-monopole	Monopole-dipole	Monopole-dipole
ρ (Å)	8	8	10	10	20
k_{∞}	8648 ($M^{-1}s^{-1}$)	9662 ($M^{-1}s^{-1}$)	1440 ($M^{-1}s^{-1}$)	1529 ($M^{-1}s^{-1}$)	25 (s^{-1})
v_{ii}	-15.55	-12.76	-9.90	-8.42	-58.16
v_{id}	---	-1.88	---	-1.82	-32.01
v_{dd}	---	-2.13	---	-0.31	-1.05

The availability of the 3D structures of these proteins allows the use of computational methods for generating a 3D model of the possible complexes. The structure of such models is important in this work for the rational design of the covalent redox chains described in the following sections. In the case of fld/hh*c* and fld/c_{553} docking simulations have been carried out [41], the results of which are generally in good agreement with the experimental data. The purported area of contact of the two proteins within both complexes show both electrostatic as well as hydrophobic interactions, with an average radius of 10 Å, well in agreement with the results obtained from the parallel plate model.

In the case of the fld/BMP complex, a model was generated by super-imposition of the 3D structure of fld on that of the truncated P450 BM3 [123]. The average radius of the contact area in this complex is 22 Å, which is considerably higher than that of the other two complexes and in good agreement with the value found from the parallel plate model. The distance between the redox centres in this complex is 18 Å, which is comparable with that found in the structure of the truncated P450 BM3 [123]. However, an alternative model is also possible, where the FMN region of fld is docked in the positively charged depression on the proximal BMP surface, around the haem ligand cysteine 400. This model brings the two cofactors at a closer distance of <12 Å. The two possible models may reflect the presence of dynamic events accompanying the formation and reorganisation of the ET competent complex that has also been postulated for the natural P450-reductase complex [50].

The models of the ET competent complexes described above were used to generate covalently linked complexes of both fld-c_{553} and fld-BMP using a flexible connecting loop by gene fusion. This method offers the advantage of keeping the two redox domains in a dynamic form. The fld-c_{553} gene fusion involves two non-physiological proteins from the same organism, *D. vulgaris*, but expressed in different cellular compartments, fld in the cytoplasm and c_{553} in the periplasm. A fld-c_{553} fusion was constructed at the DNA level, by linking the fld gene to that of c_{553} *via* a DNA sequence codifying for a flexible seven amino acid linker (GPGPGPG). The length of this peptide linker was determined by molecular modelling experiments in which the two proteins were docked and a loop was modelled to join the C-terminus of fld to the N-terminus of c_{553}, as shown in Figure 7. The resulting fusion gene expressed the chimeric protein of the correct molecular weight (25 kDa). Although the optical absorption spectrum of the partially purified chimeric protein showed the characteristic absorbencies for the oxidised haem (410 nm) and FMN (458 nm) cofactors, but the relative ratio of these cofactors was not the expected 1:1 possibly due to the expression of the fld-c_{553} protein in the cytoplasm of *E. coli* (the correct haem incorporation in c_{553} was obtained in the periplasm).

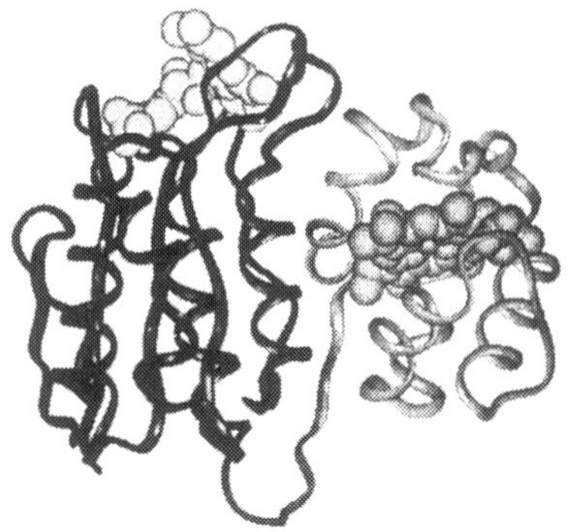

***Figure** 7. Modelled structure of the fld-*c553 *fusion protein.*

The fusion of the fld-BMP system was carried out at DNA level by linking the BMP gene (residues 1-470) with that of fld (residues 1-148) through the natural loop of the reductase domain of P450 BM3 (residues 471-479). This gene fusion was achieved by ligation of the relevant DNA sequences with engineered restriction sites. A possible model of this fusion protein is shown in Figure 8.

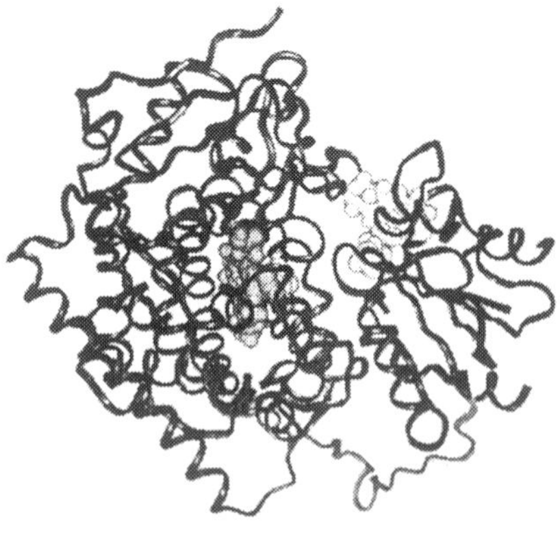

***Figure** 8. Modelled structure of the fld-BMP fusion protein.*

The fusion gene was correctly expressed in a single polypeptide chain. The absorption spectra of the purified chimeric protein indicated the incorporation of 1:1 haem and FMN. Moreover, the reduced protein was able not only to form the carbon monoxide adduct with the characteristic absorbance at 450 nm, but also to bind substrate (arachidonate) displaying the expected low- to high-spin transition from 419 nm to

397 nm, indicating that this covalent complex is indeed a functional P450. The integrity of the secondary structure of the fld-BMP fusion protein was confirmed by CD spectroscopy, with a ~2% increase in the α-helix content when compared to the BMP, probably due to the addition of the engineered loop. The spectroscopic data show that the fusion protein is indeed expressed as a soluble, folded and functional protein [121].

The presence of intra-molecular ET from the domain containing the FMN to the domain containing the haem, in the presence of substrate, was studied under steady-state conditions. The flavin domain was photo-reduced by deazariboflavin in the presence of EDTA under anaerobic conditions. The subsequent ET from the flavin domain to the haem was followed by the shift of the haem absorbance from 397 nm to 450 nm in carbon monoxide saturated atmosphere.

The kinetics of the intra-molecular ET within the fld-BMP fusion protein was studied by transient absorption spectroscopy. In the experimental set up, the FMN-to-haem ET was followed by the decrease in absorbance at 580 nm of the fld_{sq}. The ET rate measured was found to be 370 s^{-1}. This value is comparable to that measured for the intra-protein ET from FMN to haem domain of truncated P450 BM3 (250 s^{-1}) in which the FAD domain was removed [124]. These results are extremely encouraging because they demonstrate the functionality of the fld-BMP fusion protein to be equivalent to the physiological protein.

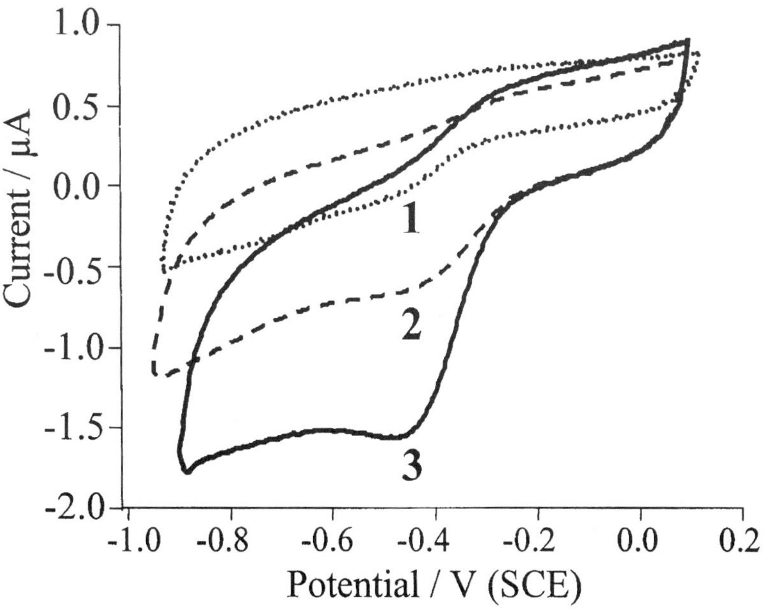

***Figure 9**. Figure 9. Cyclic voltammograms of BMP and fld-BMP fusion protein in the absence/presence of neomycin on a glassy carbon electrode; BMP (1), fld-BMP (2) and fld-BMP + neomycin (3).).*

Preliminary electrochemical experiments of the fld-BMP fusion protein, free in solution were carried out using a glassy carbon electrode [125]. The cyclic voltammograms of both the fld-BMP fusion protein and BMP are shown in Figure 9. While no current was observed for P450 BM3 enzyme on the bare glassy carbon electrode, both BMP and fld-BMP show measurable redox activities. Furthermore, the fld-BMP fusion protein interacts better with the electrode as measured by the larger current compared to that of BMP. This current is further enhanced in the presence of neomycin, a positively charged aminoglycoside which is believed to overcome the electrostatic repulsion between the negatively charged fld and the negatively charged electrode surface [126].

3. Screening methods for P450 activity

After the crucial role of P450s in the determination of the pharmacological/ toxicological properties of drugs was known, the need to develop efficient systems for the *in vitro* screening of drug-P450 interactions emerged. Thus in the recent years a great amount of effort has been put into the generation of model systems for studying *in vitro* the metabolism of pharmaceutical drugs in humans. Such model systems are aiming at obtaining valid information for the metabolism of the drug candidates *in vivo*. Additionally, it has been widely recognised that inter-individual differences in P450 content are capable of causing significant variation in pharmacokinetics, thus leading to inaccurate estimation of the toxicological and pharmaceutical action of drug candidates. As a result, the future of pharmacokinetics seems to be getting directed towards individualised drug treatment [127], where the drug therapy to be followed will be based on the genotype of the specific patient.

On the other hand the recent development of efficient non-rational protein engineering techniques [13-14] generating large combinatorial protein libraries has provided us with the capability to evolve protein function towards directions of choice. Strategies for screening protein libraries have been reviewed by Zhao and Arnold [128]. We tried to address both issues in our laboratory by developing a general assay for screening for turnover of compounds of interest by NAD(P)H-dependent oxidoreductases.

3.1. ASSAY METHODS FOR P450-LINKED ACTIVITY

A rapid investigation of the *in vitro* drug metabolism has been facilitated in the recent years by the development of high-throughput screening assays together with the advances in the creation of automated miniaturised systems for liquid handling and detection. The screens available can be categorised into

- Assays aiming at the detection of P450 inhibitors using non-specific P450 substrates which are turned over into detectable metabolites (for example fluorescent or radiolabelled) by the P450. Inhibition by a compound can thus be detected by reduction (or abolishment) of the production of the detectable metabolites [129-133].
- Assays, using an instrumental chemical analysis method, able to distinguish between the parent compound and its metabolites after turnover by the P450 [134-136].

The last category of assays provides information on the rate of metabolism and the metabolic stability of the candidate drug, and can be very important for the early prediction of drug clearance. It can also be adapted to study drug-drug interactions. *In vitro* drug metabolism screening systems have been reviewed recently by Eddersaw and Dickins [135].

Apart from the assays mentioned above for screening human P450-drug interactions, a large number of assays have been proposed in the literature for screening substrate turnover by cytochromes P450 [137-139], including two assays for P450 BM3 [140-141]. Most of these screening systems are dependent on the direct or indirect detection of the product of the turnover of certain substrates by the P450 enzyme. Although such methods can be adapted for high throughput assays, their application is limited to specific P450 enzymes with particular substrates, or they could be used for activity screening of a specific cytochrome rather than for identification of substrates within a random pool of molecules. The method proposed by Sligar's group [140], presents the advantage that it can be broadly applied to any P450 enzyme. However important limitations of the method are that it is not sufficiently sensitive to allow screening pools of mutants and that it can only be applied on cell lysates, as opposed to whole cells.

3.2. DEVELOPMENT OF A NEW HIGH-THROUGH-PUT SCREENING METHOD FOR NAD(P)H LINKED ACTIVITY

Recent developments in this laboratory have led to the development of a new high-through-put screening method for NAD(P)H-linked oxidoreductase activity. This method is applicable to any enzymatic activity that uses NAD(P)H cofactors, including the catalytically self-sufficient class II, cytochrome P450 BM3. The method has been adapted to a 96-well microtiter plate-format for screening large numbers of molecules in microlitre quantities. It is based on the spectrophotometric detection of $NA(D)P^+$ produced when a molecule of interest is being turned over by whole-cells expressing a P450 enzyme. The standard assay based on the monitoring of NADPH consumption by following the decrease of the NADPH absorbance at 340 nm could not be applied for detecting P450 activity in whole cells for the following reasons:

- light scattering by cells is interfering with the peak at 340 nm, and
- such assay would require a plate-reader able to follow enzyme kinetics. As mentioned in the introduction, assays based on expressed human P450s have actually been presented as more valid systems for the *in vitro* investigation of metabolic pathways in humans, since these systems can be adapted to take into account the genetic polymorphism, responsible for inter-individual differences in metabolic profiles.

A schematic representation of the assay can be seen in Figure 10. The principal of the assay lies on the fact that the reduced and oxidised forms of NAD(P)H have a different sensitivity to destruction in extreme pH values [142-143]. Thus by changing the pH of the reaction mixture, containing the cytochrome, a substrate, the remaining NAD(P)H and the $NAD(P)^+$ generated during catalysis, from the lower extreme of the pH scale to its other extreme, the amount of oxidised $NAD(P)^+$ can be selectively quantified. In the experimental procedure, cells expressing the cytochrome of interest are aliquoted into the wells of a microtiter plate and incubated with the pool of compounds to be screened for turnover by the enzyme. The reaction is initiated by addition of NADPH. At a certain time point after the start of the reaction, the produced $NADP^+$ is quantified: The pH is initially lowered to at least below 2.5 and subsequently risen to above 14.5. Following this procedure, $NADP^+$ is specifically quantified by measuring the absorbance of the formed product at 360 nm. To ensure that the generated $NADP^+$ is due to coupled substrate oxidation, the possible uncoupling of reducing equivalents to the formation of hydrogen peroxide is investigated. This is measured by quantification of hydrogen peroxide using the HRP-ABTS assay [144].

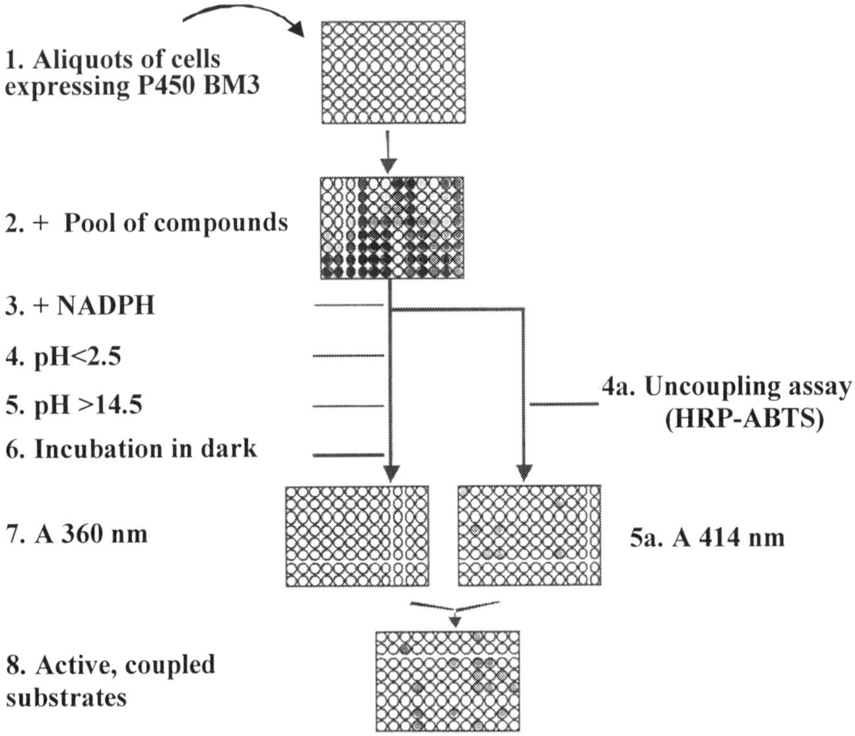

Figure 10. Schematic representation of the screening protocol for identifying novel substrates of cytochromes P450 within a random group of compounds.

Applying the assay to four known substrates of P450 BM3, the fatty acids arachidonic and lauric, the solvent 1,1,2,2-tetrachloroethane and the anionic surfactant sodium dodecyl sulphate, a 360 nm signal of only 3% was given by cells in the absence of substrate, compared to cells in the presence of substrate, under the optimum experimental conditions. Control experiments with cells non-transformed, or transformed with the same vector expressing other redox enzymes not able to turn over the substrates of interest in the presence of NADPH, were also run. They all demonstrated that the assay is specifically detecting P450 BM3 oxidising activity in whole-cells.

The proposed method presents the advantage that it can be generalised for the screening of substrates or inhibitors of any P450 enzyme, or any oxidoreductase in general, utilising NADPH or NADH (both have same behaviour in extreme basic/acidic conditions) as donor or acceptor of reducing equivalents. At the same time it can be applied for the screening of variants within a library of random mutants of a NADPH-dependent reductase for the specificity of interest.

In the work carried out in this laboratory, the viability of this new assay has been put to the test by screening

- a series of furazan derivatives, potential pharmaceuticals, against the wild type P450 BM3,
- a series of random mutants of P450 BM3 against target pollutants,
- an engineered, catalytically self-sufficient, chimeric bacterial-human P450 enzyme against some known substrates of both proteins.

3.3. VALIDITY OF THE NEW SCREENING METHOD

The validity of the assay has been demonstrated by investigating the interaction of the heterologously expressed in *E. coli* bacterial cytochrome P450 BM3 with a group of 1,2,5-oxadiazole 2-oxide (furoxan) derivatives, potential pharmaceuticals for the treatment of cardiovascular diseases, and their 1,2,5-oxadiazole (furazan) analogues (Figure 11) [145].

I.

Y	X	
$C_6H_5SO_2$	C_6H_5	+
$C_6H_5SO_2$	$SO_2C_6H_5$	−
$C_6H_5SO_2$	$S(CH_2)_2N(CH_3)_2$	+
H_5C_6	C_6H_5	+

II.

Y	X	
$C_6H_5SO_2$	C_6H_5	−
$C_6H_5SO_2$	$SO_2C_6H_5$	−

III.

Y	X	
$C_6H_5SO_2$	C_6H_5	+
$OHCH_2$	$CONH_2$	+
H_5C_6	C_6H_5	+

IV. +

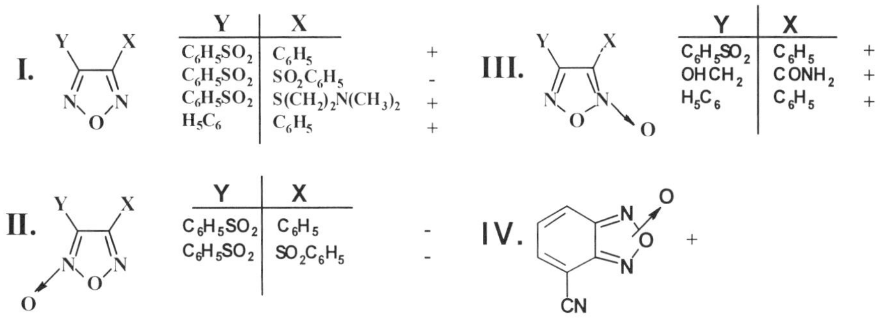

Figure 11. Structures of the ten furazan derivatives screened for turnover by wild-type cytochrome P450 BM3. The derivatives marked with + were found to cause increased consumption of NADPH, relative to the background, in contrast to analogues indicated by -, which showed NADPH-oxidising activity at background levels.

Within a group of ten analogues screened for interaction with the wild type P450 BM3, seven were identified as positives. The results from the assay in whole-cells were confirmed for all analogues, by following NADPH consumption by the purified enzyme in the presence of the same molecules on the spectrophotometer. In all cases, hydrogen peroxide formation due to uncoupling was minimal.

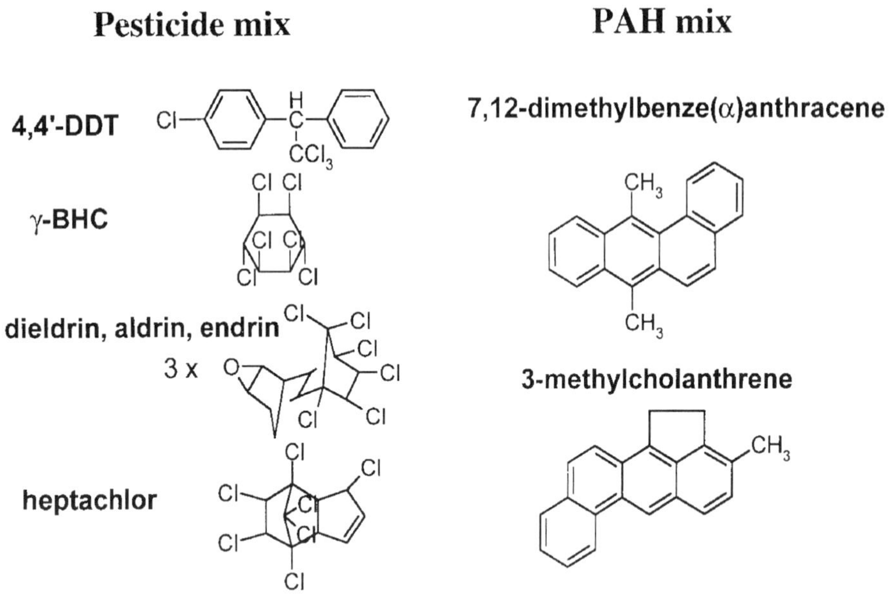

Figure 12. Structure of the pesticides and PAHs screened for turnover by random mutants of P450 BM3.

Similarly, the same assay was applied to investigate the interaction of various compounds, including a number of polycyclic aromatic hydrocarbons (PAHs) and pesticides (Figure 12), against a library of P450 BM3 variants, with random mutations in the haem domain region. Two variants have been identified among about 320 screened, showing a significantly different pattern of turnover of certain of the compounds compared to the wild type. Characterisation of the two active mutants is in progress.

4. Designing a human/bacterial 2E1-BM3 P450 enzyme

P450 2E1 is a microsomal P450 present in the liver and other tissues of many mammalian species that has been shown to catalyse the oxidation of over 80 compounds, including benzene, ethanol, acetone, chloroform, many nitrogenous compounds together with drugs such as acetaminophen and chlorzoxazone. Having as substrates ethanol and many suspect carcinogens, P450 2E1 has been considered of great interest for its possible relevance to alcoholism, chemical carcinogenesis and other diseases [84]. Its substrates have diverse structures but most of them have the common characteristic of being low molecular weight molecules [146].

In the absence of P450 2E1 crystal structure its active site conformation is not known. However, studies on its different substrates and competitive inhibitors, have provided useful information on some of the properties of its active site including:

- it binds and efficiently oxidises small molecules,
- accommodates water-soluble molecules although hydrophobicity is an important feature of the substrate-binding pocket,
- does not effectively accommodate molecules with a formal ionic charge [146].

This section reports on the modelling, construction and expression of a chimeric protein designed to confer the human P450 2E1 new and improved properties, like solubility and self-sufficiency in catalysis, hence conferring the ability of receiving electrons directly from small electron donors (NADPH), maintaining, at the same time, its quite peculiar substrate specificity. Self-sufficient and soluble cytochrome P450 systems are, in fact, essential to employ the catalytic power of these enzymes for biotechnological purposes.

The chimeric protein was obtained by fusing part of the human P450 2E1 with a portion of P450 BM3. This latter cytochrome exhibits key features, as mentioned earlier, that make it an ideal candidate for fusion with the human P450 2E1:

- it is catalytically self-sufficient due to the presence of a reductase domain within the same polypeptide chain, containing both FMN and FAD and requiring only NADPH for activity,
- it is highly homologous to the human P450 enzymes (30% homology between the P450 BM3 haem domain and 2E1), and it shares many common features leading to the classification in the same class II as the human microsomal P450 enzymes.

Furthermore, the three-dimensional structure of P450 BM3 has been solved [46, 147], its gene has been cloned [66] and its catalytic properties have been studied.

4.1. MODELLING

Previous available models of mammalian P450s were often approximate since they were based either solely on the structure of P450cam, that has a very low sequence identity with mammalian enzymes (from ca 15 to 20%) [148], or on models of the binding site which is expected to be the most variable region of the protein. Only recently, improved models were obtained after observing the necessity to generate a multiple sequence alignment of the target P450 with template proteins. These alignments should be done by looking at the structurally conserved regions of the templates and not simply relying on automated alignment procedures, since these are often not reliable, especially for some regions of the protein [148].

To help the design of a new valid chimera likely to possess the desired properties, a three-dimensional model of P450 2E1 was built in this laboratory. The P450 2E1 model was generated using a Silicon Graphics Indigo2 IRIX 6.2 workstation equipped with the Biosym/MSI software Insight. The following protocol was used for the designing of the model:

- four related proteins of known X-ray structure were chosen (P450terp, P450cam, P450eryF and the BMP domain of P450 BM3) and their sequences were aligned with P450 2E1 using the structurally conserved regions (SCRs),
- the co-ordinates of the structurally variable regions (VRs) were assigned using different templates for different VRs, depending on the degree of homology (when the VRs were of different length the co-ordinates were assigned by a loop search),
- a possible initial conformation of the side chains was searched and the co-ordinates of the more flexible N- and C- terminal regions were arbitrarily assigned,
- incorrect steric contacts (*bumps*) were corrected by manually orienting the involved rotamers, and finally
- the model was refined and energy minimised. A model of the P450 2E1-BM3 was obtained using the same procedure.

4.2. CONSTRUCTION

The information gained from the preliminary model of CYP 2E1 and previous works on isozymes [88, 149-152], was used to design a chimeric cytochrome P450, 2E1-BM3. This chimeric P450 contained the first 54 residues at the N-terminal of P450 BM3 (fragment I), the whole sequence of P450 2E1 from residue 81 to the C-terminal (fragment II) and the whole reductase domain of P450 BM3 (fragment III). The three fragments were successfully isolated from the respective cloned genes and compatible restriction sites were inserted, by PCR, at their extremities using different mutagenic oligos. The whole construct of 3150 base pairs, containing the three fragments together, was cloned back into the pT7Bm3HdZ vector [66] for the inducible expression in *E. coli.*

4.3. EXPRESSION AND FUNCTIONALITY

The 2E1-BM3 chimera with a molecular weight of 118 kDa was efficiently expressed. When reduced by sodium dithionite in a carbon monoxide saturated atmosphere the characteristic 450 nm peak was observed, giving support to a folded and functional chimeric protein.

To test the activity of the 2E1-BM3 chimeric enzyme in whole cell lysates the screening assay described earlier was used. The oxidation of NADPH by P450 2E1-BM3 in the presence of two known P450 2E1 substrates (ethanol and arachidonate) and a known 2E1 inhibitor (isoniazid) was investigated. Lysates of cells expressing P450 BM3 and non-transformed cells were used as controls. The results are shown in Figure 13, where absorbance ratio is the ratio of the $NADP^+$-alkali product given by cells expressing the P450 2E1-BM3 chimera or the control P450 BM3, against that given by non-transformed cells.

Overall these results show how the newly engineered bacterial/human P450 enzyme is active. Moreover the power of the screening method developed in this laboratory is demonstrated in its ability to identify positive, active enzymes in whole cells.

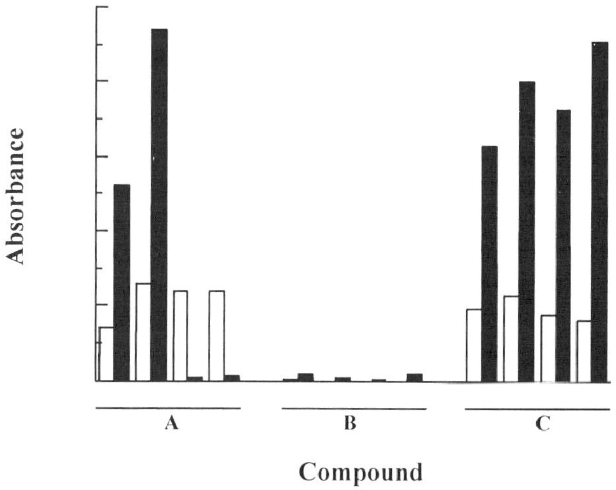

Figure 13. *Screening of P450 2E1-BM3 (white) using ethanol (A), isoniazid (B) and arachidonic acid (C) at increasing concentrations from left to right. Cytochrome P450 BM3 was used as a control (black).*

5. Conclusions

This work has shown how protein engineering methodologies can provide many useful extensions in the use of defined molecules for biosensing purposes. The experimentalist is no longer limited by the properties of the natural proteins/enzymes.

Electrochemical contact with the electrode was enhanced in the artificial chimera between flavodoxin and the haem domain of P450 BM3 (fld-BMP). Moreover, the availability of an assay able to screen for

- NAD(P)H-linked enzymatic activity towards molecules of pharmacological (new potential drugs) and biotechnological (bioremediation and biosensing) interest,
- libraries of random mutants of NAD(P)H–dependent enzymes with desired catalytic specificities,

opens many new possibilities in the engineering of novel P450 enzymes. A successful application of the assay as a method for identifying positive, active enzymes was demonstrated for the P450 2E1-BM3 chimera.

The identification of a number of active variants of P450 BM3 against several pollutants which will be used for the creation of chimeric P450 arrays with desired specificities, will follow. Finally, this approach could be seen as a step forward towards "fourth generation" of biosensors which will be based on artificial redox proteins with predicted ET pathways.

Acknowledgements

The authors wish to thank the following funding bodies for their support; EC biotechnology programme CT960413 (S.J.S.), the National Scholarship Foundation of Greece (G.E.T.) and BBSRC studentship 99/B1/E/05953 (M.F.).

References

1. Page, C. C., Moser, C. C., Chen, X., and Dutton, P. L. (1999) Natural engineering principles of electron tunnelling in biological oxidation-reduction. *Nature*, **402,** 47-51.
2. Beratan, D. N., and Onuchic, J. N. (1996) The protein bridge between redox centres in *Protein electron transfer* (Bendall, D. S., ed), pp. 23-42, BIOS scientific publishers Ltd., Oxford.
3. Canters, G. W., and Van de Kamp, M. (1992) Protein-mediated electron transfer. *Curr. Opin. Struc. Biol.*, **2,** 859-869.
4. Hill, H. A. O. (1996) The development of bioelectrochemistry. *Coord. Chem. Rev.*, **151,** 115-123.
5. Willner, I., Katz, E., and Willner, B. (1997) Electrical contact of redox enzyme layers associated with electrodes: routes to amperometric biosensors. *Electroanalysis*, **9**(13), 965-977.
6. Anne, A., Blanc, B., Moiroux, J., and Saveant, J. M. (1998) Facile derivatisation of glassy carbon surfaces by N-hydroxysuccinimide esters in view of attaching biomolecules. *Langmuir*, **14**(9), 2368-2371.
7. Heering, H. A., Hirst, J., and Armstrong, F. A. (1998) Interpreting the catalytic voltammetry of electroactive enzymes adsorbed on electrodes. *J. Phys. Chem. B*, **102**(35), 6889-6902.
8. Poulos, T. L. (1995) Cytochrome P450. *Curr. Opin. Struct. Biol.*, **5,** 767-774.
9. Kellner, D. G., Maves, S. A., and Sligar, S. G. (1997) Engineering cytochrome P450s for bioremediation. *Curr. Opin. Biotech.*, **8**(3), 274-278.
10. DiGleria, K., Nickerson, D. P., Hill, H. A. O., Wong, L. L., and Fulop, V. (1998) Covalent attachment of an electroactive sulfhydryl reagent in the active site of cytochrome P450(cam) as revealed by the crystal structure of the modified protein. *J. Am. Chem. Soc.*, **120**(1), 46-52.
11. Kurz, A., Halliwell, C. M., Davis, J. J., Hill, H. A. O., and Canters, G. W. (1998) A fullerene-modified protein. *J. Chem. Soc., Chem. Commun.*, **1,** 433-434.
12. Willner, I., Heleng-Shabtai, V., Katz, E., Rau, H. K., and Haehnel, W. (1999) Integration of a reconstituted de Novo synthesised haemoprotein and native metalloproteins with electrode supports for bioelectronic and bioelectrocatalytic applications. *J. Am. Chem. Soc.*, **121,** 6455-6468.
13. Arnold, F. H., Volkov, A.A. (1999) Directed evolution of biocatalysts. *Curr. Opin. Chem. Biol.*, **3,** 54-59.
14. Minshull, J., Stemmer, W.P.C. (1999) Protein evolution by molecular breeding. *Curr. Opin. Chem. Biol.*, **3,** 284-290.
15. Bendall, D. S. (1996) *Protein Electron Transfer* (Bendall, D. S., Ed.), BIOS Scientific Publishers Ltd., Oxford, England.
16. Davidson, V. L. (1996) Unravelling the kinetic complexity of inter-protein electron transfer reactions. *Biochemistry*, **35**(45), 14035-14039.
17. Koppenol, W. H., and Margoliash, E. (1982) The asymmetric distribution of charges on the surface of horse cytochrome *c*. *J. Biol. Chem.*, **257,** 4426-4437.
18. Margoliash, E., and Bossard, H. R. (1983) Guided by electrostatics, a textbook protein comes of age. *Trends Biochem. Sci.*, **8,** 316-320.
19. Roberts, V. A., Freeman, H. C., Olson, A. J., Tainer, J. A., and Getzoff, E. D. (1991) Electrostatic orientation of the electron-transfer complex between plastocyanin and cytochrome *c*. *J. Biol. Chem.*, **266,** 13431-13441.

20. Marcus, R. A., and Sutin, N. (1985) Electron transfers in chemistry and biology. *Biochim. Biophys. Acta*, **811**, 265-322.
21. Hoffman, B. M., and Ratner, M. A. (1987) Gated electron transfer: when are observed rates controlled by conformational interconversion. *J. Am. Chem. Soc.*, **109**, 6237-6243.
22. Nocek, J. M., Stemp, E. D. A., Finnegan, M. G., Koshy, T. I., Johnson, M. K., Margoliash, E., Mauk, A. G., Smith, M., and Hoffman, B. M. (1991) Low-temperature, cooperative conformational transition within Zn-cytochrome-c peroxidase, cytochrome-c. complexes - variation with cytochrome. *J. Am. Chem. Soc.*, **113**, 6822-6831.
23. Feitelson, J., and McLendon, G. (1991) Migration of small molecules through the structure of haemoglobin: Evidence for gating in a protein electron-transfer reaction. *Biochem.*, **30**, 5051-5055.
24. Ivkovic-Jensen, M. M., Ullmann, G. M., Young, S., Hansson, O., Crnogorac, M. M., Ejdeback, M., and Kostic, N. (1998) Effects of single and double mutatins in plastocyanin on the rate constant and activation parameters for the rearrangement gating the electron-transfer reaction between the triplet state of zinc cytochrome *c* and cupriplastocyanin. *Biochem.*, **37**, 9557-9569.
25. Walker, M. C., and Tollin, G. (1992) Laser flash photolysis study of the kinetics of electron transfer reactions of flavocytochrome *b2* from *Hansenula anomala*: Further evidence for intramolecular electron transfer mediated by ligand binding. *Biochem.*, **31**, 2798-2805.
26. Sullivan, E. P. J., Hazzard, T. J., Tollin, G., and Enemark, J. H. (1992) Inhibition of intramolecular electron transfer in sulphite oxidase by anion binding. *J. Am. Chem. Soc.*, **114**, 9662-9663.
27. Adir, N., Axelrod, H. L., Beroza, P., Isaacson, R. A., Rongey, S. H., Okamura, M. Y., and Feher, G. (1996) Co-crystallisation and characterisation of the photosynthetic reaction centre-cytochrome c(2) complex from *Rhodobacter sphaeroides*. *Biochem.*, **35**(8), 2535-2547.
28. Pelletier, H., and Kraut, J. (1992) Crystal structure of a complex between electron transfer partners, cytochrome *c* peroxidase and cytochrome *c*. *Science*, **258**, 1748-1755.
29. Chen, L., Durley, R. C. E., Mathews, F. S., and Davidson, V. L. (1994) Structure of an electron transfer complex: methylamine dehydrogenase, amicyanin, and cytochrome *c*-555i. *Science*, **264**, 86-90.
30. Ubbink, M., Ejdeback, M., Karlsson, B. G., and Bendall, D. S. (1998) The structure of the complex of plastocyanin and cytochrome f, determined by paramagnetic NMR and restrained rigid-body molecular dynamics. *Structure*, **6**(3), 323-335.
31. Moser, C. C., and Dutton, P. L. (1988) Cytochrome *c* and $c_2$binding dynamics and electron transfer with photosynthetic reaction centre protein and other integral membrane redox proteins. *Biochem.*, **27**, 2450-2461.
32. Zhou, J. S., and Hoffman, B. M. (1994) Stern-volmer in reverse: 2/1 stoichiometry of the cytochrome *c*-cytochrome *c* peroxidase electron transfer complex. *Science*, **265**, 1693-1696.
33. Peerey, L. M., Brothers, H. M., Hazzard, J. T., Tollin, G., and Kostic, N. M. (1991) Unimolecular and bimolecular oxidoreduction reactions involving diprotein complexes of cytochrome c and plastocyanin. Dependence of electron-transfer reactivity on charge and orientation of the docked metalloproteins. *Biochem.*, **30**, 9297-9304.
34. Weber, P. C., and Tollin, G. (1985) Electrostatic interactions during electron transfer reactions between *c*-type cytochromes and flavodoxin. *J. Biol. Chem.*, **260**, 5568-5573.
35. Stewart, D. E., LeGall, J., Moura, I., Moura, J. J. G., Peck, H. D. J., Xavier, A. V., Weiner, P. K., and Wampler, J. E. (1988) A hypothetical model of the flavodoxin tetrahaeme cytochrome-C3 complex of sulphate-reducing bacteria. *Biochem.* **27**, 2444-2450.
36. Poulos, T. L., and Mauk, A. G. (1983) Models for the complexes formed between cytochrome *b5* and the subunits of methaemoglobin. *J. Biol. Chem.*, **258**, 7369-7373.
37. Stayton, P. S., Poulos, T. L., and Sligar, S. G. (1989) Putidaredoxin competitively inhibits cytochrome *b5*-cytochrome P-450cam electron-transfer complex. *Biochem.* **28**, 8201-8205.
38. Geren, L., Tuls, J., O'Brien, P., Millett, F., and Peterson, J. A. (1986) The involvement of carboxylate groups of putidaredoxin in the reaction with putidaredoxin reductase. *J. Biol. Chem.*, **261**, 15491-15495.
39. Roitberg, A. E., Holden, M. J., Mayhew, M. P., Kurnikov, I. V., Beratan, D. N., and Vilker, V. L. (1998) Binding and electron transfer between putidaredoxin and cytochrome P450cam. Theory and experiments. *J. Am. Chem. Soc.*, **120**, 8927-8932.

40. Tegoni, M., White, S. A., Roussel, A., Mathews, F. S., and Cambillau, C. (1993) A hypothetical complex between crystalline flavocytochrome *b2* and cytochrome *c*. *Proteins: Struct. Funct. Genet.*, **16,** 408-422.
41. Cunha, C. A., Romao, M. J., Sadeghi, S. J., Valetti, F., Gilardi, G., and Soares, C. M. (1999) Effects of protein-protein interactions on electron transfer: docking and electron transfer calculations for complexes between flavodoxin and *c*-type cytochromes. *J. Biol. Inorg. Chem.*, **4**, 360-374.
42. Ruckpaul, K., Rein, H, and Blanck, J. (1989) Regulation mechanisms of the activity of the hepatic endoplasmic cytochrome P-450 in *Basis and mechanisms of regulation of cytochrome P-450* (Ruckpaul, K., Rein, H., ed) Vol. 1, pp. 6-29, Taylor and Francis.
43. Guengerich, F. P. (1990) Enzymatic oxidation of xenobiotic chemicals. *Crit. Rev. Biochem. Mol. Biol.*, **25**(2), 97-153.
44. Halkier, B. A. (1996) Catalytic reactivities and structure/function relationships of cytochrome P450 enzymes. *Phytochemistry*, **43**(1), 1-21.
45. Poulos, T. L., Finzel, B.C., and Howard, A.J. (1987) High-resolution crystal structure of cytochrome P450cam. *J. Mol. Biol.*, **195,** 687-700.
46. Ravichandran, K. G., Boddupalli, S.S., Hasemann, C.A., Peterson, J.A., and Deisenhofer, J. (1993) Crystal structure of haemoprotein domain of P450BM-3, a prototype for microsomal P450s. *Science*, **261,** 731-736.
47. Cupp-Vickery, J. R., and Poulos, T.L. (1995) Structure of cytochrome P450eryf involved in erythromycin biosynthesis. *Nat. Struct. Biol.*, **2,** 144-153.
48. Hasemann, C. A., Ravichandran, K.G., Peterson, J.A., and Deisenfofer, J. (1994) Crystal structure and refinement of cytochrome P450terp at 2.3 Å resolution. *J. Mol. Biol.*, **236,** 1169-1185.
49. Park, S. Y., Shimizu, H., Adachi, S., Nakagawa, A., Tanaka, I., Nakahara, K., Shoun, H., Obayashi, E., Nakamura, H., Iizuka, T., and Shiro, Y. (1997) Crystal structure of nitric oxide reductase from denitrifying fungus *Fusarium oxysporum*. *Nat. Struct. Biol.*, **4,** 827-832.
50. Williams, P. A., Cosme, J., Sridhar, V., Johnson, E.F., and McRee, D.E. (2000) Mammalian microsomal cytochrome P450 monooxygenase: Structural adaptations for membrane binding and functional diversity. *Mol. Cell.*, **5,** 121-131.
51. Graham, S. E., and Peterson, J.A. (1999) How similar are P450s and what can their differences teach us? *Arch. Biochem. Biophys.*, **369**(1), 24-29.
52. Estabrook, R. W., Hildebrandt, A.G., Remmer, H., Schenkman, J.B., Rosenthal, O., and Cooper, D.Y. (1968) Role of cytochrome P-450 in microsomal mixed function oxidation reactions in *Biochemie des Sauerstoffs* (Hess, B., Staudinger, H., ed), pp. 142-177, Springer-Verlag, Berlin.
53. White, R. E., and Coon, M.J. (1980) Oxygen activation by cytochrome P-450. *Annu. Rev. Biochem.*, **49,** 315-356.
54. Daff, S. N., Chapman, S.K., Turner, K.L., Holt, R.A., Govindaraj, S., Poulos, T.L., and Munro, A.W. (1997) Redox control of the catalytic cycle of flavocytochrome P-450 BM3. *Biochemistry*, **36,** 13816-13823.
55. Wong, L.-L. (1998) Cytochrome P450 monooxygenases. *Curr. Opin. Chem. Biol.*, **2,** 263-268.
56. Guengerich, F. P., and Schimada, T. (1998) Activation of procarcinogens by human cytochrome P450 enzymes. *Mutat. Res.*, **400,** 201-213.
57. Capdevila, J. H., Falck, J.R., and Harris, R.C. (2000) Cytochrome P450 and archidonic acid bioactivation: molecular and functional properties of the arachidonate monooxygenase. *J. Lipid Res.*, **41**(2), 163-181.
58. Estabrook, R. W., Shet, M.S., Faulkner, K., and Fisher, C.W. (1996) The use of electrochemistry for the synthesis of 17 alpha-hydroxyprogesterone by a fusion protein containing P450c17. *Endocr. Res.*, **22**(4), 665-671.
59. Hamman, M. A., Thompson, G.A., and Hall, S.D. (1997) Regioselective and stereoselective metabolism of ibuprofen by human cytochrome P450 2C. *Biochem. Pharmac.*, **54**(1), 33-41.
60. Narhi, L. O., and Fulco, A.J. (1986) Characterisation of a catalytically self-sufficient 119,000-Dalton cytochrome P-450 monooxygenase induced by barbiturates in *Bacillus megaterium*. *J. Biol.Chem.*, **261,** 7160-7169.
61. Boddupalli, S. S., Estabrook, W., and Peterson, J.A. (1990) Fatty acid monooxygenation by cytochrome P-450BM3. *J. Biol. Chem.*, **265,** 4233-4239.
62. Nakayama, N., Takemae, A., and Shoun, H. (1996) Cytochrome P450foxy, a catalytically self-sufficient fatty acid hydroxylase of the fungus *Fusarium oxysporum*. *J. Biochem.*, **119,** 435-440.

63. Fruetel, J. A., Mackman, R.L., Peterson, J.A., and de Montellano, P.R.O. (1994) Relationship of active site topology to substrate specificity for cytochrome P450terp (CYP108). *J. Biol. Chem.*, **46,** 28815-28821.
64. Alworth, W. L., Xia, Q. W., and Liu, H. M. (1997) Organochlorine substrates and inhibitors of P450 BM-3. *FASEB J.*, **11, SS,** P190.
65. Coon, M. J., McGinnity, D.F., Vaz, A.D.N, Liu, H.M., Mullin, D.A., Sato, H., and Shimizu, T. (1997) Novel substrates for mechanistic studies with cytochrome P450 BM3. *FASEB J.*, **11, SS,** 3326.
66. Darwish, K., Li, H., and Poulos, T.L. (1991) Engineering proteins, subcloning and hyperexpressing oxidoreductase genes. *Prot. Engng.*, **4,** 701-708.
67. Lewis, D. F. V. (1995) 3-Dimensional models of human and other mammalian microsomal P450s constructed from an alignment with P450 102 (P450(BM3)). *Xenobiotica*, **25(4),** 333-366.
68. Chang, Y. T., Stiffelman, O.B., Vakser, I.A., Loew, G.H., Bridges, A., and Waskell, L. (1997) Construction of a 3D model of cytochrome P450 2B4. *Prot. Engng.*, **10,** 119-129.
69. Dubois, M., Plaisance, H., Thome, J.P., and Kremers, P. (1996) Hierarchical cluster analysis of environmental pollutants through P450 induction in cultured hepatic cells- Indications for a toxicity screening test. *Ecotoxicol. Environm. Safety*, **34**(3), 205-215.
70. Harayama, S. (1997) Polycyclic aromatic hydrocarbon bioremediation design. *Curr. Opin. Biotech.*, **8,** 268-273.
71. Alworth, W. L., Mullin, D.A., Xia, Q., Kang, L., Liu, H.-M, and Zhao, W. (1995) A site specific mutant of the bacterial cytochrome-P450-102(BM-3) possessing a new capability to catalyse the hydroxylation of the polycyclic aromatic hydrocarbons pyrene and benzoapyrene. *FASEB J.*, **9,** A1491.
72. Logan, M. S., Newman, L.M., Schanke, C.A., and Wackett, L.P. (1993) Cosubstrate effects in reductive dehalogenation by *Pseudomonas putida* G786 expressing cytochrome P450cam. *Biodegradation*, **4,** 39-50.
73. Uotila, J. S., Kitunen, V.H., Saastamoinen, T., Coote, T., Haggblom, M.M., and Salkinoja-Salonen, M.S. (1992) Characterisation of aromatic dehalogenases of *Mycobacterium fortuitum* CG-2. *J. Bacteriol.*, **174,** 5669-5675.
74. Jones, J. P., O'Hare, E.J., and Wong, L.-L. (2000) The oxidation of polychlorinated benzenes by genetically engineered cytochrome P450cam: potential applications in bioremediation. *Chem. Commun.*, **3,** 247-248.
75. Wackett, L. P. (1995) Recruitment of co-metabolic enzymes for environmental detoxification of organohalides. *Environ. Health Perspect.*, **103,** 45-48.
76. WerckReichhart, D., Hehn, A., and Didierjean, L. (2000) Cytochromes P450 for engineering herbicide tolerance. *Trends Plant Sci.*, **5**(3), 116-123.
77. Erhardt, P. W. (1999) *Drug metabolism-Databases and high-throughput testing during drug design and development*, Published for International Union of Pure and Applied Chemistry by Blackwell Science Ltd, London.
78. Guengerich, F. P. (1999) Cytochrome P450: regulation and role in drug metabolism. *Annu. Rev. Pharmacol. Toxicol.*, **39,** 1-17.
79. Li, A. P. (1998) The scientific basis of drug-drug interactions: mechanism and pre-clinical evaluation. *Drug Info. J.*, **32,** 657-664.
80. Crespi, C. L., and Miller, V.P. (1999) The use of heterologously expressed drug metabolising enzymes-state of the art and prospects for the future. *Pharmacol. ther.*, **84,** 121-131.
81. Rendic, S., and DiCarlo, F.J. (1997) Human cytochrome P450 enzymes: a status report summarising their reactions, substrates, inducers and inhibitors. *Drug. Metab. Rev.*, **29,** 413-580.
82. Spatzenegger, M., and Jaeger, W. (1995) Clinical importance of hepatic cytochrome P450 in drug metabolism. *Drug Metab. Rev.*, **27,** 397-417.
83. Bertz, R. J., and Cranneman, G.R. (1997) Use of *in vitro* and *in vivo* data to estimate the likelihood of metabolic pharmacokinetic interactions. *Clin. Pharmacokinet.*, **32**(3), 210-258.
84. Gillam, E. M. J., Guo, Z., and Guengerich, F.P. (1994) Expression of modified human cytochrome P450 2E1 in *Escherichia coli*, purification, and spectral and catalytic properties. *Arch. Biochem. Biophys.*, **319,** 59-66.
85. Lieber, C. S. (1997) Cytochrome P-4502E1: Its physiological and pathological role. *Physiol. Rev.*, 77(2), 517-538.

86. Le Marchand, L., Sivaraman, L., Pierce, L., Seifried, A., Lum, A., Wikens, L.R., and Lau, A.F. (1998) Associations of CYP1A1, GSTM1, and CYP2E1 polymorphisms with lung cancer suggest cell type specificities to tobacco carcinogens. *Canc. Res.*, **58**(21), 4858-4863.
87. Beskin, M. J. (1980) Effect of combined phenylbutazone and ethanol administration on rat liver. *Exp. Pathol.*, **18,** 487-491.
88. Sibbesen, O., Devoss, J. J., and Ortiz De Montellano, P. R. (1996) Putidaredoxin reductase-putidaredoxin-cytochrome P450cam triple fusion protein. *J. Biol. Chem.*, **271**(37), 22462-22469.
89. Valetti, F., Sadeghi, S. J., Meharenna, Y. T., Leliveld, S. R., and Gilardi, G. (1988) Engineering multi-domain redox proteins containing flavodoxin as bio-transformer: preparatory studies by rational design. *Biosens. Bioelectron.*, **13,** 675-685.
90. Lamb, D. C., Kelly, D. E., Hanley, S. Z., Mehmood, Z., and Kelly, S. L. (1998) Glyphosate is an inhibitor of plant cytochrome P450: Functional expression of *Thlaspi arvensae* cytochrome P450 71B1/reductase fusion protein in *Escherichia coli. Biochem. Biophys. Res. Com.*, **244,** 110-114.
91. Schroder, G., Unterbusch, E., Kaltenbach, M., Schmidt, J., Strack, D., Luca, V. D., and Schroder, J. (1999) Light-induced cytochrome P450-dependent enzyme in indole alkaloid biosynthesis: tabersonine 16-hydroxylase. *FEBS Lett.*, **458,** 97-102.
92. Shiota, N., Nagasawa, A., Sakaki, T., Yabusaki, Y., and Ohkawa, H. (1994) Herbicide-Resistant Tobacco Plants Expressing the Fused Enzyme between Rat Cytochrome P4501A1 (CYP1A1) and Yeast NADPH-Cytochrome P450 Oxidoreductase. *Plant Physiol.*, **106,** 17-23.
93. Lacour, T., and Ohkawa, H. (1999) Engineering and biochemical characterisation of the rat microsomal cytochrome P4501A1 fused to ferredoxin and ferredoxin-NADP+ reductase from plant chloroplasts. *Biochem. Biophys. Acta*, **1433,** 87-102.
94. Friedberg, T. (2000) Recombinant *in vitro* tools to predict drug metabolism and safety. *PSTT*, **3**(3), 99-105.
95. Harlow, G. R., and Halpert, J. R. (1996) Mutagenesis study of Asp-290 in cytochrome p450 2B11 using a fusion protein with rat NADPH-cytochrome p450 reductase. *Arch. Biochem. Biophys.*, **326**(1), 85-92.
96. Schet, M. S., Fisher, C. W., Holmans, P. L., and Estabrook, R. W. (1993) Human cytochrome P450 3A4: Enzymatic properties of a purified recombinant fusion protein containing NADPH-P450 reductase. *Proc. Natl. Acad. Sci. USA*, **90,** 11748-11752.
97. Fisher, C. W., Schet, M. S., Caule, D. L., Martin-Wintrom, C. A., and Estabrook, R. W. (1992) High-level expression in *Escherichia coli* of enzymatically active fusion proteins containing the domains of mammalian cytochromes P450 and NADPH-P450 reductase flavoprotein. *Proc. Natl. Acad Sci. USA*, **89,** 10817-10821.
98. Parikh, A., and Guengerich, F. P. (1996) Expression, purification and characterisation of a catalytically active human cytochrome P450 1A2:Rat NADPH-Cytochrome P450 reductase fusion protein. *Prot. Exp. Purif.*, **9,** 346-354.
99. Estabrook, R. W., Faulkner, K. M., Shet, M. S., and Fisher, C. W. (1996) Application of Electrochemistry for P450-catalysed reactions. *Methods in Enzymology*, **272,** 44-51.
100. Eddowes, M. J., and Hill, H. A. O. (1977) A novel method for the investigation of the electrochemistry of metallo proteins: cytochrome c. *J. Chem. Soc. Chem. Commun.*, , 3154.
101. Scouten, W. H., Luong, J. H. T., and Brown, R. S. (1995) Enzyme or protein immobilisation techniques for applications in biosensor design. *TIBTECH*, **113,** 178-184.
102. Clark, L. C., and Lyons, C. (1962) Electrode systems for continuous monitoring in cardiovascular surgery. *Ann. N.Y. Acad. Sci.*, **102,** 29.
103. Cass, A. E. G., Davis, G., Francis, G. D., Hill, H. A. O., Aston, W. J., Higgins, I. J., Plotkin, E. V., Scott, L. D. L., and Turner, A. P. F. (1984) Ferrocene-mediated enzyme electrode for amperometric determination of glucose. *Anal. Chem.*, **56,** 667.
104. Cass, A. E. G., Davis, G., Green, M. J., and Hill, H. A. O. (1985) Ferricinium ion as an electron-acceptor for oxydo-reductases. *J. Electroanal. Chem.*, **190,** 117.
105. Bourdillon, C., Demaille, C., Moiroux, J., and Saveant, J.-M. (1993) New insights into the enzymatic catalysis of the oxidation of glucose by native and recombinant glucose-oxidase mediated by electrochemically generated one-electron redox cosubstrates. *J. Am. Chem. Soc.*, **115,** 2.
106. Mor, J. R., and Guamaccia, R. (1977) Assay of glucose using an electrochemical enzymatic sensor. *Anal. Biochem.*, **79,** 319.

107. Montagne, M., and Marty, J.-L. (1995) Bioenzyme amperometric D-lactate sensor using macromolecular NAD(+). *Anal. Chim. Acta*, **315,** 297.
108. Hill, H. A. O., and Higgins, I. J. (1981) Oxygen, oxidases, and the essential trace-metals. *Philos. Trans. R. Soc. London*, **302,** 267.
109. Janda, P., and Weber, J. (1991) Quinone-mediated glucose-oxidase electrode with the enzyme immobilised in polypyrrole. *J. Electroanal. Chem.*, **300,** 119.
110. Foulds, N. C., and Lowe, C. R. (1986) Enzyme entrapment in electrically conducting polymers - immobilisation of glucose-oxidase in polypyrrole and its application in amperometric glucose sensors. *J. Chem. Soc. Faraday Trans. 1*, **82,** 1259.
111. Hintz, M. J., and Peterson, J. A. (1980) The kinetics of reduction of cytochrome P-450cam by the dithionite anion monomer. *J. Biol. Chem.*, **255,** 7317-7325.
112. Contzen, J., and Jung, C. (1999) Changes in secondary structure and salt links of cytochrome P-450cam induced by photoreduction: A Fourier transform infrared spectroscopic study. *Biochem.*, **38,** 16253-16260.
113. Kazlauskaite, J., Westlake, A. C. G., Wong, L.-L., and Hill, H. A. O. (1996) Direct electrochemistry of cytochrome P450cam. *J. Chem. Soc., Chem. Commun.*, , 2189-2190.
114. Zhang, Z., Nassar, A.-E. F., Lu, Z., Schenkman, J. B., and Rusling, J. F. (1997) Direct electron injection from electrodes to cytochrome P450cam in biomembrane-like films. *J. Chem. Soc., Faraday Trans.*, **93**(9), 1769-1774.
115. Lo, K. K.-W., Wong, L.-L., and Hill, H. A. O. (1999) Surface-modified mutants of cytochrome P450(cam): enzymatic properties and electrochemistry. *FEBS Lett.*, **451,** 342-346.
116. Lvov, Y. M., Lu, Z., Schenkman, J. B., Zu, X., and Rustling, J. F. (1998) Direct electrochemistry of myoglobin and cytochrome P450(cam) in alternate layer-by-layer films with DNA and other polyions. *J. Am. Chem. Soc.*, **120,** 4073-4080.
117. Lei, C., Wollenberger, U., Jung, C., and Scheller, F. W. (2000) Clay-bridged electron transfer between cytochrome P450cam and electrode. *Biochem. Biophys. Res. Commun.*, **268,** 740-744.
118. Dryhurst, G., Kadish, K. M., Scheller, F. W., and Renneberg, R. (1982) *Biological electrochemistry*, 1, Academic press, New York.
119. Scheller, F. W., Renneberg, R., Strnad, G., Pommerening, K., and Mohr, P. (1977) . *Bioelectrochem. Bioenerg.*, **4**(500-507).
120. Sadeghi, S. J., Meharenna, Y. T., and Gilardi, G. (1999) Flavodoxin as module for transferring electrons to different c-type and P450 cytochromes in artificial redox chains in *Flavins and flavoproteins* (Ghisla, S., Kroneck, P., Marcheroux, P., and Sund, H., eds), pp. 163-166, R. Weber.
121. Sadeghi, S. J., Meharenna, Y. T., Fantuzzi, A., Valetti, F., and Gilardi, G. (2000) Engineering artificial redox chains by molecular lego. *J. Chem. Soc., Faraday Dis.*, **116,** 135-153.
122. Watkins, J. A., Cusanovich, M. A., Meyer, T. E., and Tollin, G. (1994) A "parallel plate" electrostatic model for bimolecular rate constants applied to electron transfer proteins. *Protein Science*, **3**(11), 2104-2114.
123. Sevrioukova, I. F., Hazzard, J. T., Tollin, G., and Poulos, T. L. (1999) The FMN to Heme Electron Transfer in Cytochrome P450BM-3. *J. Biol. Chem.*, **274**(51), 36097-36106.
124. Hazzard, J. T., Govindaraj, S., Poulos, T. L., and Tollin, G. (1997) Electron transfer between the FMN and haem domains of cytochrome P450BM-3. *J. Biol. Chem.*, **272**(12), 7922-7926.
125. Tsotsou, G. E., Meharenna, Y. T., Ganini, S., Fairhead, M. J., Sadeghi, S. J., and Gilardi, G. (2000) Molecular lego in generation of macromolecular assemblies of P450 enzymes for high throughput screening. *submitted.*
126. Heering, H. A., and Hagen, W. R. (1996) Complex electrochemistry of flavodoxin at carbon-based electrodes: results from a combination of direct electron transfer, flavin-mediated electron transfer and comproportionation. *J. Electroanal. Chem.*, **404,** 249-260.
127. Ingelman-Sundberg, M., Oscarson, M, and McLellan, R.A. (1999) Polymorphic human cytochrome P450 enzymes: an opportunity for individualised drug treatment. *Trends Pharmacol. Sci.*, **20,** 342-349.
128. Zhao, H., and Arnold, F.H. (1997) Combinatorial protein design: strategies for screening protein libraries. *Curr. Opin. Struct. Biol.*, **7,** 480-485.
129. Burke, M. D., and Meyer, R.T. (1983) Differential effects of phenobartitone and 3-methylcholanthrene induction on the hepatic microsomal metabolism and cytochrome P-450-binding of phenoxazone and a homologous series of its O-alkoxy ethers. *Chem.-Biol. Interact.*, **45,** 243-258.

130. White, I. N. H. (1988) A continuous assay for cytochrome P-450-dependent mixed function oxidases using 3-cyano-7-ethoxycoumarin. *Anal. Biochem.*, **172,** 304-310.
131. Rodrigues, A. D., Kukulka, M.J., Surber, B.W., Thomas, S.B., Uchic, J.T., Potert, G.A., Michael, G., Thome-Kromer, B., and Machinist, J.M. (1994) Measurement of liver-microsomal cytochrome-P450(CYP2D6) activity using o-methyl-14C. dextromethorphan. *Anal. Biochem.*, **219**(2), 309-320.
132. Crespi, C. L., Miller, V.P., and Penman, B.W. (1997) Microtiter plate assays for inhibition of human, drug-metabolising cytochromes P450. *Anal. Biochem.*, **248,** 188-190.
133. Onderwater, R. C. A., Venhorst, J., Commandeur, J.N.M., and Vermeulen, N.P.E. (1999) Design, synthesis and characterisation of 7-methoxy-4-(aminomethyl)coumarin as a novel and selective P450 2D6 substrate suitable for high-throughput screening. *Chem. Res. Toxicol.*, **12**(7), 555-559.
134. VanBreemen, R. B., Nikolic, D., and Bolton, J.L. (1998) Metabolic screening using on-line ultrafiltration mass spectrometry. *Drug Metab. Dispos.*, **26**(2), 85-90.
135. Eddershaw, P. J., and Dickins, M. (1999) Advances in *in vitro* drug metabolism screening. *Pharm. Sci. Technol. Today*, **2**(1), 13-19.
136. Yin, H., Racha, J., Li, S.Y., Olejnik, N., Satoh, H., and Moore, D. (2000) Automated high throughput human CYP isoform activity assay using SPE-LC/MS method: application in CYP inhibition evaluation. *Xenobiotica*, **30**(2), 141-154.
137. Joo, H., Lin, Z.L., and Arnold, F.H. (1999) Laboratory evolution of a peroxide-mediated cytochrome P450 hydroxylation. *Nature*, **399**(6737), 670-673.
138. Grigoryev, D. N., Kato, K., Njar, V.C.O., Long, B.J., Ling, Y.Z., Wang, X., Mohler, J., and Brodie, A.M.H. (1999) Cytochrome P450c17-expressing Escherichia coli as a first-step screening system for 17a-hydroxylase-C17, 20-lyase inhibitors. *Anal. Biochem.*, **267,** 319-330.
139. Parikh, A., Josephy, D., and Guengerich, F.P. (1999) Selection and characterisation of human cytochrome P450 1A2 mutants with altered catalytic properties. *Biochemistry*, **38,** 5283-5289.
140. Maves, S. A., Yeom, H., McLean, M.A., and Sligar, S.G. (1997) Decreased substrate affinity upon alteration of the substrate-docking region in cytochrome P450 BM-3. *FEBS Lett.*, **414,** 213-218.
141. Schwaneberg, U., Schmidt-Dannert, C., Schmitt, J., and Schmid, R.D. (1999) A continuous spectrophotometric assay for P450 BM-3, a fatty acid hydroxylating enzyme, and its mutant F87A. *Anal. Biochem.*, **269,** 359-366.
142. Kaplan, N. O., Colowick, S.P., and Barnes, C.C. (1951) Effect of alkali on diphosphopyridine nucleotide. *J. Biol. Chem.*, **191,** 461-472.
143. Lowry, O. H., Roberts, N.R., and Kapphahn, J.I. (1957) The fluorometric measurement of pyridine nucleotides. *J. Biol. Chem.*, **224,** 1047-1064.
144. Childs, R. E., and Bardsley, W.G. (1975) The steady state kinetics of peroxidase with 2'2-Azino-di(3-ethyl-benzathiazoline-6-sulphonic acid) as chromagen. *Biochem. J.*, **145,** 93-103.
145. Gasco, A., Fruttero, R., and Sorba, G. (1996) NO-donors: An emerging class of compounds in medicinal chemistry. *Farmaco*, **51**(10), 617-635.
146. Wang, M. H., Wade, D., Chen, L., White, S., and Yang, C. S. (1995) Probing the Active Site of Rat and Human Cytochrome P450 2E1 with Alcohols and Carboxylic Acids. *Arch. Biochem Biophys.*, **317,** 299-304.
147. Li, H., and Poulos, T.L. (1997) The structure of the cytochrome P450 BM-3 haem domain complexed with the fatty acid substrate, palmitoleic acid. *Nat. Struct. Biol.*, **4,** 140-146.
148. Chang, Y. T., Stiffelman, O.B., Vakser, I.A., Loew, G.H., Bridges A., and Waskell, L. (1997) Construction of a 3D model of cytochrome P450 2B4. *Prot. Engng.*, **10(2),** 119-129.
149. Shimoji, M., Yin, H., Higgins, L. A., and Jones, J. P. (1998) Design of a novel P450: A functional Bacterial- Human Cytochrome P450 chimera. *Biochem.*, **37,** 8848-8852.
150. Waterman, M. R. (1996) Cytochrome P450. *Meth. Enzymol.*, **272**(B).
151. Nelson, D. R., and Strobel, H.W. (1998) On the membrane topology of vertebrate cytochrome P-450 proteins. *J. Biol. Chem.*, **263,** 6038-6050.
152. Jenkins, C. M., and Waterman, M.R. (1998) NADPH-flavodoxin reductase and flavodoxin from *Escherichia coli* as a soluble microsomal P450 reductase. *Biochemistry*, **37,** 6106-6113.

AMPEROMETRIC ENZYME-BASED BIOSENSORS FOR APPLICATION IN FOOD AND BEVERAGE INDUSTRY

ELISABETH CSÖREGI[1*], SZILVESZTER GÁSPÁR[1], MIHAELA NICULESCU[1], BO MATTIASSON[1], WOLFGANG SCHUHMANN[2]
[1]Lund University, Centre for Chemistry and Chemical Engineering, Department of Biotechnology, P.O. Box 124, 221 00 Lund, Sweden
[2]Ruhr-Universität Bochum, Analytische Chemie - Elektroanalytik and Sensorik, Universitätsstr. 150; D-44780 Bochum, Germany

Summary

Continuous, sensitive, selective, and reliable monitoring of a large variety of different compounds in various food and beverage samples is of increasing importance to assure a high-quality and tracing of any possible source of contamination of food and beverages. Most of the presently used classical analytical methods are often requiring expensive instrumentation, long analysis times and well-trained staff. Amperometric enzyme-based biosensors on the other hand have emerged in the last decade from basic science to useful tools with very promising application possibilities in food and beverage industry. Amperometric biosensors are in general highly selective, sensitive, relatively cheap, and easy to integrate into continuous analysis systems. A successful application of such sensors for industrial purposes, however, requires a sensor design, which satisfies the specific needs of monitoring the targeted analyte in the particular application. Since each individual application needs different operational conditions and sensor characteristics, it is obvious that biosensors have to be tailored for the particular case. The characteristics of the biosensors are depending on the used biorecognition element (enzyme), nature of signal transducer (electrode material) and the communication between these two elements (electron-transfer pathway).

Therefore, the present chapter presents the different existing biosensor designs describing the possible electron-transfer pathways, discusses their advantages and disadvantages, and shows their possible application in food and beverage industry. Three practical examples are given describing biosensor designs developed in our laboratory, demonstrating their usefulness for industrial applications.

M. De Cuyper and J.W.M. Bulte (eds.), Physics and Chemistry Basis of Biotechnology, 105–129.

1. Biosensors - Fundamentals

Biosensors are selective devices that involve a communication between a biorecognition element (enzymes, microorganisms, cells, antibodies, tissues etc.) and a physical transducer, which is able to transform the chemical information i. e. the concentration of the target analyte, into a measurable signal (electrical, optical, mass, thermal, etc.) as shown in figure 1. Most biosensors use enzymes in the complementary recognition process, since their catalytic action implies a self-regeneration of the binding pocket. In addition, many enzymes are readily available and easy to couple with a large variety of transducers.

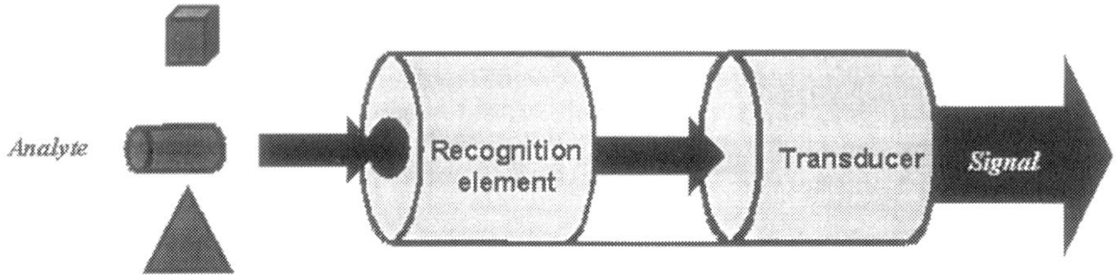

Figure 1. Schematic presentation of a biosensor.

Amperometric biosensors based on enzymes are the most studied branch, since they combine the selectivity and specificity of the enzymatic reactions with the simplicity of the electrochemical detection method. In general, the analyte is catalytically converted (oxidised or reduced) by a specific enzyme, which is usually immobilised on an electrode surface. Mostly, the redox equivalents are intermediately stored in the cofactor of the enzyme, from/to which an electron-transfer process has to occur, whereby the biocatalytic process is linked to the electrode. Thus, the chemical energy of the enzyme-catalysed reaction is transduced to an electrochemical reaction, which occurs at a certain potential determined by the nature of the compound used in the final step of the electron-transfer process. Due to a stoichiometric relation between the number of transferred electrons and the analyte, a current proportional to the concentration of the target analyte is yielded and used for quantification.

Enzyme-based biosensors attracted much attention for their possible use in food and beverage industry, representing a very promising alternative to the traditional time-consuming and often expensive analysis techniques. Due to their simplicity, low cost, and possibility of integration into on-line measurement systems needed in automated industrial technologies, enzyme-based biosensors have been widely used in this area, resulting in a large number of review articles published in the last 10 years [1-7]. As demonstrated, biosensors can be useful tools for monitoring the conversion of raw materials, the presence and concentration of possible contaminants, the product content, and product freshness [3].

2. Prerequisites for application of biosensors in food industry

A successful industrial application of biosensors in food industry, requires a sensor design assuring besides low fabrication costs the following:

- selectivity versus all possible enzymatic and electrochemical interferences
- adequate sensitivity
- appropriate response time
- stability (operational and/or storage stability)
- reliability
- simplicity of monitoring

Since the concentration of the target analytes is significantly different and each individual application requires particular operational conditions, it is obvious that biosensors have to be tailored for the considered application. E.g., long term monitoring applications, such as monitoring in fermentation processes, require biosensors, which guarantee a good mechanical and thermal stability. In addition, any source of contamination by e.g. component leakage has to be strictly avoided using an appropriate sensor design, while fast response time (<s) and high selectivity issues are often not crucial. The choice of enzyme(s), immobilisation procedure, electrode configuration, use of additional sensor elements (e.g. additional membranes for improved stability and/or interference elimination) will therefore always be determined by the considered application.

However, every enzyme-based biosensor has to take into account that its performance is determined simultaneously by the biorecognition element (the enzyme), the signal transducer (polarised electrode), and by the communication between these two elements (electron-transfer pathway). Therefore, the electrode configuration determined by the above mentioned two elements and the immobilisation procedure of the enzyme, has to be designed in such a manner, that the biochemical information is translated into a measurable electrical current with the highest efficiency (optimal electron-transfer pathway). Below, a short overview is given on the various existing biosensor designs, the electron transfer possibilities, and the advantages and disadvantages of various sensor architectures.

3. Existing biosensor configurations and related electron-transfer pathways

The simplest electron-transfer mechanism would be represented by the direct electrochemical regeneration of the active site (prosthetic group) of the enzyme at the electrode surface (see figure 2).

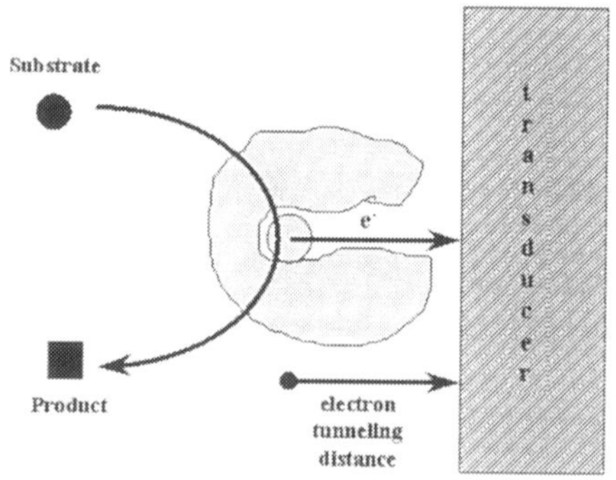

Figure 2. *Direct electron transfer pathway (via electron tunnelling) between the active site of an enzyme and the surface of a transducer.*

However, this approach can be applied only for a very few enzymes (e.g. peroxidases), which have their active site situated close to their surface, and thus allow its direct regeneration on the transducer. To insulate the enzyme-integrated active site, most of the enzymes have their prosthetic group deeply buried within the protein shell and thereby, the distance for a direct electron transfer is - according to the Marcus theory [8, 9] - too long. Therefore, in general the electron-transfer pathway has to be artificially designed, either by (i) using "electron shuttles" (e.g. redox mediators, see figure 3) or by (ii) shortening the electron-transfer distance (e.g. orientation of enzymes on electrodes, modification of the active site of an enzyme, etc).

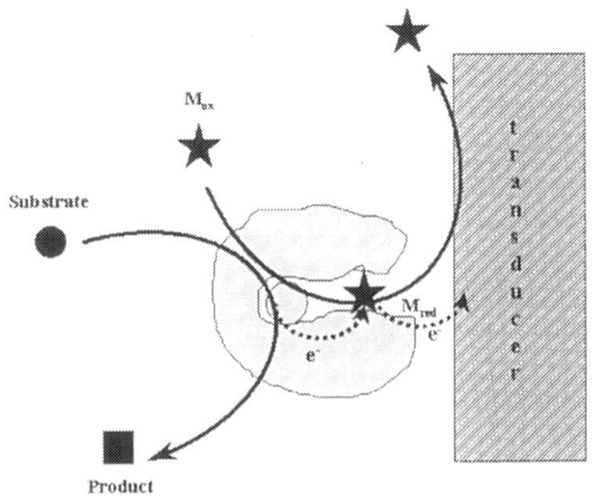

Figure 3. *Mediated electron transfer pathway using a redox mediator (M_{OX}) as the "electron shuttle" between the active site of the enzyme and the transducer exemplified for an oxidation reaction.*

3.1. BIOSENSORS BASED ON O_2 OR H_2O_2 DETECTION

The first developed enzymatic biosensors were based on the fact that in nature the active site of many enzymes is regenerated by their co-substrate (e.g. O_2 or NAD^+). Since these co-substrates can be directly oxidised or reduced on the surface of a polarised electrode, they could be successfully used as "electron shuttles". The quantification of the analyte was based in these cases on measuring either the (i) decrease of the co-substrate concentration or (ii) the increase of the co-product (e.g. hydrogen peroxide), (see scheme 1).

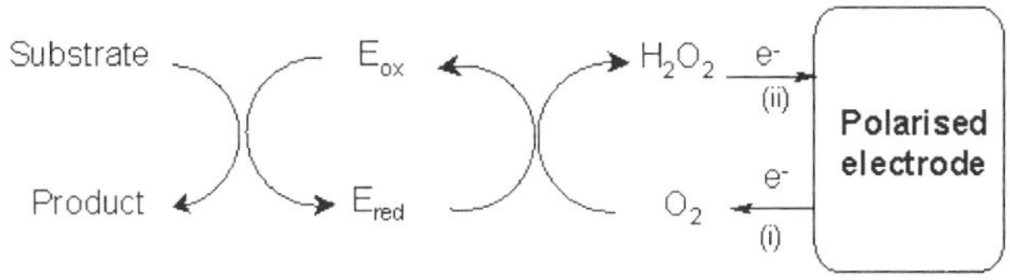

Scheme 1. Substrate detection possibilities exemplified for an oxidation reaction.

Unfortunately, both above mentioned electrochemical reactions occur at high overpotentials and thus, the signal transduction suffers of electrochemical interferences when the biosensor is used in real applications implying complex matrices. Additional problems may occur at high substrate concentrations, due to the lack of molecular oxygen. Therefore, next generation biosensors were often based on a mediated electron transfer principle, the earlier types using freely diffusing redox mediators (see section 3.2).

Despite the mentioned drawbacks, biosensors based on the direct detection of either O_2 or H_2O_2 were used to measure different analytes in food stuff and beverages, mainly due to their relative simplicity (see Table 1).

Table 1. Biosensors using direct H_2O_2 or O_2 detection with potential use in food and/or beverage industry

Analyte	Enzyme	Detection of	Characteristics*	Sample	Ref.
hypoxanthine	Xanthine oxidase	H_2O_2 or O_2	LR: 1 - 1000 µM DL: 1.3 µM	carp	[65]
lecithin	Phospholipase D; Choline oxidase	O_2	operating in organic phase	egg yolk, oil, soya flour, diet integrators	[66]
lactose	ß-galactosidase; Glucose oxidase	H_2O_2	LR: 3.5 - 2000 µM	milk	[67]
amines	Diamine oxidase	H_2O_2	LR: 1 – 50 µM DL: 0.5 µM	salted anchovy	[37]
acetaldehyde	Aldehyde dehydrogenase; NADH oxidase;	H_2O_2	LR: 0.5 - 330 µM	cider, wine, whisky, saké, champagne	[68]

Table 1. *Biosensors using direct H_2O_2 or O_2 detection with potential use in food and/or beverage industry*

Analyte	**Enzyme**	**Detection of**	**Characteristics***	**Sample**	**Ref.**
histamine, cadaverin, putrescine	Diamine oxidase	H_2O_2	LR: up to 6 mM DL: 25 μM usable for at least 60 assays	fish fillets	[39]
histamine, cadaverin, putrescine	Amine oxidase	H_2O_2	LR: 0.5 - 10 μM stable for more than 32 h of continuos operation	fish, meat homogenates	[48]
L-lysine	L-lysine α oxidase	O_2	LR: 6.7 -670 μM	wheat extracts	[69]
ethanol, methanol	Alcohol oxidase	O_2	LR: 0.5 - 15 mM (ethanol); 10 - 300 mM (methanol) response time of about 2 min stable signal for 500 assays	wine, beer	[70]
L, D-amino acids	L, D amino acid oxidase	H_2O_2	DL: 470 μM (L-leucine), 150 μM (L-glycine), 200 mM (L-phenylalanin) stable over 56 days	monitoring milk ageing effects	[71]
glucose	Glucose oxidase	H_2O_2	LR: 10 - 1000 mM	lemon juice, milk, soda Seven Up, orangeade, tonic water , grapefruit	[72]
glucose	Glucose oxidase	H_2O_2	LR: 0 - 11 mM	honey, sugar pancake syrup, punch, sweet potato, apple juice	[73]

* *LR denotes the linear range and DL the detection limit of the biosensors*

3.2. BIOSENSORS BASED ON FREE-DIFFUSING REDOX MEDIATORS

Considering any real application, it is of great importance that the biosensors are operated within an optimal potential window (approximately between -0.10 to +0.05 V vs. SCE), where electrochemical interference is minimal [10]. This can be practically realised in two different sensor architectures.

The first sensor types use artificial, free-diffusing, electron-transfer mediators (redox couples with a formal potential lying within the optimal potential range), replacing the natural electron acceptor the hydrogen peroxide/oxygen, couple (see scheme 2).

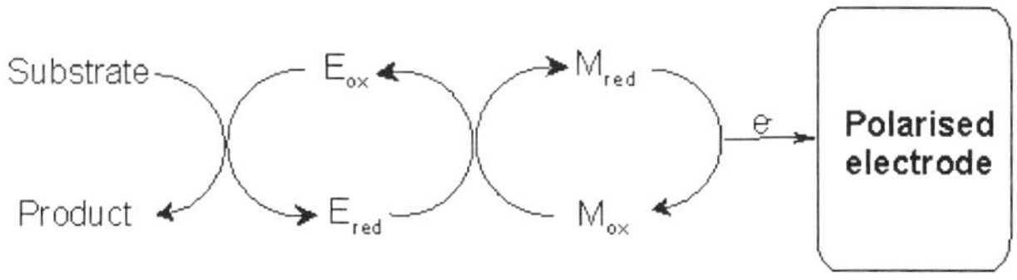

Scheme 2. *Detection of a substrate using an artificial redox mediator.*

Despite assuring detection at low potentials, these biosensor configurations often cause problems by contaminating the sample. Thus, the application of these biosensors in food and/or beverage industry is almost impossible, unless extra protection membranes are applied in the particular sensor design. Moreover, a competition between the natural cofactor of the enzyme (e.g. O_2) and the artificial mediator occurs. Therefore, the search for and use of, oxygen independent enzymes was and is obvious (e.g. measurement of D-fructose in food samples based on PQQ-dependent D-fructose dehydrogenase [11-13]).

The second type of sensors is based on coupled enzymes (oxidase-peroxidase). These bi-enzyme electrodes are mostly applied for the detection of the substrates of H_2O_2-producing oxidases and make use of the selectivity of peroxidases (POD) towards H_2O_2. Since peroxidases are able to directly exchange electrons with the electrode (via electron tunnelling), many of the reported electrode designs make use of a direct electron-transfer pathway [14] as shown in scheme 3a. The most often used peroxidase in this context is horseradish peroxidase (HRP). Since only 48 % of the randomly immobilised HRP molecules were reported to be able to undergo a direct electron transfer [15], often an orientation of the peroxidase molecule is required to improve the rate of the electron transfer. Besides the accessibility of the enzyme's redox centre, the glycosylation degree of the enzymes is also playing an important role. However, even when using an orientated binding of peroxidases, (improved electron-transfer reaction rate [16]), the major drawback of this sensor design is a small current response.

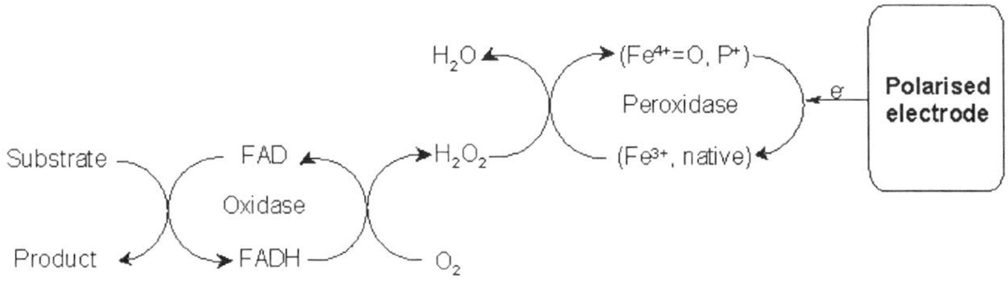

Scheme 3a. *Electron-transfer pathway in coupled enzyme electrodes, the final electron transfer step is via direct electron tunnelling.*

Since the direct electron transfer between the commonly used horseradish peroxidase (HRP) and the electrode was shown to be sluggish, often the use of an additional mediator is required, as shown in scheme 3b.

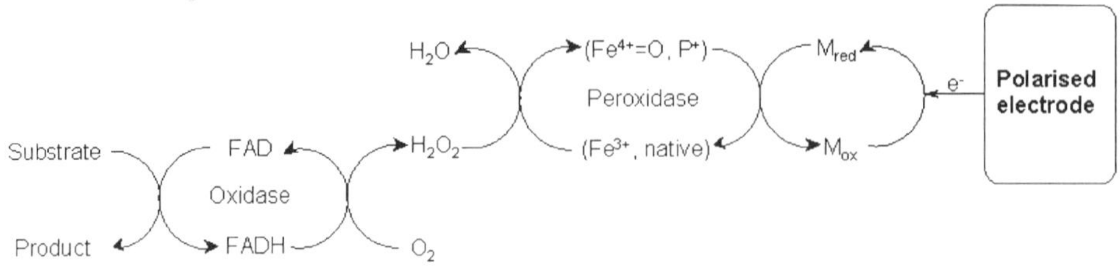

Scheme 3b. *Electron-transfer pathway in coupled enzyme electrodes, the final electron transfer step is via mediated electron transfer.*

Moreover, HRP is characterised by a low selectivity vs. reducing substrates, thus, highly motivating search for new, more selective, and/or stable peroxidases. Use of lactate peroxidase [17-19], tobacco peroxidase [20, 21], microperoxidase [17, 18, 22, 23], peroxidase from *Arthromyces ramosus* [19, 24, 25], soybean peroxidase [19, 26, 27] or sweet potato peroxidase [21] in biosensor designs was already reported.
Examples of biosensor architectures based on electron transfer principles mentioned in this section with possible application in food and/or beverage industry are given in Table 2.

Table 2. *Biosensors using artificial mediators with potential use in food and/or beverage industry*

Analyte	Enzyme	Detection of	Characteristics*	Sample	Ref.
D-fructose	D-fructose dehydrogenase	Hexacyanoferrate (in solution)	LR: 0.05 - 10 mM selectivity against ascorbate stable for 6 months	cherry jam, floral honey, milk chocolate, orange juice, wines	[11]
D-fructose	D-fructose dehydrogenase	Hexacyanoferrate (in solution)	LR: 0.01 - 1 mM selectivity against glucose and other sugars	apple juice, orange juice, pear juice	[74]
D-fructose	D-fructose dehydrogenase	$Os(bpy)_2Cl^{2+}$	LR: 0.2 - 20 mM DL: 35 µM	cola, apple juice, honey pineapple juice,	[12]
ethanol	Alcohol dehydrogenase; NADH oxidase;	Hexacyanoferrate (in solution)	LR: 0.3 - 200 µM	cider, whisky	[13]
ethanol	Alcohol dehydrogenase;	Poly(phenylene diamine)	LR: 0.03 - 3 µM response time 20 s	cider, wine, whisky	[75]

Table 2. *Biosensors using artificial mediators with potential use in food and/or beverage industry*

Analyte	Enzyme	Detection of	Characteristics*	Sample	Ref.
ethanol	Alcohol dehydrogenase;	Meldola's blue	LR: up to 35 mM 90% activity after 49 days	gin	[76]
L-lactate	Lactate oxidase, Peroxidase	Ferrocene	DL: 0.9 - 1.4 μM no loss of the activity after 6 months of storage at 4 °C	red wine, shaken yoghurt	[77]
sulphite	Sulphite oxidase	Tetrathiafulvalen; tetracyanoquinodimethane	DL: up to 5 mM 55 repeated analyses during 26 h of continuos operation with no loss of activity	wine, beer, dried fruit samples	[78]
fructose	Fructose dehydrogenase	Coenzyme ubiquinone-6	DL: 10 μM no significant ascorbic acid interferences	apple, orange juice	[79]
methylcarbamates	Cholinesterase	Cobalt phthalocyanine	LR: $5x10^{-5}$ - 50 mg/Kg DL: $1x10^{-4}$ - 3.5 mg/Kg	potato, carrot, sweet pepper	[80]
Histamine putrescine cadaverine	Amine oxidase	Poly(1-vinylimidazole) modified with Os(4,4'dimethylbipyridine)$_2$Cl$^{+/2+}$	DL:0.33 μM (histamine), 0.17 μM (putrescine) 90 % activity observed after 10 days of storage at 4 °C	turbot fish muscle	[31]
D-lactate	D-lactate dehydrogenase; NADH oxidase;	Hexacyanoferrate (in solution)	LR: 0.01 -1 mM stable for 4 months when stored at 4 °C	yoghurt, milk, cheese	[81]
glucose	Glucose oxidase	Tetrathiafulvalene	LR: 1 - 3 mM	wine, orange and apple juice	[82]
glucose	Glucose oxidase Horseradish peroxidase	Ferrocene	LR: 10 - 800 μM DL: 1.9 μM	must, wine	[83]

* *LR denotes the linear range and DL the detection limit of the biosensors*

3.3. INTEGRATED SENSOR DESIGNS (REAGENTLESS BIOSENSORS)

An improved communication between enzyme(s) and the electrode, meeting the requirements for real sample applications is respected in the greatest extent by using integrated ("reagentless") biosensors. These electrodes contain all needed components integrated in a sensing layer without any leakage possibility, since there is no need of addition of free-diffusing mediators, co-factors, etc. Most of these electrode designs make use of a mediated electron-transfer mechanism (as shown in scheme 2), the mediating molecules being covalently immobilised either on conducting (e.g. polypyrrole) or non-conducting (e.g. poly vinyl imidazole) polymeric backbones (see figure 4).

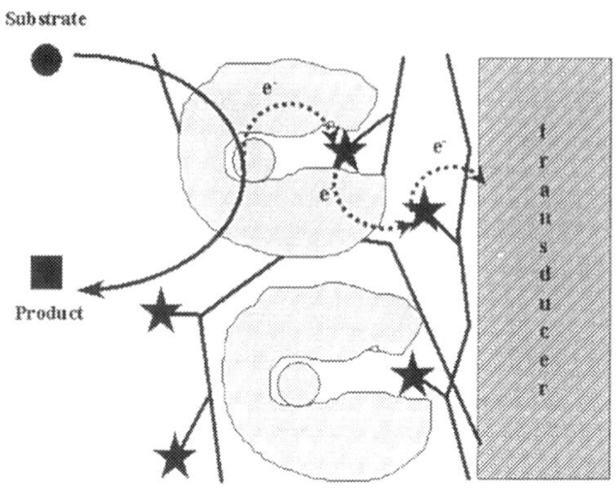

Figure 4. *Mediated electron-transfer pathway with both the enzyme and the redox mediator (stars) immobilised either on conducting or non-conducting polymeric chains.*

These constructions offer an efficient electron transfer and improved stability, mainly due to the lack of any free-diffusing components. One of the most promising approaches makes use of enzymes entrapped in redox polymers (non-conducting polymers modified with highly efficient Os- or Ru-complexes as mediators, often called "wires"). The redox polymer integrated enzymes result in highly permeable, efficient, and stable redox hydrogels [28, 29]. In this design, the mediator is retained in the close proximity of the enzyme's redox centre and electrons are rapidly exchanged (via electron hopping between the relaying redox centres along the wire and/or occasionally crossing between undulating segments of the wires) assuring an efficient electron-transfer pathway.

Practical examples of this type of biosensor (developed in our laboratories) with potential application in food and/or beverage industry are presented in the following section. Detection of hydrogen peroxide (as above mentioned) constitutes the base of sensing many other analytes, therefore one of the hereby presented examples is considering development of hydrogen peroxide sensors based on different, newly isolated, purified, and characterised plant peroxidases (4.1) [21]. The other two examples focus on the detection of biomarkers (biogenic amines) using amine oxidase entrapped in redox hydrogels (4.2) [30-32] and monitoring of alcohol using a newly isolated and characterised PQQ-dependent (oxygen independent) alcohol dehydrogenase (4.3) [33] (adsorbed on graphite, glassy carbon or platinum electrodes and entrapped in conducting or redox polymers, respectively).

4. Selected practical examples

4.1. REDOX HYDROGEL INTEGRATED PEROXIDASE BASED HYDROGEN PEROXIDE BIOSENSORS

The importance of finding new and more efficient peroxidases, with improved bioelectrochemical characteristics, has been outlined in section 3.2. In this example, development of hydrogen peroxide sensors has been targeted, using two newly purified peroxidases extracted from tobacco (TOP) and sweet potato (SPP), comparing their characteristics to the ones obtained for similarly constructed electrodes based on the generally used HRP. Electrodes were prepared according to a previously published protocol [21]. Briefly: the enzymes were cross-linked to poly(vinylimidazole) complexed with Os(4,4'dimethylbipyridine)$_2$Cl (PVI$_7$dmeOs) using poly (ethyleneglycol) diglycidyl ether (PEGDGE) as the cross-linker. Premixed hydrogels were made using stock solutions of peroxidases (4.78 mg ml^{-1} of TOP in 0.15 M Tris-HCl, pH 6.0; 5 mg ml^{-1} of HRP in 0.1 M phosphate buffer pH 7.0; 1.86 mg ml^{-1} of SPP in 0.005 M Tris-HCl, pH 8.1), PVI$_7$dmeOs (3.3 mg ml^{-1}) and PEGDGE (2.5 mg ml^{-1}). Defined amounts of hydrogels (film thickness) were applied on spectrographic graphite rods and the electrodes were cured at room temperature for 20 h in desiccated conditions. The electron-transfer pathway is outlined in scheme 4.

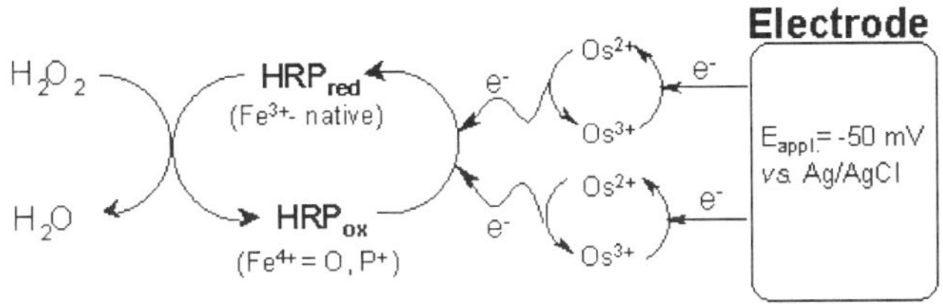

Scheme 4. Electron-transfer pathway in peroxidase containing redox hydrogels.

The comparative study concerning the bioelectrochemical characteristics of the developed biosensors showed that, irrespective of studied peroxidase, the biosensors' sensitivity was strongly influenced by the composition of the redox hydrogel, curing procedure, film thickness and applied potential. Therefore, all hydrogels were optimised with regard to these parameters. The optimal enzyme and cross-linker content was found to be of 15 - 40 % POD and 10 - 25 % PEGDGE, respectively. A loading of 22 μg per electrode was yielding the best results (adhesion, stability, and sensitivity). The current response increased with the applied potential and have shown a levelling off tendency at 0, -50 and around -100 mV vs. Ag/AgCl, KCl 0.1 M for TOP, SPP and HRP, respectively. Calibrations curves obtained using the optimal electrodes in a single manifold flow-injection system (see figure 5) with a flow-through wall-jet electrochemical cell [34] are presented in figure 6. The optimised SPP biosensor (48 %

PVI_7dmeOs, 23 % PEGDGE and 29 % SPP, w/w %) displayed the highest sensitivity for H_2O_2 (3.2 A $M^{-1}cm^{-2}$), a linear range up to 220 μM, a detection limit of 25 nM (calculated as twice the signal-to-noise ratio, 2S/N) and a response time ($t_{95\%}$) of about 2 min (see Table 3).

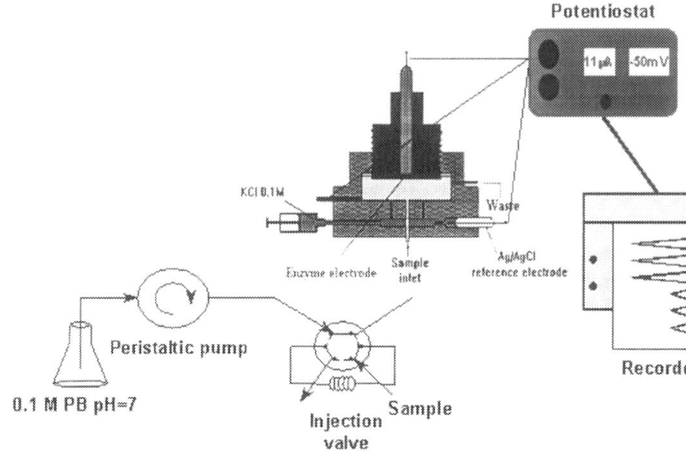

Figure 5. *Flow Injection Analysis system with bioelectrochemical detection.*

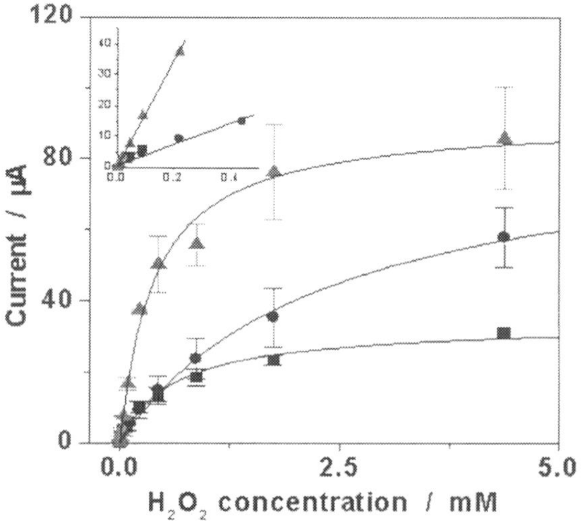

Figure 6. *Calibration curves obtained with HRP (■), TOP (●) and SPP (▲) modified electrodes of optimal composition. Experimental conditions: flow injection system, applied potential -50 mV vs. Ag/AgCl, KCl 0.1 M, flow rate of 0.5 ml/min; carrier 0.1M PB at pH 7.0.*

***Table 3**. Bioelectrochemical characteristics of peroxidase modified electrodes.*

Enzyme	Sensitivity ($\mu A\ mM^{-1}$)[a]	Linear range (μM)	Detection limit (nM)[b]
HRP	55 ± 12	0.5 - 130 R = 0.9981	35 ± 15
TOP	37 ± 6.2	0.5 - 470 R = 0.9958	81 ± 68
SPP	236 ± 38	0.5 - 220 R = 0.9988	25 ± 10

[a]calculated as the ratio between I_{max} and $K_M{}^{app}$ [b]estimated for 2S/N

The obtained results indicated the possibility of using some of the newly isolated peroxidases (e.g. SPP) instead of HRP in bi-enzyme sensor designs, especially in cases when the concentration of the target analyte is low, and therefore low detection limit and high sensitivities are crucial.

4.2. AMINE OXIDASE-BASED BIOSENSORS FOR MONITORING OF FISH FRESHNESS

Biogenic amines can act as possible biomarkers for control of food products [35-37]. Putrescine, cadaverine, tyramine and histamine are the most known compounds in this class, their concentration being a good indicator of fish, meat, and cheese freshness [38-42]. Biogenic amines are generally produced by microbial decarboxylation of corresponding amino acids and their toxicological significance in food products is still unclear. However, they can cause severe effects, such as headache and facial flushing, even when consuming very small amounts of infested fermented beverages and/or food [43, 44]. Many enzymatic methods have been developed for measuring biogenic amines in blood, biological tissues and food products [45-51], most of them being based on amine oxidase (AO) which catalyses the following reaction (1):

$$R\text{-}CH_2\text{-}NH_2 + H_2O + O_2 \rightarrow R\text{-}CHO + H_2O_2 + NH_3 \qquad (1)$$

We recently reported on the development, characteristics, and application of amperometric graphite electrodes based on a newly isolated and characterised AO, both in a mono- [30] and bi-enzymatic (co-immobilised AO and HRP) design [31], either in the presence or in the absence of an electrochemical mediator (Os-based redox polymer). The grass-pea AO used during this work is a newly described copper-containing enzyme, which besides the metal ions also contains an organic cofactor with a quinoid structure (topa quinone) in its catalytic site [52, 53], showing the possibility of transferring electrons directly to the graphite electrode [32]. The working principle of the mediated mono- and bi-enzyme biosensor architectures is presented in schemes 5 and 6, respectively.

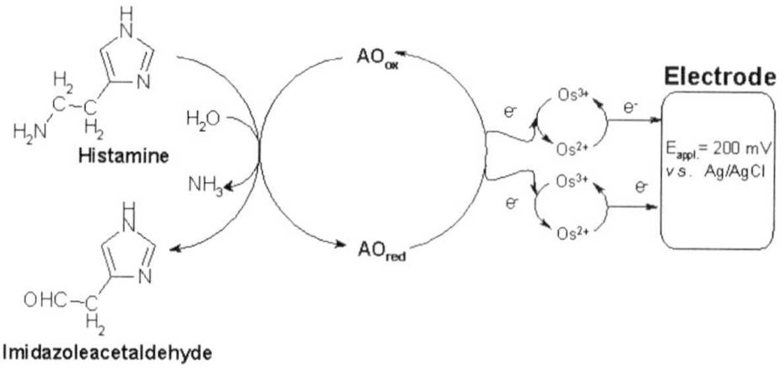

Scheme 5. Electron-transfer pathway for a mono-enzyme histamine electrode.

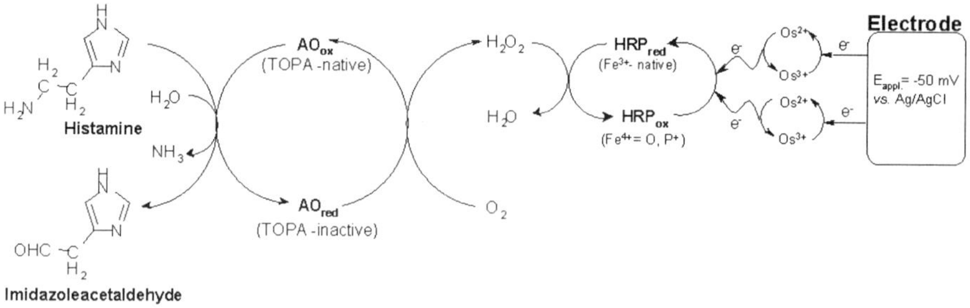

Scheme 6. Electron-transfer pathway for a bi-enzyme histamine electrode.

Mono- and bi-enzyme electrodes were prepared following previously published protocols [30, 31]. Briefly, unmediated electrodes were prepared by placing a defined amount of AO (6 μl of 5 mg/ml in phosphate buffer at pH 7.2) or 6 μl of a premixed solution of AO and HRP (80 % AO and 20 % HRP; AO, as above, HRP 1.25 mg/ml in phosphate buffer pH 7.2). When integrating these enzymes in redox polymers, a premixed solution of AO, PVI_{13}dmeOs and PEGDGE (66.3 % AO, 27 % PVI_{13}dmeOs and 6.7 % PEGDGE, w/w %) was added on the top of the electrodes. Electrodes were cured overnight at 4 °C. The working potential of the mono-enzyme electrodes was +200 mV vs. Ag/AgCl while bi-enzyme electrodes were operated at -50 mV vs. Ag/AgCl, regardless using a direct or a mediated electron-transfer mechanism.

Generally, the bi-enzymatic electrodes showed considerably improved sensitivity (e.g. 0.073 A/Mcm^2 for histamine) and lower detection limits (e.g. 0.33 μM for histamine, 2S/N) as compared with the mono-enzymatic ones (0.007 A/Mcm^2 and 2.2 μM, respectively), especially in the presence of the electrochemical mediator (optimum composition of the sensing film; 49 % AO, 12 % HRP, 19.5 % PVI_{13}dmeOs and 19.5 % PEGDGE, w/w %).

The optimised redox hydrogel integrated bi-enzyme biosensor could be also applied for the measurement of biogenic amines in extracts of fish samples stored in different conditions (at 4 and 25 °C, respectively).

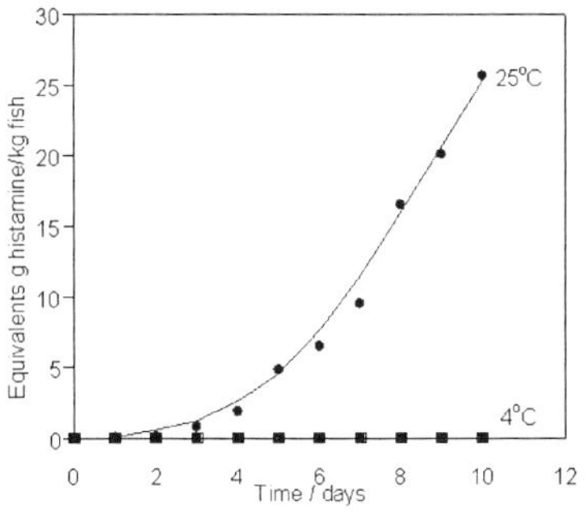

***Figure** 7. Monitoring of biogenic amines in fish sample extracts stored in different conditions. The total amine concentration is expressed in histamine equivalents.*

Internationally established regulation accept a maximum level of histamine in fish of 200 - 400 mg/kg [36]; as seen, the sample stored at room temperature already attained this value after 4 days of storage, showing thus, that the developed electrode is effective to screen food samples (detection limit: 0.33 μM for histamine, 0.17 μM putrescine and cadaverine, respectively at a sample throughput of 30 samples/hour). Interestingly, the different sensor designs display different selectivity patterns [30, 31]. The selectivity of the mono-enzymatic electrode for the biogenic amines decreased according to the row: histamine >> agmatine > spermidine > ethylenediamine > putrescine > cadaverine > Z, E-2-butene-1,4-diamino dihydrochloride. While the selectivity of the bi-enzymatic electrodes was following the order: putrescine > spermidine > cadaverine > histamine > agmatine > Z, E-2-butene-1,4-diamino dihydrochloride > ethylenediamine. By comparing these selectivity patterns it is obvious that using both types of electrodes it is not only possible to determine the sum of the biogenic amines, but also histamine alone with high specificity.

4.3. ALCOHOL BIOSENSORS BASED ON ALCOHOL DEHYDROGENASE

The importance of aliphatic alcohols and especially ethanol determination in clinical, industrial, food and environmental analysis, coupled with the generally high selectivity displayed by biosensors, resulted in the development of redox enzyme-based amperometric alcohol electrodes [54-56]. Alcohol oxidase [56-59], NAD^+-dependent alcohol dehydrogenase [60-62], PQQ-dependent alcohol dehydrogenase [33, 63] are the most studied enzymes in this context, all displaying inherent advantages and disadvantages. Alcohol oxidase is catalytically active for a range of short chain aliphatic alcohols including methanol, whereas NAD^+-dependent alcohol dehydrogenase is more specific for primary aliphatic and aromatic alcohols other than methanol [64] but requires the addition of its soluble cofactor that complicates the analysis system.

The present examples describes the development of biosensors based on a newly isolated quinohaemoprotein alcohol dehydrogenase from *Gluconobacter sp. 33* which contains several cofactors (haem and pyrroloquinolin quinone /PQQ, denoted QH-ADH). Three different biosensor designs have been tested and optimised: (i) Type I electrodes were based on the enzyme simply adsorbed on the electrode surface (direct electron transfer) (see scheme 7a), (ii) Type II electrodes integrated the enzyme into an Os-modified redox polymer (mediated electron transfer) (see scheme 7b), and (iii) Type III electrodes entrapped the same enzyme into a conducting polymer network (polypyrrole) (see scheme 8).

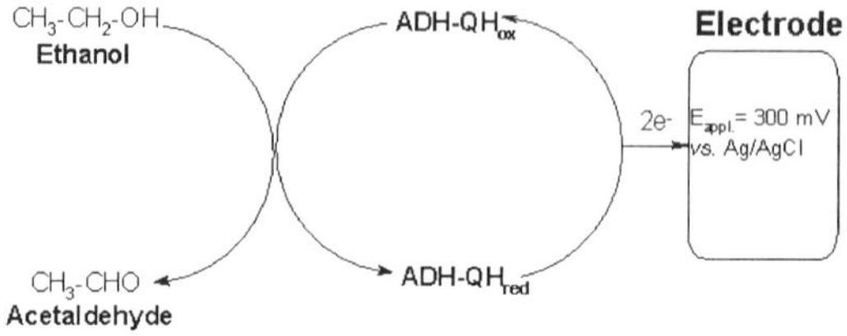

Scheme 7a. *Electron-transfer pathway for Type I QH-ADH electrodes.*

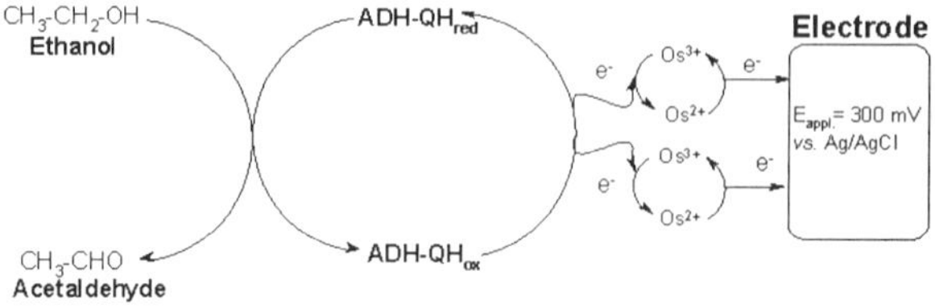

Scheme 7b. *Electron-transfer pathway for Type II QH-ADH electrodes.*

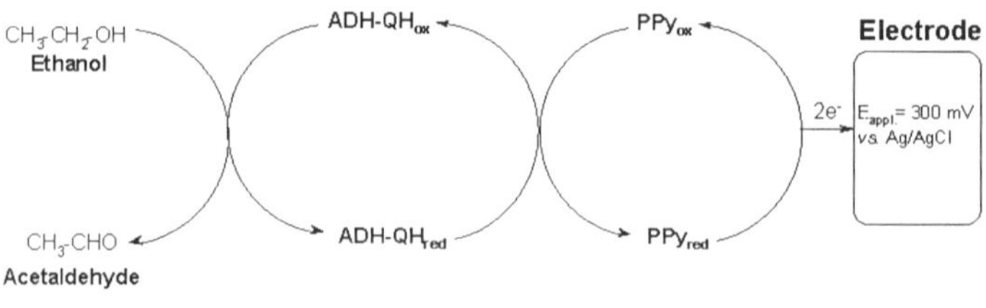

Scheme 8. *Electron-transfer pathway for Type III QH-ADH electrodes.*

The ethanol sensors were prepared as follows: Type I, screen printed graphite electrodes were coated with 1 μL of QH-ADH (2.5 mg/ml in 50 mM sodium acetate buffer, at pH 6.0 and 100 mM KCl) and dried overnight at 4 °C in refrigerator. Type II, screen printed graphite electrodes were prepared by placing the same amount of a premixed solution of QH-ADH, redox polymer and crosslinker/PEGDGE (36 % QH-ADH, 57 % $PVI_{13}dmeOs$ and 7 % PEGDGE, w/w %). When preparing type III electrodes, QH-ADH has been entrapped into a polypyrrole film during its electrochemical-induced formation following a potential-pulse profile as previously described [33]. A solution containing 2.5 mg/ ml QH-ADH, 100 mM pyrrole and 100 mM KCl was used for the electrochemical formation of the conducting film by applying 30 potential pulses from 950 mV (1s) to 350 mV (10s).

The calibration plots for ethanol obtained for the Type I and Type II electrodes operated at +300 mV *vs.* Ag/AgCl showed typical Michaelis Menten profiles (see figure 8). However, Type II electrodes were characterised by increased sensitivity due to a more efficient electron-transfer pathway in the presence of the polymer-bound Os-mediator. The response of the optimised redox hydrogel-based biosensor was linear in the range 5 - 100 μM and displayed a detection limit of 0.97 μM ethanol (defined as 2S/N ratio).

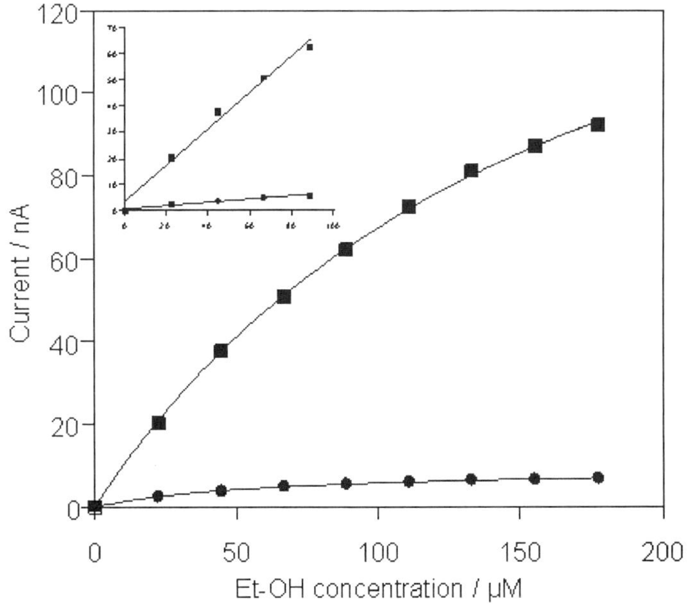

Figure 8. *Calibration curves for ethanol obtained for Type I (QH-ADH, ●) and II (QH-ADH-PVI13dmeOs-PEGDGE, ■) electrodes. Experimental conditions: batch system, applied potential +300 mV* vs. *Ag/AgCl.*

Type III electrodes were based on a direct electron-transfer between the polypyrrole (PPy) entrapped QH-ADH and a platinised Pt-electrode or glassy carbon *via* the conducting-polymer network [33]. These electrodes displayed very different bioelectrochemical characteristics as compared with Type I and II ones, as discussed below more in details.

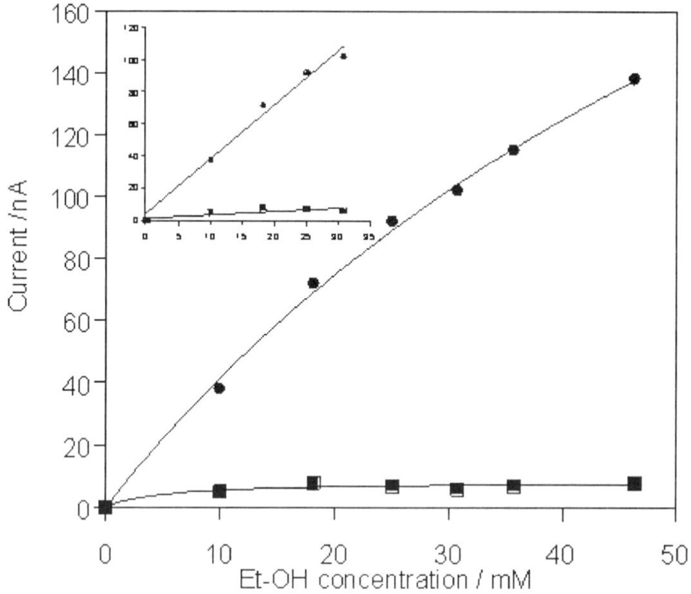

Figure 9. *Calibration curves for ethanol obtained for Type III electrodes in the presence (●) or absence (■) of QH-ADH. In the absence of QH-ADH, bovine serum albumin was used. Experimental conditions: batch system, applied potential +300 mV* vs. *Ag/AgCl.*

The multi-subunit enzyme has been integrated into the polymer film in an active conformation, demonstrated by the current generated in the presence of ethanol and phenazine methosulphate (PMS) as a free-diffusing redox mediator (constant-potential amperometry). Although the diffusion of the mediator into the QH-ADH/PPy film may be slow due to the properties and morphology of the conducting-polymer network, a current-concentration curve could be obtained saturating at about 5-6 mM of ethanol. Addition of ethanol - in the absence of any free-diffusing redox mediator - up to concentrations of 100 mM, gave rise to an unexpected increase of the steady-state current. This observed current was attributed to a possible internal electron-transfer pathway involving the different enzyme-integrated redox sites (PQQ and haem), located in the different subunits of the enzyme. It was therefore anticipated that the alcohol is primarily oxidised via the PQQ-site, which might be regenerated in a subsequent step by heam reduction of the haem units located in subunits I and II. Control experiments using the enzyme adsorbed at a platinised Pt-surface did not show a significant catalytic current and hence clearly demonstrated the efficient electron-transfer pathway via the conducting polymer chains. Since a direct electron transfer implies by definition, that the redox reaction should occur close to the formal potential of the involved active site, cyclic voltammograms have been recorded in oxygen-free 50 mM acetate buffer containing different ethanol concentrations. The increase of a redox wave at potentials of +190 mV (haem oxidation) supported this hypothesis. The co-operative action of the enzyme-integrated prosthetic groups - PQQ and haem- is assumed to allow this electron-transfer pathway from the enzyme's active site to the conducting polymer backbone. This unusual electron-transfer pathway leads to an accentuated increase of the K_M^{app}-value (up to about 100 mM in dependence from the polymer-film thickness and the electrode material [33]) and hence to a significantly increased linear detection range of an ethanol sensor based on this enzyme.

By changing the electrode material from Pt to glassy carbon a similar electron-transfer pathway via the conducting polymer chains could be obtained. In figure 9, the related calibration plot recorded on glassy carbon type III electrodes for increasing ethanol concentrations is shown, characterising the described ethanol sensor by a linear range of up to about 25 mM, a sensitivity of 4.54 μA/M, and an apparent K_M^{app} of 80.86 mM.

Comparing the calibration curves obtained with the studied electrode types one can see the great flexibility regarding the linear range offered by the different immobilisation methods. The presented alcohol biosensors fulfil requirements for very different K_m values, from 62 μM (electrodes type I) and 170 μM (electrodes type II) up to 80 mM (electrodes type III based on glassy carbon) and 100 mM (electrodes type III based on platinum). The three different ethanol sensors clearly demonstrated the different electrochemical characteristics one can obtain using various sensor architectures, and thus the possibility of their tailoring for a particular application.

5. Enzyme-based amperometric biosensors for monitoring in different biotechnological processes

Although, as mentioned, biosensors are not yet widely spread in food and beverage industry, a number of enzyme-based electrodes have already been successfully applied to monitor fermentation processes, the biosensors being an essential part of a process control system. It has to be accentuated that the biosensors themselves, are only parts of the entire analysis system, where other components for sampling, for elimination of contamination, for sample transport, etc. are of equal importance. A detailed discussion of this subject is, however, beyond the scope of this chapter.

Biosensors have been considered for this application mainly because of their versatility offering the possibility of easy and automated on-line monitoring, replacing thus, off-line analysis which involved manual sampling and sample handling steps. Biosensors proved to be the right tools and therefore have been also used to develop feedback control strategies [84-90]. The implementation of the biosensor in the technological line is made usually in one of the following ways:

- the target substrate is directly-detected (in-situ) in the fermenter. The practical requirements for an in-situ biosensor, such as: sterilisation possibility, adequate measuring range, resistance to membrane fouling, have so far not been entirely met, which has precluded the widespread application of this approach [84]. Due to these requirements, measurements in an external flow stream or in on-line systems are more often applied. However, a mediated amperometric glucose biosensor for the in-situ monitoring of a pulse-fed baker's yeast cultivation on defined medium was already reported [85]. The biosensor displayed an improved stability (4 days of continuous use) and extended working range (up to 20 g/l). Also, an autoclavable glucose biosensor was used to monitor in-situ the fed-batch fermentation of *Escherichia coli* [84].
- the target substrate is sampled via a flow-injection system and is detected using an amperometric biosensor. A split-stream flow-injection analysis system was described for the simultaneous determination of glucose and L-glutamine in serum-free hybridoma bioprocess media. In this approach the system assayed 12 samples/h with a linear response to glucose in the range of 0.03 to 30 mM [86].
- the target substrate is sampled from the fermenter using a microdialysis system and is detected subsequently with a biosensor housed in a flow-injection system. The microdialysis system provides a cell-free dialysate while the flow-injection system permits a high sampling rate. Such a system was used to monitor glucose and lactate (up to 70 mM) in lactic acid fermentation of *Lactobacillus delbrueckii*. The sensor system monitored glucose and lactate concentrations during a 24 h long fermentation process, without any interfering signals, as confirmed with a conventional (colorimetric) method [87]. The same fermentation process was also monitored by coupling the microdialysis sampling with a flow-through electrochemical cell housing both a glucose and a lactate biosensor. The system was characterised by a sampling frequency of 15 h-1 and a delay between sampling and detection of less than 3 minutes. Obtained results were confirmed with a standard off-line analysis using HPLC [88]. An interesting study compared the characteristics of such analytical systems with those obtained for an off-line system, based on manual sampling and clean-up, and column liquid chromatography in combination with refractive index detection [89].
- the target substrate is sampled via an automated analyser, passed through an oxidase-immobilised mini-reactor, monitoring the produced hydrogen peroxide by amperometry. Such a system was combined with a column switching valve downstream from the injector for monitoring of glucose, ethanol and glutamate during the fermentation of aged fish sauce in a fermenter loaded with *Torulopis versatiles*-immobilised beads [90].

An example from our laboratory illustrates the monitoring of glucose and ethanol during the fermentation process of Tokay wine (see Fig. 10). Commercially available glucose and ethanol biosensors were purchased from SensLab (Leipzig, Germany), integrated into an on-line sampling and detection system (OLGA, Institut für

Bioanalytik, Göttingen, Germany [91]) and their characteristics were evaluated and compared with those obtained using reagentless biosensors developed in our laboratories (see section 4.3).
As seen from the figure 10 the used sequential-injection analyser with integrated biosensors was able to follow the decrease of glucose and simultaneous increase of the ethanol concentration in the expected way.

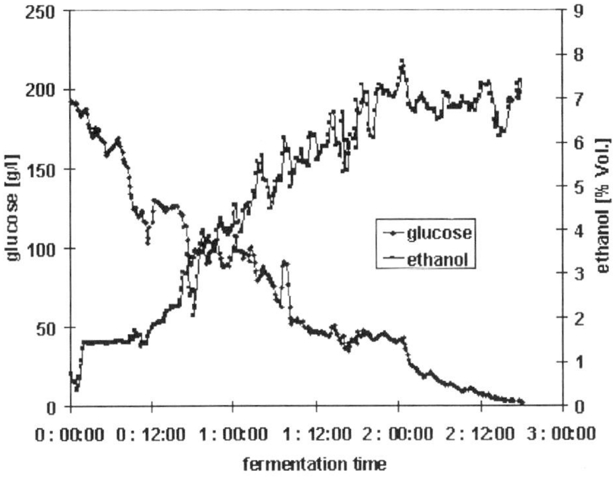

Figure 10. *Monitoring of glucose and ethanol during the fermentation process of Tokay wine (see text for more details).*

6. Conclusions

Many integrated biosensors fulfil the requirements for their analytical applications in food and beverage industry. Their bioelectrochemical characteristics (sensitivity, selectivity, and stability) combined with their simplicity in use and relative cheapness forecast a wide spread of these analytical tools in the field of production and control of various foodstuff and beverages.

Acknowledgements

The authors thank the following organisations for financial support: the European Commission (Contract No. IC15CT96-1008 and IC15-CT98-0907), the Swedish Council for Forestry and Agricultural Research (SJSF), and Swedish National Board for Industrial and Technical Development (NUTEK).

References

1. Nunes, G.S. and Barcelo, D. (1998) Electrochemical biosensors for pesticide determination in food samples, *Analysis* 26, M156-M159.
2. Turner, A.P.F. and Newman, J.D. (1998) An introduction to biosensors, Spec. Publ. -R. Soc. Chem 167, pp.13-27.
3. Loung, J.H.T., Bouvrette, P. and Male, K.B. (1997) Developments and applications of biosensors in food analysis, *Trends Biotechnol.* 15, 369-377.
4. Maines, A., Ashworth, D. and Vadgama, P. (1996) Enzyme electrodes for food analysis, *Food Technol. Biotechnol.* 34, 31-42.
5. Warsinke, A. (1997) Biosensors in food analysis, in Scheller, F.W., Schubert, F. and Fedrowitz, J. (eds.), Frontiers in Biosensorics II Fundamental aspects, Birkhäuser Verlag, Basel, pp.121-139.
6. Campanella, L. and Tomassetti, M. (1996) Biosensors for food analysis in aqueous and non-aqueous media, *Food Technol. Biotechnol.* 34, 131-141.
7. Wagner, G. and Schmid, R.D. (1990) Biosensors for food analysis, *Food Biotechnol.* 4, 215-240.
8. Marcus, R.A. and Sutin, N. (1985) Electron transfer in chemistry and biology, *Biochim. Biophys. Acta* 811, 265-322.
9. Marcus, R.A. (1993) Electron-transfer reactions in chemistry. Theory and experiment (Nobel lecture), *Angew. Chem. Int. Ed. English* 32, 1111-1121.
10. Marko-Varga, G., Emnéus, J., Gorton, L. and Ruzgas, T. (1995) Development of enzyme-based amperometric sensors for the determination of phenolic compounds, *Trends Anal. Chem.* 14, 319-328.
11. Stredansky, M., Pizzariello, A., Stredanska, S. and Miertus, S. (1999) Determination of D-fructose in foodstuffs by an improved amperometric biosensor based on a solid binding matrix, *Anal. Commun.* 36, 57-61.
12. Parades, P.A., Parellada, J., Fernandez, V.M., Katakis, I. and Domínguez, E. (1997) Amperometric mediated carbon paste biosensor based on D-fructose dehydrogenase for the determination of fructose in food analysis, *Biosens. Bioelectron.* 12, 1233-1243.
13. Leca, B. and Marty, J.L. (1997) Reusable ethanol sensor based on a NAD^+ - dependent dehydrogenase without coenzyme addition, *Anal. Chim. Acta* 340, 143-148.
14. Ruzgas, T., Csöregi, E., Emnéus, J., Gorton, L. and Marko-Varga, G. (1996) Peroxidase-modified electrodes: Fundamentals and application, *Anal. Chim. Acta* 330, 123-138.
15. Lindgren, A., Tanaka, M., Ruzgas, T., Gorton, L., Gazaryan, I., Ishimori, K. and Morishima, I. (1999) Direct electron transfer catalysed by recombinant forms of horseradish peroxidase: insight into the mechanism, *Electrochem. Commun.* 1, 171-175.
16. Zimmermann, H., Lindgren, A., Schuhmann, W. and Gorton, L. (2000) Anisotropic orientation of horseradish peroxidase by its reconstitution on a thiol-modified gold electrode, *Chemistry - A European Journal,* 6, 592-599.
17. Csöregi, E., Jönsson-Petersson, G. and Gorton, L. (1993) Mediatorless electrocatalytic reduction of hydrogen-peroxide at graphite electrodes chemically-modified with peroxidases, *J. Biotechnol.* 30, 315-317.
18. Gorton, L., Bremle, G., Csöregi, E., Jönsson-Petersson, G. and Persson, B. (1991) Amperometric glucose sensors based on immobilised glucose-oxidising enzymes and chemically modified electrodes, *Anal. Chim. Acta* 249, 43-54.
19. Vreeke, M.S. and Heller, A. (1994) Hydrogen peroxide electrodes based on electrical connection of redox centres of various peroxidases to electrodes through a three-dimensional electron-relaying polymer network, in Usmani, M.U. and Akmal N. (eds.), Diagnostic Biosensor Polymers, ACS Symposium Series 556, Washington, pp. 180-192.
20. Munteanu, F.D., Lindgren, A., Emnéus, J., Gorton, L., Ruzgas, T., Csöregi, E., Ciucu, A., Huystee, R.B., Gazaryan, I.G. and Lagrimi, L.M. (1998) Bioelectrochemical monitoring of phenols and aromatic amines in flow injection using novel plant peroxidases, *Anal. Chem.* 70, 2596-2600.
21. Gáspár, S., Popescu, I.C., Gazaryan, I.G., Bautista, A.G., Sakharov, I.Y., Mattiasson, B. and Csöregi, E. (2000) Biosensors based on novel plant peroxidases; a comparative study, *Electrochim. Acta* in press.
22. Razumas, V., Kazlauskaite, J. and Kulys, J. (1992) Bioelectrochemistry of microperoxidases, *Bioelectrochem. Bioenerg.* 28, 159-176.

23. Tatsuma, T. and Watanabe, T. (1991) Peroxidase model electrodes: haem peptide modified electrodes as reagentless sensors for hydrogen peroxide, *Anal. Chem.* 63, 1580-1585.
24. Csöregi, E., Gorton, L., Marko-Varga, G., Tüdös, A.J. and Kok, W.T. (1994) Peroxidase-modified carbon-fiber microelectrodes in flow-through detection of hydrogen-peroxide and organic peroxide, *Anal. Chem.* 66-71, 3604-3610.
25. Kulys, J. and Schmid, R.D. (1990) Mediatorless peroxidase electrode and preparation of byenzyme sensors, *Bioelectrochem. Bioenerg.* 24, 305-311.
26. Kenausis, G., Chen, Q. and Heller, A. (1997) Electrochemical glucose and lactate sensors based on "wired" thermostable soybean peroxidase operating continuously and stable at 37 °C, *Anal. Chem.* 69, 1054-1060.
27. Lindgren, A., Emnéus, J., Ruzgas, T., Gorton, L. and Marko-Varga, G. (1997) Amperometric detection of phenols using peroxidase-modified graphite electrodes, *Anal. Chim. Acta* 357, 51-62.
28. Heller, A. (1990) Electrical wiring of redox enzymes, *Acc. Chem. Res.* 23, 128-134.
29. Heller, A. (1992) Electrical connection of enzyme redox centers to electrodes, *J. Phys. Chem.* 96, 3579-3587.
30. Niculescu, M., Frébort, I., Peč, P., Galuska, P., Mattiasson, B. and Csöregi, E. (2000) Amine oxidase based biosensors for histamine detection, *Electoanalysis,* 5, 369-375.
31. Niculescu, M., Nistor, C., Frébort, I., Peč, P., Mattiasson, B. and Csöregi, E. (2000) Redox hydrogel based amperometric bienzyme electrodes for fish freshness monitoring, *Anal. Chem., 72, 1591-1597.*
32. Niculescu, M., Ruzgas, T., Nistor, C., Frébort, I., Šebela, M., Peč, P., Mattiasson, B. and Csöregi, E. (2000) Electrooxidation mechanism of biogenic amines at amine oxidase modified graphite electrode, *Anal. Chem.,* in press.
33. Ramanavicius, A., Habermüller, K., Csöregi, E., Laurinavicius, V. and Schuhmann, W. (1999) Polypyrrole entrapped quinohaemoprotein alcohol dehydrogenase. Evidence for direct transfer via conducting polymer chains, *Anal. Chem.* 71, 3581-3586.
34. Appelqvist, R., Margo-Varga, G., Gorton, L., Torstensson, A. and Jönsson, G. (1985) Enzyme determination of glucose in a flow system by catalytic oxidation of the Nicotinamide coenzyme at a modified electrode, *Anal. Chim. Acta* 169, 237-239.
35. Yang, X. and Rechnitz, G.A. (1995) Dual Enzyme Amperometric Biosensor for Putrescine with Interference Suppression, *Electroanalysis* 7, 105-108.
36. Chemnitius, G.C. and Bilitewski, U. (1996) Development of screen-printed enzyme electrodes for the estimation of fish quality, *Sens. Actuators* B 32, 107-113.
37. Draisci, R., Volpe, G., Lucentini, L., Cecilia, A., Frederico, R. and Palleschi, G. (1998) Determination of biogenic amines with an electrochemical biosensor and its application to salted anchovies, *Food Chem.* 62, 225-232.
38. Chemnitius, G.C., Suzuki, M. and Isobe, K. (1992) Thin-film polyamine biosensor: substrate specificity and application to fish freshness determination, *Anal. Chim. Acta* 263, 93-100.
39. Male, K.B., Bouvrette, P., Luong, J.H.T. and Gibbs, B.F. (1996) Amperometric biosensor for total histamine, putrescine and cadaverine using diamine oxidase, *J. Food Sci.* 61, 1012-1016.
40. Volpe, G. and Mascini, M. (1996) Enzyme sensors for determination of fish freshness, *Talanta* 43, 283-289.
41. Yano, Y., Yokoyama, K., Tamiya, E. and Karube, I. (1996) Direct evaluation of meat spoilage and a progress of ageing using biosensors, *Anal. Chim. Acta* 320, 269-276.
42. Bouvrette, P., Male, K.B., Luoung, J.H.T. and Gibbs, B.F. (1997) Amperometric biosensor for diamine using diamine oxidase purified from porcine kidney, *Enz. Microb. Technol.* 20, 32-38.
43. Taylor, S.L., Hiu, J.Y. and Lyons, D.E., (1984)Toxicology of scombroid poisoning, in Ragelis, E.P. (ed.), Seafood Toxins, ACS Symposium Series 262, Washington, pp 417-430.
44. Stratton, J.E., Hutkins, R.W. and Taylor, S. (1991) Biogenic amines in cheese and other fermented foods: a review, *J. Food Protect.* 54, 460-470.
45. Karube, I., Satoh, I., Araki, Y. and Suzuki, S. (1980) Monoamine oxidase electrode in freshness testing of meat, *Enzyme and Microbial Technology* 2, 117-120.
46. Matsumoto, T., Suzuki, O., Katsumata, Y., Oya, M., Suzuki, T., Nimura, Y. and Hattori, T. (1981) A New Enzymatic Assay for Total Diamines and Polyamines in Urine of Cancer Patients, *J. Cancer Res. Clin. Oncol.* 100, 73-84.
47. Stevanato, R., Mondovi, B., Sabatini, S. and Rigo, A. (1990) Spectrophotometric assay for total polyamines by immobilised amine oxidases, *Anal. Chim. Acta* 273, 391-397.

48. Gasparini, R., Scarpa, M., Di Paolo, M.L., Stevanato, R. and Rigo, A. (1991) Amine oxidase amperometric biosensor for polyamines, *Bioelectrochem. Bioenerg.* 25, 307-315.
49. Gasparini, R., Scarpa, M., Vianello, F., Mondoví, B. and Rigo, A. (1994) Renewable miniature enzyme-based sensing devices, *Anal. Chim. Acta* 294, 299-304.
50. Xu, C.X., Marzouk, S.A.M., Cosofret, V.V., Buck, R.P., Neuman, M.R. and Sprinkle, R.H. (1997) Development of a diamine biosensor, *Talanta* 44, 1625-1632.
51. Tombelli, S. and Mascini, M. (1998) Electrochemical biosensors for biogenic amines: a comparison between different approaches, *Anal. Chim. Acta* 358, 277-284.
52. Šebela, M., Luhová, L., Frébort, I., Hirota, S., Faulhammer, H.G., Stužka, V. and Peč, P. (1997) Confirmation of the presence of a Cu(II)/topa quinone active site in the amine oxidase from fenugreek seedlings, *J. Exp. Bot.* 48, 1897-1907.
53. Šebela, M., Luhová, L., Frébort, I., Faulhammer, H.G., Hirota, S., Zajoncová, L., Stužka, V. and Peč, P. (1998) Analysis of active sites of copper/topa quinone-containing amine oxidases from *Lathyrus odoratus* and *Lathyrus sativus* seedlings, *Phytochem. Anal.* 9, 211-222.
54. Nanjo, M. and Guibault, G.G. (1975) Amperometric determination of alcohols, aldehydes and carboxylic acids with an immobilised alcohol oxidase enzyme electrode, *Anal. Chim. Acta* 75, 169-180.
55. Baratti, J., Courdec, R., Cooney, C.L. and Wang, D.I.C. (1978) Preparation and properties of immobilised methanol, *Biotechnol. Bioenerg.* 20, 333-348.
56. Vijayakumar, A.R., Csöregi, E., Heller, A. and Gorton, L. (1996) Alcohol biosensors based on coupled oxidase-peroxidase systems, *Anal. Chim. Acta* 327, 223-234.
57. Kulys, J. and Schmid, R.D. (1991) Bienzyme sensors based on chemically modified electrode, *Biosens. Bioelectron.* 6, 43-48.
58. Gorton, L., Jönsson-Pettersson, G., Csöregi, E., Johansson, K., Domínguez, E. and Marko-Varga, G. (1992) Amperometric biosensors based on an apparent direct electron-transfer between electrodes and immobilised peroxidases, *Analyst* 117, 1235-1241.
59. Johansson, K., Jönsson-Pettersson, G., Gorton, L., Marko-Varga, G. and Csöregi, E. (1993) A reagentless amperometric biosensor for alcohol detection in column liquid-chromatography based on co-immobilised peroxidase and alcohol oxidase in carbon-paste, *J. Biotechnol.* 31, 301-316.
60. Miyamoto, S., Murakami, T., Saiti, A. and Kimura, J. (1991) Development of an amperometric alcohol sensor based on immobilised alcohol-dehydrogenase and entrapped NAD^+, *Biosens. Bioelectron.* 6, 563-567.
61. Wang, J., Gonzalez-Romero, E. and Reviejo, A.J. (1995) Improved alcohol biosensor based on ruthenium-dispersed carbon-paste enzyme electrodes, *J. Electroanal. Chem.* 353, 113-120.
62. Wang, J. and Liu, J. (1993) Fumed-silica containing carbon-paste dehydrogenase biosensors, *Anal. Chim. Acta* 284, 385-391.
63. Ikeda, T., Kobayashi, D., Matsushita, S., Sagara, T. and Niki, K. (1993) Bioelectrocatalysis at electrodes coated with alcohol-dehydrogenase, a quinohaemoprotein with haem-c serving as a built-in mediator, *J. Electroanal. Chem.* 361, 221-228.
64. Woodward, R. (1990) Advances in autotrophic microbiology and one-carbon metabolism, in G.A. Codd et al. (eds.), Kluwer Academic, Dordrecht, pp. 193-225.
65. Niu, J. and Yang, L.J. (1999) Renewable-surface graphite-ceramic enzyme sensors for the determination of hypoxanthine in fish meat, *Anal. Commun.* 36, 81-83.
66. Campanella, L., Pacifici, F., Sammartino, M.P. and Tomassetti, M. (1998) New organic phase bienzymatic electrode for lecithin analysis in food products, *Bioelectrochem. Bioenerg.* 47, 25-38.
67. Leochel, C., Chemnitius, G.C. and Borchardt, M.Z. (1998) Amperometric bi-enzyme based biosensor for the determination of lactose with an extended linear range, *Z. Lebensm.-Unters. Forsch. A* 207, 381-385.
68. Nouguer, T. and Marty, J.L. (1997) Reagentless sensors for acetaldehyde, *Anal. Lett* 30, 1069-1080.
69. Vrbova, E., Marek, M. and Ralys, E. (1992) Biosensor for the determination of L-lysine, *Anal. Chim. Acta* 279, 131-136.
70. Belghith, H., Romette, J.L. and Thomas, D. (1987) An enzyme electrode for on-line determination of ethanol and methanol, *Biotechnol. Bioenerg.* 30, 1001-1005.
71. Sarkar, P., Tothill, I.E., Setford, S.J. and Turner, A.P.F. (1999) Screen-printed amperometric biosensors for the rapid measurement of L- and D- amino acids, *Analyst* 124, 865-870.

72. Amine, A., Patriarche, G.J., Marrazza, G. and Mascini, M. (1991) Amperometric determination of glucose in undiluted food samples, *Anal. Chim. Acta* 242, 91-98.
73. Wei, D., Lubrano, G.J. and Guilbault, G.G. (1995) Dextrose sensor in food analysis, *Anal. Lett.* 28, 1173-1180.
74. Xie, X., Kuan, S.S. and Guilbault, G.G. (1991) A simplified fructose biosensor, *Biosens. Bioelectron.* 6, 49-54.
75. Castano, M.J.L., Ordieres, A.J.M. and Blanco, P.T. (1997) Amperometric detection of ethanol with poly-(o-phenylenediamine)-modified enzyme electrodes, *Biosens. Bioelectron.* 12, 511-520.
76. Sprules, S.D., Hartley, I.C., Wedge, R., Hart, J.P. and Pittson, R. (1996) A disposable reagentless screen-printed amperometric biosensor for the measurement of alcohol in beverages, *Anal. Chim. Acta* 329, 215-221.
77. Serra, B., Reviejo, A.J., Parrado, C. and Pingarron, J.M. (1999) Graphite-Teflon composite bienzyme electrode for the determination of L-lactate: application to food samples, *Biosens. Bioelectron.* 14, 505-513.
78. Groom, C.A., Luong, J.H.T. and Masson, C. (1993) Development of a flow injection analysis-mediated biosensor for sulphite, *J. Biotechnol.* 27, 117-127.
79. Kinnear, K.T. and Monbouquette, H.G. (1997) An amperometric fructose biosensor based on fructose dehydrogenase immobilised in a membrane mimetic layer on gold, *Anal. Chem.* 69, 1771-1775.
80. Nunes, G.S., Skladal, P., Yamanaka, H. and Barcelo, D. (1998) Determination of carbamate residues in crop samples by cholinesterase-based biosensors and chromatographic techniques, *Anal. Chim. Acta* 362, 59-68.
81. Montagne, M. and Marty, J.L. (1995) Bi-enzyme amperometric D-lactate sensor using macromolecular NAD^+, *Anal. Chim. Acta* 315, 297-302.
82. Bilitewski, U., Chemnitius, G.C., Ruger, P. and Schmid, R.D. (1992) Miniaturised disposable biosensors, *Sensors and Actuators B* 7, 351-355.
83. Cerro, M.A., Cayuela, G., Raviejo, A.J. and Pingarron, J.M. (1997) Graphite-Teflon-peroxidase composite electrodes. Application to the direct determination of glucose in musts and wines, *Electroanalysis* 9, 1113-1119.
84. Phelps, M.R., Hobbs, J.B., Kilburn, D.G. and Turner, R.F.B. (1995) An autoclavable glucose biosensor for microbial fermentation monitoring and control, *Biotechnol. Bioeng.* 46, 514-524.
85. Bradley, J. and Schimd, R.D. (1991) Optimisation of a biosensor for in-situ fermentation monitoring of glucose-concentration, *Biosens. Bioelectron.* 6, 669-674.
86. Meyerhoff, M.E., Trojanowicz, M. and Palsson, B. (1993) Simultaneous enzymatic electrochemical determination of glucose and L-glutamine in hybridoma media by flow-injection analysis, *Biotechnol. Bioeng.* 41, 964-969.
87. Suzuki, M., Kumagai, T. and Nakashima, Y. (1999) On-line monitoring system of lactic acid fermentation by using integrated enzyme sensors, *Kagaku Kogaku Ronbunshu* 25, 177-181.
88. Min, R.W., Rajendran, V., Larsson, N., Gorton, L., Planas, J. and Hahn-Hagerdal, B. (1998) Simultaneous monitoring of glucose and L-lactic acid during a fermentation process in an aqueous two-phase system by on-line FIA with microdialysis sampling and dual biosensor detection, *Anal. Chim. Acta* 366, 127-135.
89. Buttler, T., Lidén, H., Jönsson, J.A., Gorton, L., Marko Varga, G. and Jeppsson, H. (1996) Evaluation of detection and sample clean-up techniques for on- and off-line fermentation monitoring, *Anal. Chim. Acta* 324, 103-113.
90. Chen, R.L.C. and Matsumoto, K. (1995) Sequential enzymatic monitoring of glucose, ethanol and glutamate in bioreactor fermentation broth containing a high-salt concentration by a multichannel flow-injection analysis method, *Anal. Chim. Acta* 308, 145-151.
91. Schuhmann, W., Wohlschläger, H., Huber, J., Schmidt, H.-L., Stadler, H. (1995) Development of Fermentation Processes an Extremely Flexible Automatic Analyzer with Integrated Biosensors for on-line Control of, *Anal. Chim. Acta.* 315, 113-122.

SUPPORTED LIPID MEMBRANES FOR RECONSTITUTION OF MEMBRANE PROTEINS

BRITTA LINDHOLM-SETHSON
Department of Chemistry, Analytical Chemistry, Umeå University, SE - 901 87 Umeå, Sweden

Abstract

Various methods for creation of supported lipid membranes suitable for incorporation of membrane proteins are described, including Langmuir-Blodgett techniques, self-assembly of thiolipids and/or phospholipids and fusion of vesicles. Practical applications that are discussed include ligand-receptor binding, immunosensing devices, membrane fluidity, ion-selective sensors and signal transduction from reconstituted membrane proteins.

1. Introduction

Scientists have been fascinated by the delicate structure of the cell membrane ever since the first indications of a bilayer structure were demonstrated [1]. This is not surprising since the plasma membrane that embraces all living cells is essential for life itself. Firstly, it serves as a selective filter that controls the entry of nutrients into the cell and the exit of waste products out of the cell. Moreover, the intra- and extra cellular fluids are generally quite dissimilar although the separating membrane is only a few nanometer thick. The concentration difference gives rise to an electrostatic trans-membrane potential that plays an essential role in a variety of biological processes including transport, bioenergetics and the propagation of nerve impulses.

The delicate balance is regulated by membrane proteins solvated in the phospholipid bilayer. Some of the membrane proteins have enzymatic functions whereas others serve as specific receptors or transporters. The integral membrane proteins are tightly bound to the hydrophobic part of the lipid bilayer and in most cases span the whole membrane, whereas the peripheral membrane proteins are bound only to one or the other face of the membrane.

Little is known about the function of many of these membrane proteins, whereas others are well known for their high selectivity. The challenge to the scientific community is therefore to find methods to create artificial plasma membranes, where

M. De Cuyper and J.W.M. Bulte (eds.), Physics and Chemistry Basis of Biotechnology, 131–165.

membrane proteins can be incorporated without loss of activity with the aim of studying their specific functions or in the development of biosensing devices.

2. Objective

2.1. THE PLASMA MEMBRANE

The lipid bilayer in the plasma membrane is composed mainly of three types of lipids where the phospholipids are the most abundant, but there are also significant amounts of cholesterol and glycolipids. The mixture of lipids in the inner and outer monolayers of the plasma membrane are different and a large variation is also prevalent in the lipid composition in membranes of different types of cells. The reason for the large variation in lipid composition is in most cases not well understood. The lipid bilayer is fluid and it is known that the presence of cholesterol increases the fluidity, elasticity and mechanical stability of the lipid bilayer and moreover it is believed to decrease the permeability of small water-soluble molecules.

One family of the glycolipids is the gangliosides, which contributes with up to 10% of the total lipid mass in the nerve cell membrane. The ganglioside, G_{M1}, for instance, binds bacterial toxins and has been used as a model receptor in supported lipid membranes to signal for cholera toxins [2-7]. The main function for the gangliosides is obviously not to signal for cholera toxin, but probably to serve as receptors in the signalling between cells. The surface of a biological cell membrane is covered with a layer of charges, that might be up to 20 nm thick. The contribution to this surface charge layer comes from a non-uniform ion distribution near the membrane surface that is governed by ordinary coulombic surface interactions, but also other ion-surface affinities that are not electrostatic to their nature. The dipole potential of zwitterionic amphiphiles must also be taken into account when considering the electrostatic potentials of the surface region [8]. The net charge of the biological membrane is most often negative and the charge density is rather low, typically $-(0.02\text{-}0.2)C{*}m^{-2}$. The electrostatic membrane surface potential plays an important role in the processes of membrane interaction, recognition and solute binding.

2.2. The artificial cell membrane

The discussion above underlines the complexity of the biological membrane and the problems the architect of an artificial lipid membrane faces. The quest to build a structure that totally mimics the plasma membrane in all its details is overwhelming. Instead one has to focus on the creation of an overall structure, which possesses the most important characteristics of the living cell membrane. The artificial membrane should therefore consist of a continuous bilayer of phospholipids where the inner core is composed of the hydrocarbon chains and the hydrophilic head groups in both leaflets are facing an aqueous environment. Furthermore, the phospholipids should be able to diffuse freely within the two monolayers. This is of vital importance for the long term stability of the artificial bilayer since it would impose self-healing properties to the bilayer. A fluid membrane is also of greatest significance for the successful incorporation of an active membrane protein.

It has been observed that the composition of the lipid layers is not of crucial importance for a successful incorporation of an active membrane protein. This is rather surprising considering the great effort Nature has taken to vary the lipid composition in the biological membrane. In practice, artificial cell membranes rarely consist of more than a few different phospholipids and in many cases only one has been used. Therefore they are much more homogeneous than the real plasma membrane and in some respects they differ significantly from the forerunner. The net surface charge of the artificial bilayer membrane is for instance often rather large and can vary between -0.4 and + 0.4 $C*m^{-2}$ and the effective interfacial width is only around 0.6 nm [8]. Furthermore, in many cases only one side of the biomimetic membrane faces an aqueous phase whereas the other one faces a solid support.

2.2.1. Unsupported artificial bilayer membranes

Two types of unsupported synthetic cell membranes have been used successfully for decades in experimental studies: Firstly, it has long been recognised that a solution of phospholipids under special conditions spontaneously forms uni- or multi-lamellar vesicles consisting of one or several hundred concentric lipid bilayer membranes. The size of the vesicles varies within large limits and they are sometimes classified according to their size. Thus D.D. Lasic distinguishes between large multilamellar vesicles (MLV´s), and large and small unilamellar vesicles (LUV´s and SUV´s) [9] and this terminology is also adopted in this paper. Vesicle studies have found many applications both in theoretical and experimental sciences, i.e.; topology investigations of two-dimensional surfaces in three-dimensional space, phase transition studies in two dimensions, artificial photosynthesis, drug delivery and medical diagnostics, etc. In biochemistry and biology the focus is on reconstitution of membrane proteins into artificial membranes and the study of model biological membranes. The long term stability of the vesicles is good but since it is an unsupported membrane the bioactivity of incorporated proteins cannot always be investigated in a straightforward way.

Secondly, so called black lipid membranes can easily be formed by painting a lipid solution across an aperture in a hydrophobic septum that separates two aqueous phases. In the pioneering work by Mueller et al variations in the dielectric properties of such an artificial membrane caused by spontaneous adsorption of various water-soluble macromolecules was reported [10]. Although these types of BLM provide excellent models for biological membranes they suffer from some severe drawbacks. The total surface area is small, typically less than 1 mm^2, most often only low protein densities can be reconstituted into them. Furthermore, they rarely last more than a few hours. Less fragile painted lipid membranes can be formed on polycarbonate ultrafiltration membranes for use in FIA-systems [11-13].

2.2.2. Supported artificial bilayer membranes (s-BLMs)

Originally, s-BLMs were introduced in investigations of the immune system particularly in studies of cell-cell and cell-membrane interactions and in an early review H.M. McConnell and co-workers point out the bright future for supported lipid membranes in cellular recognition events. The authors underline that the most important assignment is to find a technique to assemble fluid lipid bilayers, where integral membrane-spanning proteins can be incorporated with retained mobility and activity [14].

The stabilised biomembranes on solid supports permit long-termed investigations in contrast to the classical BLM. This is essential for fundamental membrane research but also in many practical applications [15,16]. Various techniques to create s-BLMs, with or without reconstituted membrane proteins, have been suggested in the literature and the details will be discussed in the forthcoming sections. However, firstly a brief outline is given.

2.2.2.1. Formation of s-BLMs. One suggested method is the building up of the biomembrane step by step on the electrode surface with either Langmuir-Blodgett, LB, techniques or self-assembly of amphiphiles or a combination of both [17-20]. Another method is to fuse small unilamellar vesicles to either supported phospholipid monolayers [21] or alkylated surfaces [22] or to incubate them with a hydrophilic surface [23]. When the correct thermodynamic conditions prevail, i.e. an attractive surface and/or the vesicles are under tension, a supported lipid bilayer results.

Different methods are proposed for creation of supported lipid membranes contacted by an aqueous phase on both sides to meet the requirement for a successful incorporation of membrane proteins. One suggestion involves the self assembly of ordered lipidic monolayers separated from the planar gold support with a hydrophilic spacer. This monolayer can serve as a support for consecutive Langmuir-Blodgett transfers, vesicle fusion [24] and/or formation of the second layer by detergent dilution [25]. Tien et al. have reported the self assembly of agar-gel supported bilayer membranes [26,27]. Sackmann et al have recently reported the combined LB and Langmuir-Schäfer, LS, transfer of phospholipid monolayers to a solid substrate covered with a water - swellable polymer film covalently linked to the surface [28].

2.2.2.2. Reconstitution of membrane proteins into the membrane. Langmuir-Blodgett techniques are not well suited for reconstitution of membrane proteins into artificial membranes, because of the risk for denaturation of the protein at the air/water interface. An alternative method is fusion to a solid support of unilamellar vesicles containing proteins [29]. Other methods include the prefabrication of a biomimetic structure and subsequent detergent dilution of a detergent solubilised protein [30] or to contact the preformed bilayer membrane with proteo-vesicles [31].

2.2.3. Various methods of investigation

Deposition of the lipid-protein film on a planar support enables a number of different methods for investigation of the lipid membrane and/or the biological activity of reconstituted membrane proteins. Here I will give only a brief introduction to a few of them and others will be introduced in the forthcoming chapters.

X-ray diffraction techniques have been widely used in determination of the structure of biological membranes. Hence, X-ray diffraction was employed to reveal details on the nature of the binding between an integral membrane protein, i.e.: a photoreaction center and an anionic lipid [32]; and the thickness and structure of films prepared from membrane lipids extracted from Archaebacteria [33]. Furthermore, this technique in combination with neutron reflectivity [34] made an estimation possible on the lipid head group hydration of a floating surface biotin-functionalised lipid monolayer that was bound to a monomolecular layer of the protein streptavidin [35]. Infrared spectroscopy is another appealing method for structural investigation of these systems. Accordingly, in a phosphatidic acid multilayer film the tilt angle of the alkyl chains was determined and from observation of dichroism in the infrared spectra it was indicated that the incorporated peptide gramicidin was in a channel forming conformation [36]. A single supported planar bilayer carries too little material to give a measurable signal in Solid State NMR. However, with the phospholipid bilayers deposited on silica beads in an aqueous suspension this problem was circumvented and it was possible to use NMR to investigate the dynamics of the partly structured water film between the solid support and the phospholipid headgroups [37].

With the Surface Forces Apparatus the interactions between two lipid layers can be measured with a method suggested for development of suitable lipid compositions for liposomes in drug delivery systems [38] and in Atomic Force Microscopy one could get a "snapshot" of the surface morphology of for instance an arachidic acid Langmuir-Blodgett (LB) film with incorporated glucose oxidase [39]. In biospecific interaction analysis surface plasmon resonance [40] and the shear acoustic waveguide [41] are two techniques that have both attracted attention in these systems.

Nevertheless, electrochemical techniques are considered more convenient and also more simple then the other techniques. This is particularly correct for signal transduction in biosensing devices. With an artificial membrane residing on an electrode surface the bioactivity of reconstituted proteins can in many cases easily be detected and transformed to an electrical signal. This is obvious for an oxidase, provided the electrons from the reduced protein find their way to the electrode and the oxidation current can be distinguished from the background. Moreover, a conformational change of a membrane protein located close to the electrode surface causes capacitive charging currents that can be detected with amperometry [42]. Another example is activation of an ion channel located in a highly resistant supported membrane. This imposes a change in the permeability of the membrane that can be detected for instance with electrochemical impedance spectroscopy, EIS [43].

EIS is acknowledged as an attractive technique for the characterisation of surface processes and is therefore frequently employed. It is particularly suitable for judging the quality of a deposited lipid bilayer since it provides a means to estimate the membrane resistance and capacitance, even in the presence of an ongoing electrochemical reaction. However, with respect to EIS some words of warning are called for. The method involves the application of a small-amplitude alternating voltage in a wide frequency range. The current response is characterised by the different time constants of the relaxation processes that are prevailing. This might involve the mass transport in an ion channel, the charging of the double layer capacitance in the electrolyte/electrode interface or the diffusion of an electroactive ion to the electrode surface. Two different methods are available to interpret the impedance spectra, both of them rely on a proper identification of exactly what is going on. In the first case a physical model is evaluated and the corresponding flux equations are solved [44]. Thereafter, the impedance data are fitted to the resulting equation. In the other case an equivalent circuit is proposed, most often consisting of resistances and capacitances in series and/or in parallel, reflecting the different suggested processes. Provided, the assumptions are correct, data fitting in both cases gives information on the electrochemical events in the interface. However, if the wrong equivalent circuit is chosen, data fitting yields numbers of little or no value, even in cases with good correlation between fitted and measured data. A general rule is: the more elements in the equivalent circuit the larger is the risk for misinterpretation. The best practice is to perform various experiments on the system, with different conditions [45].

3. s-BLMs in close contact with the solid support

Supported bilayer membranes that are deposited directly on solid support, glass or metal, are described in this section.

3.1. LANGMUIR-BLODGETT FILMS ON SOLID SUPPORTS

One of the most attractive methods to manipulate thin films at the molecular level is the Langmuir-Blodgett technique [18, 46-48]. Briefly, the technique involves spreading of water insoluble molecules at the surface of an aqueous subphase that fills a Teflon basin to the brim. On top of the tray one or more movable and in most cases hydrophilic barriers reside. The surface confined molecules are trapped inside a surface area that is defined by the walls of the trough and the position of the barriers. Thus it is possible to exactly control the mean molecular area of a single molecule to Ångström precision. In commercial troughs it is possible to slowly compress the monolayer to a predetermined surface pressure. The electronic circuits hold the target surface pressure at a constant level by small movements of the barriers back and forth. The compressed monolayer can then be transferred to a suitable substrate simply by moving it slowly and perpendicularly through the air/water interface as in the Langmuir-Blodgett, LB, techniques or by a horizontal dip as in the Langmuir-Schaefer, LS, techniques.

A monolayer consisting of phospholipids, or any other type of amphiphile, spontaneously orients itself at the air/water interface with the hydrophilic head groups solvated in the aqueous phase and the hydrophobic carbon chains pointing out into the air. If a hydrophilic substrate is withdrawn from the subphase, the phospholipid monolayer is transferred head down to the substrate and one half of an artificial cell membrane is obtained on the solid support. By dipping the substrate up and down through the compressed monolayer it should be possible to build multilayer films, at least theoretically. Two consecutive strokes should yield a complete phospholipid bilayer with a precisely controlled composition and surface concentration (Fig. 1).

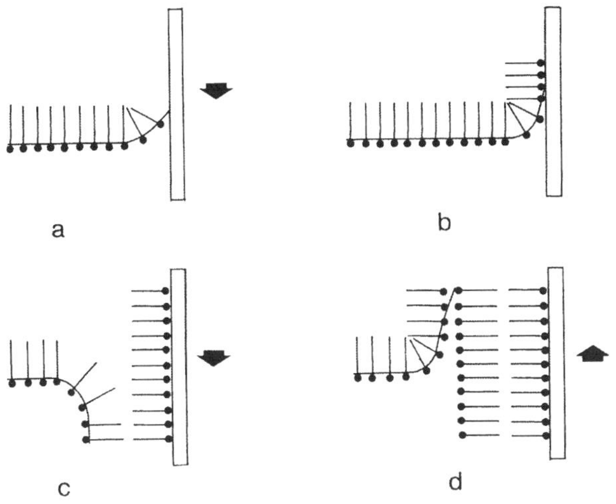

Figure 1. Deposition of multilayers by the Langmuir-Blodgett technique: a) first immersion, b) first withdrawal, c) second immersion, d) second withdrawal. Reprinted with permission from: Langmuir-Blodgett films, Roberts, G.G.(ed), Plenum press, New York, 1990, figure 2.10. page 28.

3.1.1. Pure phospholipid films

Thus, almost thirty years ago Procairone and Kauffman describes the electrical properties of bilayers of distearoyl phosphatidylcholine, DSL, in vacuum [49]. The bilayers were produced with Langmuir - Blodgett techniques and this was the first study on basic electrical properties of thin phospholipid films on solid support. They found a room temperature capacitance for the DSL bilayer within the range 0.35 - 0.40 $\mu F*cm^{-2}$ and could also observe phase transitions.

The close resemblance to biological membranes has made LB films very attractive to bioscientists. However, reality does not always agree with theory and it is unfortunately found that phospholipids in most cases do not deposit as bilayer structures. Usually, there is no problem with transfer of the first monolayer. However when attempting to transfer the second layer to complete the bilayer, the first monolayer transfers back to the air/water interface [18]. The problem is recognised as the poor adhesion of the first monolayer of phospholipids to the substrate. In some instances this has been circumvented by a combination of LB and LS techniques [50,51].

Factorial experimental design was used in an ambitious endeavour to improve the adhesion of the first phospholipid monolayer to glass, platinum and chromium support. The aim of the work was to find optimum conditions for subsequent deposition of the next phospholipid monolayer in the formation of stable bilayers on solid supports composed of biologically relevant phospholipids [19,20,52]. An appropriate composition of lipids was found with significant amounts of phosphatidylcholine that showed good adhesion to platinum surfaces. Furthermore, the presence of cholesterol was found to greatly improve the stability of the monolayer. The reason for this is not clear but the authors speculate on a condensing effect of cholesterol on the monolayer or that cholesterol is less hydrated than any phospholipid and thus less water is co-deposited. Another suggested explanation is that cholesterol counterbalances phase separation in the mixed phospholipid monolayer.

3.1.2. s-BLM as receptor surface

A monolayer consisting of a 1:1 mixture of dipalmitoylphosphatidylcholine and dimyristoylphosphatidic acid was transferred to a chromium electrode covered with cadmium arachidate and the membrane capacitance was recorded as polylysine was added to the subphase [53]. A significant decrease in the membrane capacitance indicates the adsorption of the protein to the membrane surface and demonstrates the function of the phospholipid head groups as a receptor surface.

3.1.3. s-BLM with ion channels and/or ionophores

The ionophore valinomycin is a mobile ion carrier that transports potassium ions across a lipid bilayer down its electrochemical gradient, by picking up the ion on one side of the membrane and delivering it at the other. Valinomycin-phospholipid multilayer LB-films consisting of three bilayers were transferred onto platinum wires [54]. With potassium in the subphase a significant decrease in the film impedance was obtained, which was interpreted as the presence of an active potassium binding valinomycin within the film.

Similarly, a mixed lipid/ionophore monolayer transferred to a silanised silicon electrode has shown good potential in development of admittance based sensors with selective response to various cations. Thus, a potassium ion sensor has been reported [55] and also a calcium selective ion sensor both with detection limits of ca. $1*10^{-6}$ M [56]. In the latter work the sensor showed a competitive response manifested in a decreasing calcium signal when other cations were present. The selective response of membranes with several ionophores was also investigated, where the different cations had their "own" ionophore. The response was additive, indicating a non-competitive response.

3.1.4. s-BLM with other integral membrane proteins

Microsomal cytochrome b_5 is a ubiquitous electron transport protein supplying electrons to oxidative enzymes. The anchor sequence penetrates only half the thickness of the membrane, which makes it suitable for LB-transfer. Thus in an early work, the LB-transfer of a proteolipid monolayer consisting of a mixture of dipalmitoylphosphatidyl-choline, and cytochrome b_5 to a platinum wire was reported [57]. Identical cyclic voltammograms revealed that the transferred film was stable for at least 48 h and a selective response was found for electron transfer to certain acceptor dyes. A plausible reason for the selectivity is the electrostatic interaction between the charged dye molecules and the negatively charged functional groups on the protein surface.

3.2. VESICLE FUSION

Langmuir-Blodgett techniques are in most cases not suitable for simultaneous deposition of lipid monolayers and integral proteins. The method has been applied successfully only with small ionophores and integral proteins that only span half the cell membrane as in the articles cited in the previous section. The obvious reason is that spreading mixtures of large integral proteins and phospholipids at the air/water interface imposes a risk for denaturation of the protein.

An alternative and very attractive method was presented by Brian and McConnell, comprising the incubation of small unilamellar vesicles, SUV, with alkylated or unalkylated glass coverslips to form planar membranes [58]. Lateral diffusion coefficients for the phospholipids were obtained from fluorescence recovery after photobleaching and showed that the membrane was fluid. When the glass slides were incubated with vesicles containing the transmembrane protein H-2K^k a planar membrane was formed on the surface where the reconstituted protein clearly was recognised by precursor cytotoxic T cells. However, the integral membrane protein did not show any measurable mobility which indicates that the protein interacts strongly with the supporting substrate. Thus, a facile way to investigate immunological responses utilising supported lipid bilayer membranes was demonstrated for the first time. However, a qualitative evaluation could not be carried out since the protein was immobilised. The paper soon acquired successors and the progress of the work is discussed in the forthcoming sections.

3.2.1. LB/vesicle method and/or direct fusion

3.2.1.1. Structure, fluidity and formation of s-BLMs. A monolayer of dipalmitoyl phosphatidic acid was transferred to an ATR-plate head-down with LB-techniques and was subsequently incubated with a SUV solution. The formation of a planar bilayer membrane and the subsequent adsorption of a local anaesthetic, oxybutylprocaine were monitored in situ with Attenuated Total Reflection, ATR Spectroscopy [59]. It was concluded that the adsorption of the anaesthetic did not involve significant penetration, because of the small lipid loss, 10%, upon desorption. In later work the same group describes the structure and stability as determined with FTIR-ATR of the same type of supported bilayer. The mean molecular area of the phospholipids, the hydrocarbon chain order, and in some cases the chain tilt were reported [60].

In an original paper on a new application of neutron reflectivity, SUVs were prepared from dimyristoylphosphatidylcohline with or without perdeuterated hydrocarbon chains. These were fused on quartz surfaces to form stable bilayer membranes. Specular reflection of neutrons was used for the first time to investigate the structure of a single lipid bilayer. Selective deuteration of the lipids provided the necessary conditions for an estimation of the thickness of the total bilayer, the head group and the hydrocarbon chain. In addition, the chain tilt angle and the degree of lipid hydration could be calculated and the presence of a thin water film between the lipid membrane and the quartz surface was verified and its thickness was determined to 30 ±10Å [61].

This is in good agreement with the later findings by Kalb and Tamm, who employed vesicles containing a fluorescent lipid and fused them on a LB monolayer. The photobleaching experiments revealed that both halves of the bilayer were at least partly mobile [21]. This was interpreted as indicating the presence of a thin aqueous film in between the bilayer and the solid support, which is good news for successful incorporation of membrane proteins. With cytochrome b_5 in the vesicles the formation rate of the bilayer increased and the formation of multilayers was suppressed. Furthermore, a significant fraction of the membrane protein was mobile but with a very small diffusion coefficient, i.e.:100 times smaller than expected [29]. In addition, in a recent NMR study of supported lipid bilayer on silica beads in an aqueous suspension it has been shown that the lipids in the inner leaflet diffuse about 50% slower than those in the outer layer [62].

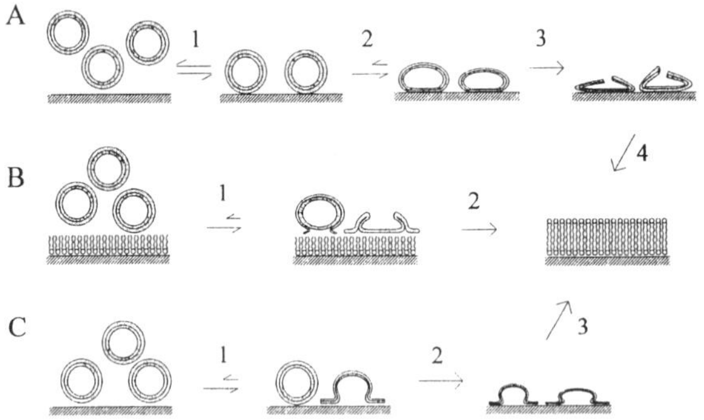

Figure 2*. Schematic representation of possible mechanisms for planar bilayer formation from vesicles on hydrophilic and hydrophobic surfaces. Reprinted from: Puu, G. and Gustafson, I. (1997) BBA Biomembr. 1327, 149-161 with permission from Elsevier Science.*

The observation of the increased fusion rates for vesicles with incorporated proteins as compared to pure vesicles, conforms well with studies on forces and molecular mechanisms linked to bilayer adhesion and fusion made with the surface force apparatus [63]. The authors concluded that the major force leading to direct bilayer fusion is hydrophobic interaction and they also suggest that one of the causes for fusion activation is local stress in the bilayer. The reconstitution of proteins into the vesicle probably causes an disclosure of the internal hydrophobic carbon chains and also locally stressed defects which both will assist fusion.

Vesicle fusion kinetics were recently investigated by G. Puu et al and accordingly an enhanced fusion rate was observed with proteins in the vesicles. The investigation comprised factorial experimental design with the option to identify factors important for adhesion and stability of the supported lipid bilayer [64]. Different mixtures of membrane forming lipids were employed in combination with proteins. Important factors for a successful formation of stable bilayer films were the underlying solid support and the presence of saturated phospholipids in the vesicles. It was also found that addition of calcium ions to the electrolyte had a crucial influence on fusion kinetics (Fig. 2).

Most membrane proteins carry a net charge or have an inhomogeneous charge distribution and are therefore expected to impose an electric field on adjacent phospholipids in the bilayer. In order to explore the effect of an electric field on a fluid lipid membrane, mixed bilayers containing charged and uncharged fluorescent probes were deposited onto microslides. In a two-dimensional microelectrophoresis experiment the mobility of the probe molecules in an electric field was estimated and also the frictional forces due to electroosmosis [65]. In a similar experiment, SUVs were produced from lipids with head groups that were either neutral or carried a net negative charge. The SUVs were incubated with a glass coverslip and a fluid bilayer was obtained consisting of a homogeneous mixture of phospholipids of different charges. When the model membrane is subjected to an electric field the charged molecules start to drift in the membrane until an equilibrium is reached. A thermodynamic model for the concentration distribution in the steady state is developed and confirmed with experiments. The resulting concentration profile that develops in the electric field is determined by, for instance, the number of carbon chains in the migrating phospholipid or the effect of lateral electrostatic interactions within the lipid layer [66].

3.2.1.2. s-BLM with membrane proteins.

Lactose permease is a transmembrane polypeptide, that carries lactose into the cell. The active transport is driven by the proton gradient across the membrane and one proton is co-transferred with every lactose molecule. Protein lactose permease was reconstituted into vesicles of Escherichia coli lipids, which afterwards were incubated with the Ta2O5/SiO2/Si heterostructures on an ISFET to form supported bilayer membranes with reconstituted lactose permease on the pH-sensitive surface [67]. When lactose was injected in the electrolyte facing the outer leaflet of the membrane, a rapid response was obtained, that was interpreted as originating from the active co-transport of lactose and protons across the membrane, thus changing the pH in the interface.

In a number of papers G. Puu and co-workers have demonstrated the usefulness of fusion of SUV on solid substrates [4,23,42,64,68]. Thus, stable bilayers could be formed from SUV with one of the following reconstituted proteins: bacteriorhodopsin, acetylcholinesterase, and the nicotinic acetylcholine receptor or cytochrome c oxidase. The vesicles were incubated with either a plain platinum support or a platinum surface covered with an LB-transferred monolayer of phospholipids. All reconstituted proteins possessed initial activity and bacteriorhodopsin and acetylcholine esterase retained their biological activity for several weeks [42]. In a recent publication, lipidic membranes containing biotinylated bacteriorhodopsin and the nicotinic acetylcholine receptor were formed on silicon support with vesicle fusion. The biotinylated proteins were labelled with streptavidin conjugated to colloidal gold. The subsequent atomic force microscopy in air showed that neither of the proteins forms aggregates within the membrane structure [68].

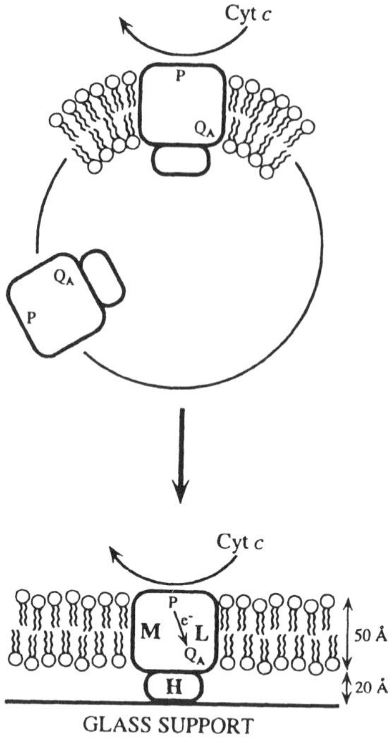

Figure 3. Schematic diagram of the bacterial photosynthetic reaction center in a bilayer membrane. The three protein subunits (L, M, and H) and the relevant functional components, the special pair primary electron donor, P, and the primary quinone acceptor QA are indicated. The RC is depicted in an orientation that is consistent with a mechanism in which the vesicles fuse to the glass support by opening out. Reprinted from: Salafsky, J., Groves, J.T. and Boxer, S.G. (1996) Biochemistry 35, 14773-14781 with permission from American Chemical Society.

A photosynthetic reaction center, RC, is a small transmembrane protein that accomplishes the initial photoinduced charge separation steps in photosynthesis. Light energy is transformed into a charge separation across the cell membrane with a resulting electro-chemical gradient that stores the energy. The internal charge separated state, P^+QA^-, is created by electron transfer from an electron donor site, P^+, to a quinone acceptor, QA^-, that is subsequently reduced by reduced cytochrome c. In the development of preparation strategies for supported fluid membranes containing orientated proteins, vesicle fusion was performed on planar glass coverslips with SUVs containing oriented RC, [69] (Fig. 3). In photobleaching experiments it was revealed that the phospholipids in the supported membrane were mobile, but the RC was immobilised. The recombination kinetics of P^+QA^- were monitored after a laser pulse in the presence or absence of ferrous cytochrome c. It was revealed that the protein was active and that more than 90 % of the protein had the desired orientation.

Three proteins, each with a glycan-phosphatidyl inositol, GPI, linkage were tethered to phospholipid vesicles which were fused with clean glass coverslips. The aim was to investigate the possibilities to fabricate surfaces consisting of laterally separated membrane patches of different and precisely controlled compositions Y. Proteins, bound to supported membranes via a GPI linkage are not immobilised. Hence, when the bilayer was subjected to an electric field, migration of the charged proteins commenced and concentration profiles were created. In some regions a protein density approaching close-packing was achieved and the experiments clearly showed that a precise manipulation of the spatial distribution was possible [70].

High resistance supported bilayer membranes were obtained by fusion of positively charged unilamellar vesicles and subsequent annealing to 60°C onto optically transparent semiconductor indium- tin oxide (ITO) planar electrodes. The surface of the electrode is negatively charged at the current pH, which promotes the fusion. Gramicidin or porin were reconstituted into the preformed supported planar membrane with a novel technique by incubation with a solution of vesicles containing the polypeptide. The transfer rate of gramicidin from the proteo-vesicles into the supported membrane was monitored with impedance spectroscopy. It was found that the membrane resistance decreased as a function of time, indicating a successful reconstitution of the protein into the supported bilayer [31].

The annexins constitute a class of ubiquitous membrane proteins that are known to bind strongly to anionic phospholipids in cell membranes in a calcium-dependent manner. The cellular role of this peripheral protein is not well understood and in a recent study supported lipid membranes were used to shed some light on the protein function. Model membranes consisting of phosphatidylserine and phosphatidylcholine were formed by fusion of SUV on glass coverslips and were later exposed to calcium ions and annexin V. The effect of annexin V binding to the fluidity of a planar supported lipid bilayer was investigated with photobleaching experiments and it was found that both the protein and phosphatidylserine were totally immobilised whereas the lateral mobility of phosphatidylcholine was significantly slowed down. This is in good accordance with a proposed model for the protein interaction with the cell membrane, i.e.: that annexin V forms a proteolipid complex, comprising an extended two-dimensional crystalline network that interacts strongly with phosphatidylserine. Thus, one of the functions of this particular membrane protein is to diminish the mobility of other membrane proteins and thus modulate their activities [71].

3.2.2. Hybrid bilayer membranes (HBMs)

The other route to form bilayer systems on planar supports with vesicle fusion as suggested by Brian and McConnell has been employed in a number of papers [58]. The method comprises fusion of phospholipid vesicles onto alkylated supports, as for instance gold electrodes modified with alkanethiols. In these cases a hybrid bilayer membrane, HBM, is formed where one half of it is immobile and anchored to the gold surface via a covalent bond. The obvious drawback is that only the outermost layer is mobile and the biomimetic surface is probably not optimal for incorporation of membrane-spanning proteins. However HBM is useful in studies of biological events that concern interactions at only one side of the lipid membrane, such as cellular immune responses or specific receptor-ligand bindings.

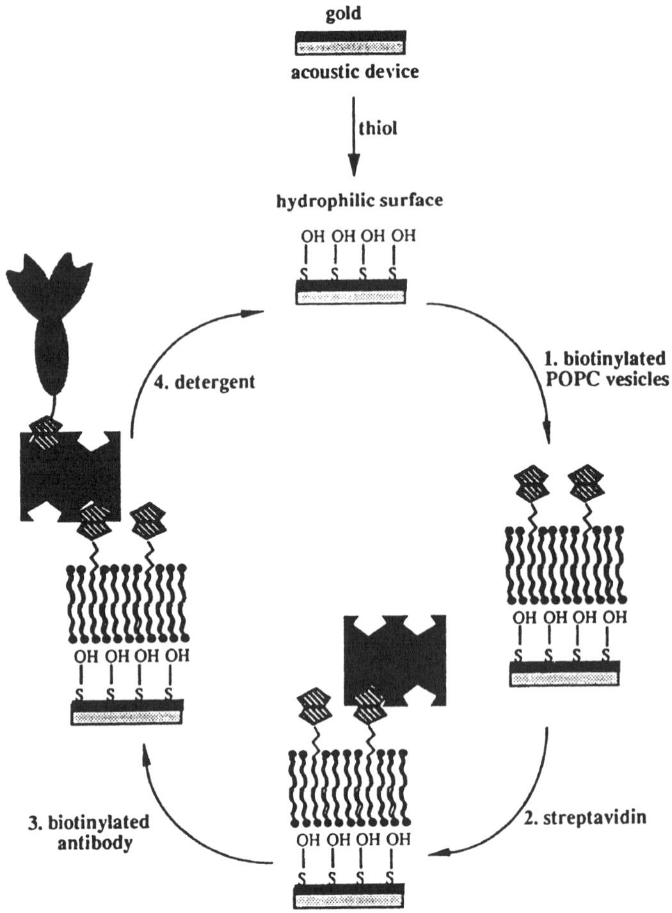

Figure 4. *Schematic representation of the modification of the surface. The gold coated device surface is rendered hydrophilic by self-assembly of a thiol monolayer. 1) Addition of biotinylated vesicles results in the spontaneous formation of a lipid bilayer on the hydrophilic surface. 2) streptavidin is attached to the surface through binding to biotin molecules. 3) a biotinylated antibody is coupled to streptavidin via one of the remaining streptavidin binding sites and 4) the surface is regenerated on addition of detergent. Reprinted from: Gizeli, E., Liley, M., Lowe, C.R. and Vogel, H. (1997) Anal. Chem. 69, 4808-4813 with permission from American Chemical Society.*

3.2.2.1. HBM as receptor surface and in immunological responses

HBM is a good candidate for investigations of ligand binding to receptors, particularly since the structure of the membrane allows the receptors to be mobile in the outer leaflet. Thus, Stelzle et al reported on molecular recognition of biotin by the bacterial protein streptavidin [72] where biotin was bound to the outermost membrane of the

HBM with a lipid anchor. In a similar system specific adsorption of neutravidin to the biotinylated membrane was reported [73].

As mentioned previously, the gangliosides represent receptor functions at the cell surface and among them, G_{M1} has been used as a model in many studies. With the ultimate goal of developing lipid-based receptors on solid supports, HBM were formed by fusion of unilamellar vesicles consisting of a mixture of phospholipids and G_{M1} on gold surfaces modified with alkanethiol [3]. The specific binding of cholera toxin to the membrane was monitored with surface plasmon resonance and a binding constant in good agreement with literature values was estimated. Later and in similar systems the ganglioside-peanut agglutinin [6], the ganglioside-lectin and ganglioside-toxin interactions were investigated [5].

Self-assembly of mercaptoundecanol on a gold surface produced a hydrophilic interface where vesicles were readily fused to form a lipid bilayer containing 3 mol% biotinylated phospholipids [74]. The specific binding of streptavidin and the subsequent binding of biotinylated antibodies were monitored with a shear acoustic waveguide. The device constitutes an acoustic immunosensor that successfully detects rabbit anti-goat I_gG in the concentration range $3*10^{-8}$ - 10^{-6} M (Fig. 4).

To study specific interactions of various lipophilic ligands with their analytes HBMs were formed by fusion of SUV on an octadecane-thiol self-assembled monolayer on a gold surface [75]. Thereafter, small acylated ligands (<1000 Da) were inserted into the preformed membrane simply by injection of dilute solutions across the preformed HBM. For larger molecules the vesicles had to be loaded with the receptor before incubation with the hydrophobic gold surface. The specific bindings of cry toxin and glycopeptide antibiotics were observed with surface plasmon resonance and kinetic analysis of the interactions at the model membrane was performed.

Another interesting investigation where HBM have been utilised involves the specific interactions of a lipid-anchored antigenic peptide with antibodies[76]. The current lipopeptide carried a peptide sequence analogous to one of the capsid proteins of the picorna virus causing foot- and-mouth decease in cattle. The specific binding of the monoclonal antibody to the peptide part of the lipopeptide was investigated and was reasonably well fitted to a Langmuir isotherm. Non-specific interactions are suppressed due to the lateral mobility of the lipids and lipopeptides in the outer leaflet of the HBM.

3.2.2.2. HBM and membrane proteins.

Melittin is a polypeptide consisting of 26 amino acids and constitutes the dominant component in bee venom. When it binds to the lipid bilayer it can effect cell death through pore formation. Phospholipid vesicles were fused on thiolated gold electrode surfaces and the permeability of the HBM was probed by performing cyclic voltammetry with ferrocyanide in solution. It was shown that the presence of melittin in solution decreases the insulating properties of HBM [77].

Bacteriorhodopsin, BR, is a light driven proton pump where a single photon induces a conformational change that results in the transport of a proton from the inside to the outside of the cell. HBMs were formed by incubation of BR containing purple membrane fragments with a thiolated gold surface. The activity of BR was certified as a light induced electrical current with a maximum coinciding with the adsorption maximum of bacteriorhodopsin. Moreover, in the same paper, the activity of three different types of reconstituted cationic ion pumps, i.e.: P-type ATPase was investigated. The ion pumps were activated by the creation of a swift concentration change of ATP, whereas transient electrical currents were observed [22].

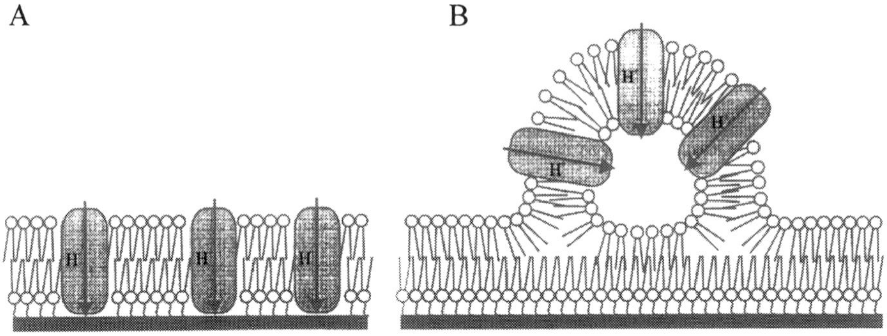

***Figure 5**. Schematic diagram of the proposed model of the lipid structures after the fusion of proteo-vesicles onto a thiolipid monolayer. A) BR molecules are inserted in the hybrid bilayer membrane. B)Vesicles are not fused completely; BR is mainly located in the unfused areas of the vesicles. Reprinted from:Steinem, C., Janshoff, A., Galla, H.J. and Sieber, M.(1997)Bioelectrochem. Bioenerg. 42, 213-220 with permission from Elsevier Science.*

In a later work bacteriorhodopsin was again employed as a model protein, when SUVs containing the membrane protein were fused on gold electrodes modified with the thiolipid: 1,2-dimyristoyl-sn-glycero-3-phosphothioethanol [78]. The objective was to construct a hybrid bilayer where the transmembrane protein could be inserted (Fig. 5). Upon irradiation of the supported lipid layer at various pH's photocurrents were detected with a maximum at pH 6.4 in good agreement with what is expected from literature values. However, the authors conclude that a perfect hybrid bilayer was not formed and BR is probably located in unfused vesicles as indicated in Fig. 5B. The important message is that a successful incorporation of transmembrane proteins in supported bilayer membranes, probably requires that both the upper and lower leaflets of the membrane are mobile.

The prime function of the membrane protein acetylcholinesterase, ACTHase, is to enzymatically hydrolyse acetylcholine to acetate and choline. It is a suitable membrane protein to incorporate in HBM systems since it does not span the membrane. It can be found in red blood cells were it is linked to the outer leaflet of the lipid bilayer. Hence, a membrane suspension of red blood cell ghosts was incubated with a hydrophobic alkanethiol gold electrode surface and the fusion was monitored with ellipsometry, surface plasmon resonance and AFM [79]. An increase in the thickness of the hybrid structure of 30 - 40 Å was obtained, indicating that a single layer of cell membrane was formed on the hydrophobic surface. The activity of ACTHase was confirmed with standard optical adsorbance measurements and it was shown that the protein retained its activity for several days.

The peripheral membrane protein, pyruvate oxidase, Pox, was inserted in a HBM formed on a thiolated gold surface by lipid vesicle fusion. The reconstitution was accomplished by incubation with a reaction medium containing the protein, via its small amphiphatic α-helix [80]. Ferricinium methanol was used as an artificial electron acceptor and the apparent Michaelis Menten constants of Pox both for pyruvate and ferricium were estimated, in fair agreement to what can be deduced from the literature [81].

A series of papers by F.M. Hawkridge and co-workers are attributed to studies of cytochrome c oxidase incorporated in an HBM. The biomimetic membrane is formed by immersing a thiolated gold surface in a buffer solution of deoxycholate, phospholipids and cytochrome c oxidase. The deoxycholate is removed in a dialysis procedure and a lipid bilayer with a controlled orientation of the reconstituted cytochrome c oxidase is suggested to form spontaneously on the gold surface. Cyclic voltammetry shows that the enzyme can be oxidised and reduced directly at the electrode surface and the mediated oxidation of reduced cytochrome c was reported [82]. However, in a later publication [83] the modified electrode was incorporated in a FIA system and the amperometric response from enzymatic oxidation of reduced cytochrome c was recorded with increasing concentration. A continuous increase in current is clearly recognised, surprisingly without any observation of saturation even at concentrations far above the reported Michaelis Menten constant [84]. The enzyme kinetics was investigated further and a turn-over rate was found, that was significantly lower than literature values [85]. The poor accordance with literature values on enzyme kinetics indicates that an intact lipid bilayer with an incorporated protein probably is not prevalent. This is not surprising, and again it is highlighted that HBM probably does not provide the ideal conditions for large integral membrane proteins. In the living cell membrane this particular protein extends its extramembraneous parts 35 Å into the aqueous phase on either side of the bilayer [86] which is not allowed for in this configuration.

3.3. SELFASSEMBLED BILAYERS ON SOLID OR GEL SUPPORTS

A phospholipid bilayer spontaneously forms on a solid support provided the right conditions are chosen. Thus, Hongyo and colleagues report on the formation of a self-assembled phospholipid/cholesterol bilayer on an agar support [87]. The incorporation of valinomycin into the membrane gave a large increase in conductance without affecting the capacitance. Oppositely, on adsorption of phloretin to the membrane surface the capacitance was affected more than the conductance.

An alkylated gold film was covered with a drop of a lipid solution and the electrode compartment was filled with an electrolyte solution. The thinning of the film to a hybrid bilayer membrane was monitored with surface plasmon microscopy. The formation of the biomimetic membrane was shown to dramatically decrease the conductivity of the system and also to effectively suppress the oxidation of ferrocyanide [88].

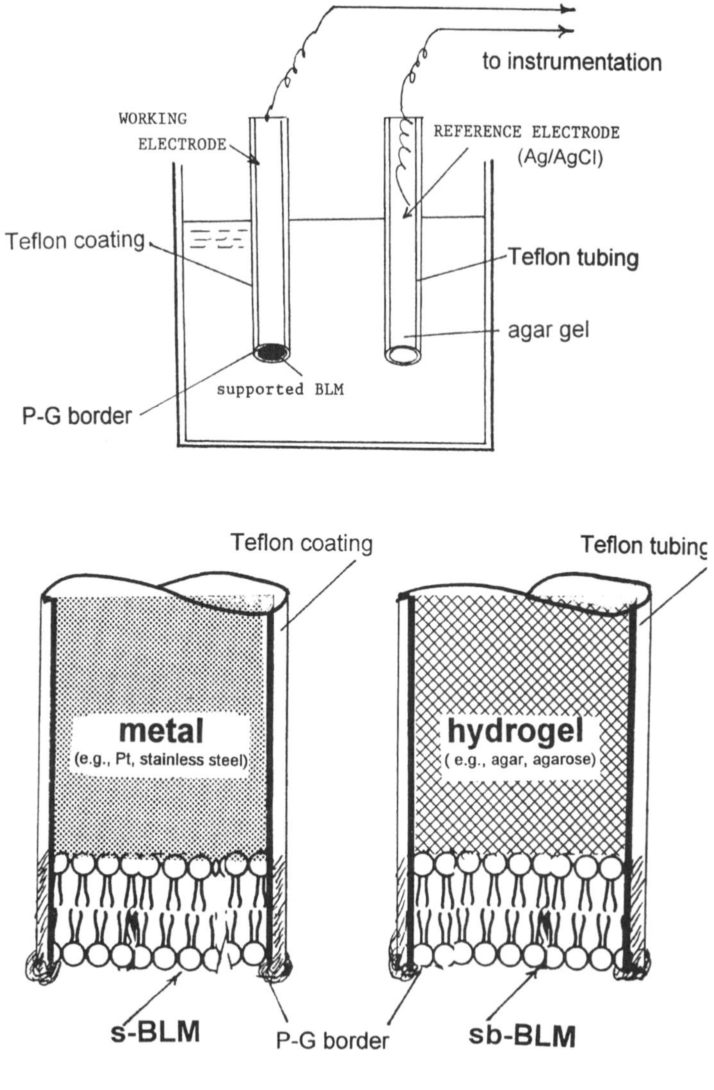

Figure 6. *Upper: Schematic illustration of a cell assembly used for determining the electrical properties of supported planar lipid bilayers. A Plateau-Gibbs border, a torous of lipid solution surrounds the BLM. Lower: Enlarged views of supported BLM-based probes. Showing on the left is a supported BLM on a metal substrate and on the right on agar gel. Reprinted from: Tien, H.T. and Ottova, A.L. (1998) Electrochim. Acta, 43, 3587-3610 with permission from Elsevier Science.*

Similar methodology as in the two preceding paragraphs has been adopted by a number of groups. Here I have chosen only two of them. Firstly: A small amount of lipid bilayer forming solution was spread on the surface of an indium oxide electrode and the binding of cytochrome c was investigated. Cyclic voltammetry indicated three different modes of electron transfer related to three types of protein-membrane interactions [89]. Secondly: Si/SiO2 surfaces were silanised and a lipid monolayer containing various surface concentrations of crown-ether was formed on the hydrophobic surface by a self-assembly procedure [90]. The crown-ether was provided with a carbon chain that links it to the lipid layer and the amount can easily be estimated with admittance measurements. Since the incorporation of the ionophore is slow, it can be monitored in real time and interrupted when the surface concentration has reached a predestined value. Thus the analytical range of the cation sensitive membrane can be precisely controlled.

An extremely simple method to produce lipid bilayers on metal supports was introduced by Tien and Salamon [91]. Firstly, a Teflon-coated metal-wire is introduced into a phospholipid solution where it is cut with a scalpel. Now, the phospholipid head groups spontaneously assemble on the exposed hydrophilic metal surface with the carbon chains pointing out into the aqueous phase. This renders the gold surface hydrophobic and a second layer is self-assembled on top of the first but with the opposite orientation (Fig. 6). The lipid coated metal wire is now transferred to an electrolyte solution where the lipid layer thins to a bilayer.

Later the same group extended the method to agarose-supported membranes, resembling the earlier work by Hongyo et al. [87]. The procedure is roughly the same as the one described above. That is, a Teflon tubing was filled with a hot electrolyte solution containing 2.5% agarose and after cooling it was cut with a scalpel. The freshly cut surface was dipped into a BLM-forming solution for 30 s thereafter it was moved to an aqueous solution were the lipid layer was thinned to a bilayer. These salt bridge supported BLMs are stable with no significant change in the electrical properties for at least 48h [27,26]. The self-assembled bilayers formed on a hydrogel are better models for real biological membranes, than those supported directly on the rough metal surface. This is of special importance for biosensors that rely on incorporation of bulky transmembrane proteins.

There are many interesting articles on biosensing devices based on the technology described in the two latest paragraphs. These are thoroughly reviewed by H.T. Tien and colleagues and the interested reader should consult them [92-96]. Thus there are investigations concerning: pH sensing, ion sensing, molecular sensing, immunological reactions, urea detection etc. Particularly I want to mention a cyanide minisensor consisting of s-BLMs on metal support containing methaemoglobin with low detection limits for cyanide, 4.9 nM and fast response time [97]; a metal-supported s-BLM operative as an acetylcholine minisensor for determinations of environmental pollutants, such as organophosphorous Trichlorphon and eserine, by biotin/avidin technology [98]; and finally an investigation concerning the effect on membrane conductance on an agar supported s-BLM, imposed by the presence of neuropeptides in the surrounding electrolyte [99].

4. s-BLMs with an aqueous reservoir trapped between the solid support and the membrane

For supported lipid bilayers intended as a model systems for natural biological membranes where membrane proteins can be incorporated with retained activity it is beneficial that the membrane is surrounded with an aqueous phase. When the phospholipid layers are transferred directly onto a metal support, the membrane is only separated from the electrode surface with an ultrathin layer of water. This water film is sufficient to give mobility to the phospholipids in the inner leaflet of the bilayer.

However, incorporated large membrane spanning proteins are immobilised which has an unfavourable effect on their activity. One elegant solution is the self-assembly of lipid bilayers on agarose as developed by H.T. Tien and described in the preceding section. However, although this method is superior in its simplicity and gives bilayer lipid membranes facing aqueous solutions on both sides, the technology does not allow precise manipulation of the composition of the bilayer. Moreover, the total membrane area is small, typically with a diameter of 1 mm.

In this section I will focus on the attempts to build lipid bilayers linked to a solid support through an aqueous phase of significant thickness. The work presented below is divided in two sections: tethered lipid membranes and polymer cushioned lipid membranes.

4.1. TETHERED LIPID MEMBRANES

A new class of lipids was introduced by H. Lang et al., [24,25] comprising the attachment of a hydrophilic spacer to the phosphate group of dipalmitoylglycero-phosphatidic acid. Three ethylene glycol spacers of different lengths were employed, all terminated with a disulfide group. Thus, these so-called "thiolipids" can be covalently linked to a gold substrate in a simple self-assembling procedure. Various types of monolayers and bilayers were assembled and investigated with surface plasmon resonance and electrochemical techniques. It was found that the longer the spacer, the higher was the molecular integrity and also the optical density of the first monolayer. A monolayer of conventional phospholipids was spontaneously formed on top of the first thiolipid monolayer simply by exposing it to a lipid/detergent solution which was diluted stepwise below the critical micelle concentration of the detergent. Thus, supported phospholipid bilayers could be formed which were mechanically and chemically stable for several weeks. The capacitance of the planar gold-supported membrane was in good accordance with solvent free planar unsupported black lipid membranes and therefore these systems were proposed as most suitable aspirants for incorporation of membrane proteins with large extramembraneous parts. In an ensuing paper the same group reports on the formation of 2D structured surfaces based on the combination of Langmuir-Blodgett and self-assembly techniques [100, 101] Briefly, the proposed method involves Langmuir-Blodgett transfer of a mixed monolayer consisting of phase separated palmitic acid and thiolipids to a planar gold surface. The fatty acid was washed away after transfer, exposing bare gold domains. These 2D structured surfaces were further modified as described in the two forthcoming paragraphs.

In the first example the bare gold spots were modified with an antigen carrying an N-terminal cysteine. The thiol group in the cysteine links the peptide to the gold surface via a covalent bond. The structured surface was exposed to either bovine serum albumin or the monoclonal antibody against the chemisorbed peptide. Surface plasmon resonance clearly indicated that the protein was specifically adsorbed on the hydrophobic thiolipids whereas the antibody interacted only with the peptide-covered domains. This structured surface is suggested for development of multichannel biosensors.

In a second application, 21-mercaptoheneicosanol was self-assembled on the gold domains and then the surface was incubated with vesicles. Surface plasmon micrographs indicate that a phospholipid monolayer is deposited on top of the thiolipid monolayer and that a phospholipid bilayer is formed on the hydroxyl thiols. The domains of phospholipid bilayers are suggested as suitable hosts for incorporation of transmembrane proteins since the phospholipids in both leaflets are most likely mobile. This is not the case for the mixed bilayers with thiolipids in one plane and conventional phospholipids in the other.

The formation of a 2D structure was also obtained in a different approach where a squared micropattern on a gold electrode surface made it possible to self-assemble lipid "anchor" molecules in small wells 20x20 μm large and separated by 200 μm. The remaining part of the gold surface was modified with mercaptoethanol and the surface was incubated with LUV, whereby a lipid membrane was formed with a membrane capacitance of 0.9 μF cm-2. Two types of ionophores were incorporated in the membrane and their expected selective responses for various cations were verified [102,103].

A paper following essentially the same line as the work by Vogel and co-workers, was recently published where the first leaflet of the supported bilayer membrane consists of thiolipids supplied with a hydrophilic spacer composed of three ethoxy groups and terminated by a thiol group. A sulphur-gold bond links the thiolipid to the gold substrate and the second monolayer was formed by vesicle fusion. With valinomycin incorporated in the bilayer the electrode shows a selective response to potassium ions as compared to sodium ions [104].

R. Naumann and co-workers [105-107] presented another variant on the same theme. Firstly, a helix forming hydrophilic peptide was functionalised with a terminal sulphur group. Then it was covalently attached to a planar gold surface forming an almost complete monolayer. In the next step, the free terminal carboxylic group of the immobilised peptide was activated and coupled to a phospholipid, whereby an imperfect monolayer of thiolipids was formed, with only 70% coverage. In the last step the thiolipid monolayer was incubated with vesicles with or without reconstituted ATP synthase and a complete monolayer of phospholipids was formed. With proteins in the vesicles, surface plasmon resonance measurements indicate an increased membrane thickness as compared to the pure bilayer.

ATP synthase, also called F_0F_1ATPase, works in two reversible directions and couples ion movement to ATP synthesis or hydrolysis. Square wave voltammetry was employed to investigate the activity of the incorporated protein. The lipid layer is exposed to ATP at increasing concentrations and an increasing peak is observed at negative potentials which is interpreted as reduction of protons to hydrogen at the gold electrode. The growing height reflects an increasing concentration of hydrogen ions in the thin aqueous layer adjacent to the electrode surface due to ATP-dependent transport of protons. It is not explained, however, why the peak does not shift to more positive potentials as the pH is decreasing (Fig. 7).

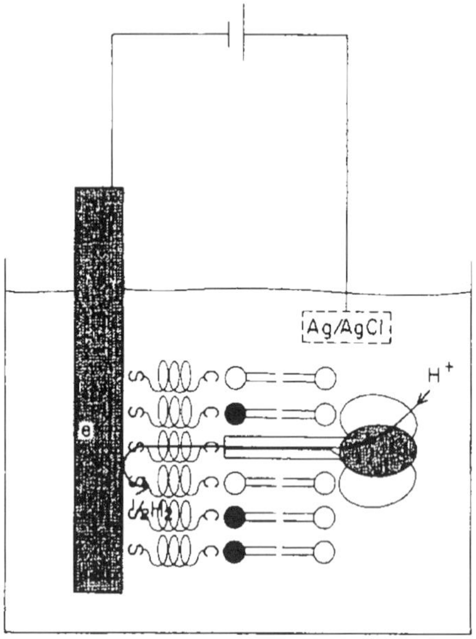

Figure 7. Schematic representation of the coupling between the translocation of protons across the lipid bilayer catalysed by the reconstituted ATPase and their discharge at the gold electrode. Reprinted from: Naumann, R., Jonczyk, A., Hampel, C., Ringsdorf, H., Knoll, W., Bunjes, N. and Graber, P. (1997) Bioelectrochem. Bioenerg. 42, 241-247 with permission from Elsevier Science.

Similarly, the dimer species of the nicotinic acetylcholine receptor was incorporated in the peptide supported lipid membrane and studied with surface plasmon resonance Q and fluorescence spectroscopy [108]. The activity of the receptor was verified with the observed binding of firstly: primary monoclonal and secondary polyclonal antibodies and secondly: a-bungarotoxin. Thus, the receptor was successfully incorporated in the lipid bilayer in its active form and at least a fraction of the receptor molecules were oriented with the extracellular domain pointing outwards.

Finally, the same group has incorporated cytochrome c oxidase into a preformed peptide tethered lipid bilayer [30]. Cytochrome c oxidase is one of the major membrane-bound proteins in the respiratory chain, accepting electrons from cytochrome c and passing them on to oxygen. The protein was incorporated into the lipid layer by exposing it to a protein/detergent solution and diluting it below the critical micellar concentration of the detergent. During incorporation of the protein the total thickness of the layer including the spacer, increased from 6.2 nm to 12.2 nm. The protein was activated by introducing cytochrome c in the reduced form in the air-saturated buffer and square wave voltammetry and impedance spectroscopy was performed. Also this time a peak was observed at negative potentials which was attributed to reduction of hydrogen ions. This time the peak was decreasing with increasing concentration of cytochrome c. The authors suggest that the reason for the decreasing peak is that the active protein transports protons away from the electrode surface. Another peak was observed at 0.3 V vs. Ag/AgCl, increasing with increasing concentration of cytochrome c in the concentration range 15 μM to 300 μM. The peak is attributed to enzymatic oxidation of the cytochrome c, but surprisingly, no saturation of the enzyme is observed at these concentrations. Finally, the electrode system was probed with impedance spectroscopy and an attempt was made to fit the response to various equivalent circuits. Based on the proposed equivalent circuit a saturation of the electrochemical response was obtained at a cytochrome c concentration of 12 μM in one set-up and 28 μM in another. This is in fair agreement with literature values on the Michaelis Menten constant [84].

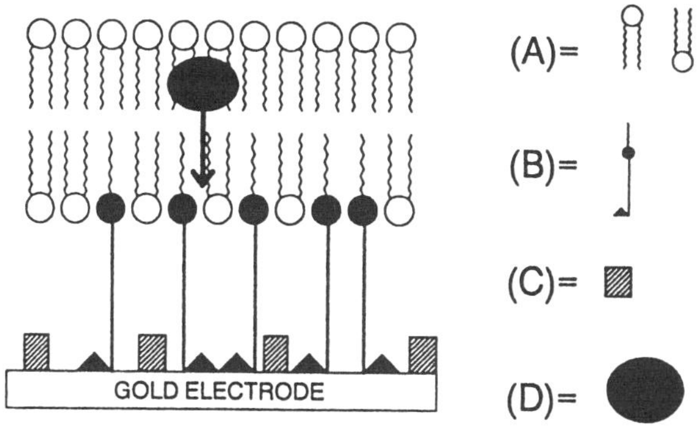

Figure 8. Schematic representation of the tethered bilayer membrane where A) is the mobile lipid that makes up the bulk of the membrane, B) is the reservoir lipid that defines the ionic reservoir and tethers the membrane for the gold surface, C) is the spacer molecules used to laterally space the reservoir lipids and D) is the valinomycin ionophore used to modulate the membrane conductivity. Reprinted from: Raguse, B., Braach-Maksvytis, V., Cornell, B.A., King, L.G., Osman, P.D.J., Pace, R.J. and Wieczorek, L. (1998) Langmuir, 14, 648-659 with permission from American Chemical Society.

An elegant approach was recently presented on an electrode system, seemingly well suited for use with most types of receptors. A fluid lipid bilayer membrane containing the ionophore gramicidin [43] or valinomycin [109] is linked to a planar gold electrode via a spacer molecule. The disulfide bearing spacer molecule, the so-called reservoir lipid, is the critical component in the novel electrode configuration. In Ref. 109 and as can be seen in Fig. 8, a hydrophilic headgroup is linked through a long hydrophilic chain to the gold electrode surface, whereas another hydrocarbon chain is extended from the headgroup into the hydrophobic core of the tethered bilayer. Firstly, the reservoir lipids were self-assembled to the gold electrode surface and then the bilayer was formed in situ with a solvent/dilution technique. The length of the hydrophilic spacer precisely determines the size of the reservoir. For a number of lipids the capacitance and thickness of these bilayer membranes are in good agreement with literature values. A potassium sensor with an analytical range over physiologically relevant concentrations was obtained with valinomycin incorporated in this bilayer.

In the other paper the channel-forming ionophore was exploited as an essential part of the biosensor [43]. Gramicidin spans only half a cellmembrane, and when a dimer is formed with a gramicidin in the opposite part of the membrane a channel is opened that passively transports monovalent cations. In this work two types of reservoir lipids were employed, one of them a synthetic archaebacterial membrane-spanning lipid possessing antibody fragments (Fab′) and one half-membrane spanning lipid. Both of them were attached to the gold surface via a hydrophilic spacer terminated with a disulfide group. Gramicidin was inserted in the lower membrane and immobilised with a similar spacer as the anchor lipids. Mobile gramicidin linked to antibodies with biotin/streptavidin technology was incorporated in the upper leaflet. When a targeted analyte is added to the surrounding electrolyte it crosslinks the Fabs on the membrane spanning lipid and the gramicidin. Thus the gramicidin in the upper leaflet is immobilised and prevented from forming a dimer with its partner in the lower leaflet. This is registered as a decreasing conductivity in the admittance spectrum. The receptor in this sensor can be varied and the authors claim that the technology has been applied successfully in the detection of bacteria, virus particles, DNA, antibodies and electrolytes.

4.2. POLYMER CUSHIONED BILAYER LIPID MEMBRANES

The benefits of using a polymer cushioned lipid bilayer as a support for artificial cell membranes were clearly pointed out by Sackmann [15], who in earlier work evaluated frictional coefficients between the lipids in a Langmuir-Blodgett transferred film and a supporting polyacrylamide film [110]. The bilayer was not continuously closed but the structural, dynamic and thermodynamic properties were preserved after deposition.

The first contribution to this class of supported membranes came several years earlier and involved mixed monolayers composed of phosphatidylcholine and cholesterol that were transferred to a polyacrylamide hydrogel by Langmuir-Blodgett techniques [111]. Only a fraction of the attempts to produce low conductance bilayers was successful, but when they were transient ion currents were observed upon addition of valinomycin or phloretin close to the membrane.

In a similar but improved configuration spincoated polyanhydride was used as a support for Langmuir-Blodgett transfer of phospholipids to form supported bilayers [112]. In both studies the stability problems with fragile black lipid membranes were recognised, but also the need for an ionic reservoir on both sides of the artificial membrane.

In a novel approach a copolymer consisting of a hydrophilic main chain, hydrophobic lipidic parts and finally a disulfide group was self-assembled on a gold surface with gold/sulphur bonds. Thus a polymer supported monolayer of a lipid-like phase was formed and after fusion with phospholipid vesicles a bilayer was obtained surrounded with an aqueous phase on both sides [113]. A successful streptavidin binding experiment with a biotinylated lipid layer was performed. The same group extended the investigations on the self-assembled amphiphilic polymer in a later publication [114] to improve the barrier properties of the first monolayer and to confirm that a water layer exists between the support and the lipid membrane.

In a later work a reactive polymer was chemisorbed on a functionalised glass slide thus forming a thin polymer film on the solid support [28]. Hydrophilic amino groups were linked to the polymer and a phospholipid bilayer was transferred to the ca. 80 Å thick hydrated polymer film. The first layer was obtained with Langmuir-Blodgett techniques and the second leaflet with Langmuir-Schäfer techniques. The fluidity of the lipids was investigated with photobleaching techniques and it was found that the lipid bilayer was fluid and stable for several days.

A hydrophilic polymer cushion with a smooth outer surface, is easily self-assembled at the electrode surface by alternate adsorption of polycations and polyanions [115]. Such a surface was employed as a support for deposition of phospholipid multilayers with LB-techniques and impedance spectroscopy was employed to investigate if the polyelectrolyte was suitable as a support for an artificial cell membrane. A bilayer lipid membrane was obtained with a membrane capacitance of ca 0.60 μF cm-2 that was linked to the polyelectrolyte film via a calcium bridge provided the outermost layer was negatively charged [116]. A similar polymer cushion was prepared on a gold electrode, consisting of three layers of polyelectrolyte with an electroactive polycation sandwiched between thin layers of polystyrenesulphonate. Vesicles containing cytochrome c oxidase were fused on this surface and a biomembrane containing the active enzyme was formed resting on the polyelectrolyte surface. The activity of the enzyme was confirmed with amperometry in a FIA system by monitoring the transient current from oxidation of a pulse of reduced cytochrome c at anaerobic conditions. The fused lipid layer did not block the cytochrome c from being oxidised directly, either at the gold electrode or via the osmium complex in the film resulting in high background currents. However, inhibition of the enzyme with sodium azide resulted in a temporarily decreased cytochrome c signal, which is a clear indication of an active membrane protein. Moreover, only biomembranes with a mean thickness corresponding to a transmembrane-containing lipid membrane, i.e.: ca 60 Å resulted in reproducible results and the enzyme kinetics estimated from a fitted Michaelis - Menten relationship was in good agreement with literature values [117].

Another and rather amazing way to form a polymer-cushioned lipid bilayer was recently reported and involves firstly the fusion of small unilamellar dimyristoylphosphatidyl-choline vesicles to form an intact bilayer on a quartz substrate. Secondly, when the cationic polyethyleneimine, PEI; is added to the solution it creeps beneath the bilayer and forms a 40 Ångström thick soft cushion between the lipid layer and the solid support. The process is monitored with Neutron Spectroscopy and interestingly an attempt to fuse vesicles on a solid support already covered with the polyelectrolyte failed [118]. In a subsequent paper it was found that if the PEI-coated slide was allowed to dry before it was incubated with vesicles, the fusion was successful and a continuous bilayer could be formed on top of the polyelectrolyte [119]. Instead of building a polymer-cushioned lipid monolayer on a solid support step by step, the whole template can be formed in the air/water interface and then transferred to a freshly cleaved mica plate with Langmuir-Blodgett techniques. This was accomplished by spreading a reactive lipid in the air/water interface in the Langmuir trough. The lipid was provided with an isothiocyanate function that reacts with the amino groups in the polyethylenimine dissolved in the water subphase. After transfer a lipid bilayer could be formed on the mica support by vesicle fusion [120].

The lateral mobility of phospholipids in mono- and bilayers supported on silane-, dextran-, and crystalline bacterial surface layer proteins on planar silicon substrates was investigated with fluorescence recovery after photobleaching [121]. It was shown that the phospholipids in a monolayer on a silane supported membrane, i.e.: a hybrid bilayer, had significantly lower mobility than the phospholipids in the bilayers on either dextran or S-layer protein. The bilayer supported on S-layer proteins was in a fluid state with a lateral diffusion coefficient as high as 2.5 - 3.1 $*10^{-6}$ m^2s^{-1}. Furthermore, when an S-layer protein lattice was formed as a cover for the lipid layer the fluidity in almost all cases was even higher.

Soft polymer cushions formed from Langmuir-Blodgett transferred hairy-rod multilayers onto indium-tin-oxide electrodes provide a promising support for bilayer membranes. The multilayer is hydrophobic and highly insulating directly after transfer since the hairy-rod molecules consist of amphiphilic cellulose derivatives with hydrophobic side-chains. However, by cleavage of trimethylsilyl groups, cellulose is regenerated and the film becomes ionically conducting. When vesicles are fused to this surface lipid bilayers were formed with a resistance of 0.44 $M\Omega cm^2$, and a membrane capacitance of 0.57 μFcm^{-2}. This indicates very high quality lipid bilayer films and its potential as a template for incorporation of active membrane proteins is elegantly demonstrated with reconstituted gramicidin channels. The experiment clearly shows a selective response for sodium and potassium ions with the expected preference for potassium ions [122].

5. Phospholipid monolayers at the mercury/water interface, "Miller -Nelson films"

It is of vital importance to create a full lipid bilayer when the interest is focused on studies of membrane-spanning proteins. However, in investigations concerning biological processes at the cell membrane, where the two leaflets of the bilayer are not jointly involved it is not necessary to build a full lipid bilayer. In many applications it is therefore sufficient to have a fluid lipid monolayer, as has been shown in the many successful investigations on the hybrid bilayer membranes.

The original articles of Miller et al.[123,124] described how a phospholipid monolayer can be deposited on a hanging mercury drop electrode, HMDE, simply by passing it through a spread lipid film residing at the air/water interface. The mercury electrode is hydrophobic in a wide potential range and certifies that the lipid molecules keep their orientation with the hydrocarbon chains pointing toward the electrode surface. The monolayer is fluid and rests on a perfectly flat surface and the dielectric properties can be measured readily with for instance impedance spectroscopy, as one side of the membrane is directly contacted with the mercury electrode (Fig. 9).

The method has been utilised [125] to exploit melittin, alamethicin and protein kinase C interactions with a lipid monolayer on a mercury surface. A lipid/protein mixture was spread in the air/water interface. Thereafter, a proteolipid monolayer was formed in the same way as previously described for a pure lipid monolayer. The changes in film capacitance and permittivity were investigated with ac polarograms and small increments were found in the ionic permeability. It was concluded, however, that neither of these proteins could form channels in the "half a bilayer" lipid film.

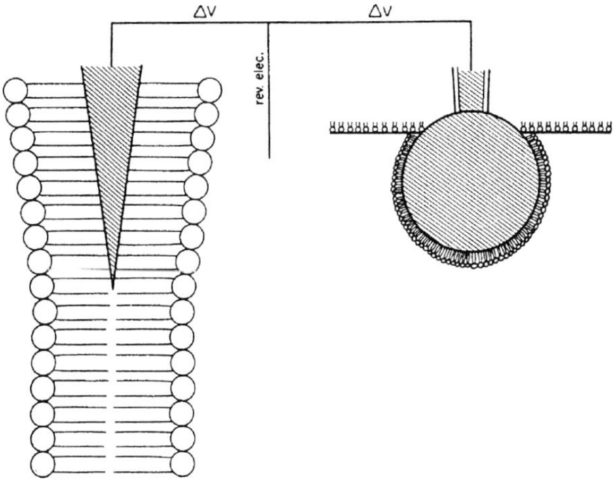

Figure 9. *Equivalence of a hypotethical electrode inserted into the middle of a lipid bilayer with a mercury electrode contacting the hydrocarbon side of a lipid monolayer. Reprinted from: Miller, I.R., Doll, L. and Lester, D.S. (1992) Bioelectrochem. Bioenerg. 28, 85-103 with permission from Elsevier Science.*

A. Nelson early recognised this highly reproducible system as a versatile tool for investigations of biomimetic lipid layers [126]. The induced phase transitions and the final complete desorption of the monolayer at a voltammetric scan from -0.2 V to -1.8 V vs. Ag/AgCl (sat'd KCl) have been thoroughly studied [127,128] The spontaneous respreading of the monolayer at reversal of the potential provides evidence for the high mobility of the lipids in the supported layer.

The gramicidin ionophore was successfully incorporated in the mercury-supported lipid monolayer and different aspects of the transport of thallium ions through the channel have been investigated in a series of papers. Tl^{+} is isoelectric with K^{+} and has similar size and hydration number. Moreover, the ion has a reduction potential within the voltage range where the supported lipid monolayer is stable and was therefore chosen as a suitable probe for the protein activity. In the first paper it was concluded that the presence of Mg^{2+} enhances the transport as opposed to K^{+} ions and that chloride ions suppress the reduction current of Tl^{+} compared to that of nitrate ions [129]. In one of the subsequent papers the gramicidin-mediated permeability of the film was investigated with respect to added retinal and the pesticide DDT [130]. Various models for the monomolecular channel function have also been discussed [44,131-133]. Other biological relevant investigations from the research of Nelson include: insertion of the antibiotic 23187 into the lipid layer and the investigation of the ionophore assisted translocation of heavy metal ions from the solution through the film [134]; reduction of vitamin A aldehyde with various mixtures of phospholipids in the monolayer [135]; reduction of ubiquionone-10 at various pH's [136].

Many research groups have become inspired by this research and in a recent contribution the selective interaction of alkali metal cations with azocrown molecules assembled in the lipid monolayer was reported [137]. Other papers of particular biological importance concern determination of intrinsic pKa of various phospholipids in the monolayer configuration [138,139] and still another the extent of penetration of various hydrophobic molecules into the monolayer [140].

In recent work a hybrid bilayer membrane was prepared on the sessile mercury drop and utilised in an investigation concerning bacteriorhodopsin. In this case, a hexadecanethiol monolayer was first assembled on the mercury electrode and a phospholipid monolayer was then added to the surface according to the "Miller-Nelson techniques". Bacteriorhodopsin was subsequently adsorbed on the surface and the photoactivity of the protein was confirmed [141].

6. Conclusions

During the last decade a completely new research area has developed concerning the fabrication of supported lipid membranes for reconstitution of integral membrane proteins. The challenging task has inspired many research groups to ingenious solutions as described in this review. Although many authors describe the vesicle fusion method as a universal method for this purpose it must be underlined that it is not always so. The technique is very sensitive to many different factors [64] and sometimes multilayers or stacks of unfused vesicles are formed [142,143]. Furthermore when proteo-vesicles are fused on solid substrates the protein is often immobilised albeit the lipids in both leaflets are mobile.

However, a universal method for building the lipid membrane is not feasible and is maybe not even desirable. The method of choice should be adopted to what the biomimetic system is intended for. Thus, because of the ease to fabricate BLMs according to the self-assembling technique developed by Tien and co-workers [27,91] those systems seem attractive for commercial production of membrane based biosensors to be used for instance in large scale screening devices and combinatorial chemistry.

The hybrid bilayer membranes, BLMs from direct fusion of vesicles and the Miller-Nelson films have already qualified for investigations of cellular immune responses, specific receptor-ligand bindings and various membrane interactions. Peripheral proteins or membrane proteins that are not spanning the whole membrane are also suitable for investigations in these simpler systems. In the near future much is expected to merge from these approaches, for instance in pharmaceutical approaches such as drug interaction with a biomembrane surface.

For reconstitution of large membrane spanning proteins in supported membranes more sophisticated solutions are called for. The dimensions of the enzyme must be considered, which in most cases demands an aqueous phase on both sides of the supported bilayer. A measurement technique must be available to register the bioactivity correctly. In section 4 the many creative ventures are presented that have emerged during the last decade in order to fulfil these requirements. One of the most promising approach concerns deposition of lipid bilayers on polymer cushions from multilayers of hairy rods. Here the polymer support is the equivalent of the cytoskeleton in the real cell. Stable bilayer membranes have been formed on these cushions with electrical properties similar to those of as black lipid membranes [122]. In the first decade of this millennium a substantial break-through is expected, which will open up excellent possibilities to study specific functions of integral proteins.

References

1. Gorter, E. and Marsch, D. (1925) On bimolecular layers of lipids on the chromocytes of the blood., J. Exp. Med. 41, 439-443.
2. Reed, R.A., Mattai, J. and Shipley, G.G. (1987) Interaction of Cholera-Toxin With Ganglioside Gm1 Receptors in Supported Lipid Monolayers, Biochemistry 26, 824-832.
3. Terrettaz, S., Stora, T., Duschl K, C. and Vogel, H. (1993) Protein binding to supported lipid membranes: Investigation of the cholera toxin-ganglioside interaction by simultaneous impedance spectroscopy and surface plasmon resonance., Langmuir 9, 1361-1369.
4. Ohlsson, P.-Å., Tjärnhage, T., Herbai, E., Lövås, S. and Puu, G. (1995) Liposome and proteoliposome fusion onto solid substrates, studied using atomic force microscopy, quartz crystal microbalance and surface plasmon resonance. Biological activities of incorporated components, Bioelectrochem. Bioenerg. 38, 137-148.
5. Steinem, C., Janshoff, A., Wegener, J., Ulrich, W.P., Willenbrink, W., Sieber, M. and Galla, H.J. (1997) Impedance and shear wave resonance analysis of ligand-receptor interactions at functionalised surfaces and of cell monolayers, Biosensors and Bioelectronics 12(8), 787-808.
6. Janshoff, A., Steinem, C., Sieber, M. and Galla, H.J. (1996) Specific binding of peanut agglutinin to G_{M1}-doped solid supported lipid bilayers investigated by shear wave resonator measurements, Eur. Biophys. J. Biophys. Lett. 25, 105-113.
7. Cooper, M.A., Hansson, A., Löfås, S. and Williams, D.H. (2000) A vesicle capture sensor chip for kinetic analysis of interactions with membrane-bound receptors, Anal. Biochem. 277, 196-205.
8. Cevc, G. (1990) Membrane electrostatics, Biochim. Biophys. Acta 1031, 311-382.
9. Lasic, D.D. (1995) Liposomes: from physics to applications, Elsevier Science, Amsterdam.
10. Mueller, P., Rudin, D.O., Tien, H.T. and Wescott, W.C. (1962) Reconstitution of cell membrane structure in vitro and its transformation into an excitable system, Nature 194, 979-980.
11. Nikolelis, D.P. and Siontorou, C.G. (1995) Bilayer lipid membranes for flow injection monitoring of acetylcholine, urea and penicillin, Anal. Chem. 67, 936-944.
12. Nikolelis, D.P. and Siontorou, C.C. (1996) Flow injection monitoring and analysis of mixtures of simazine, atrazine, and propazine using filter-supported bilayer lipid membranes (BLMs), Electroanalysis 8, 907-912.
13. Siontorou, C.G., Nikolelis, D.P. and Krull, U.J. (2000) Flow injection monitoring and analysis of mixtures of hydrazine compounds using filter-supported bilayer lipid membranes with incorporated DNA, Anal. Chem. 72, 180-186.
14. McConnell, H.M., Watts, T.H., Weis, R.M. and Brian, A.A. (1986) Supported planar membranes in studies of cell-cell recognition in the immune system., Biochim. Biophys. Acta 864, 95-106.
15. Sackmann, E. (1996) Supported Membranes: Scientific and practical applications, Science, 271, 43-48.
16. Tiede, D.M. (1985) Incorporation of membrane proteins into interfacial films: model membranes for electrical and structural characterisation, Biochim. Biophys. Acta 811, 357-379.
17. Ulman, A. (1991) An introduction to. "Ultrathin Organic Films" from Langmuir-Blodgett to Self-Assembly., Academic Press, Inc., San Diego.
18. Roberts, G.G. (Ed) (1990) Langmuir-Blodgett films, Plenum press, New York.

19. Sellström, Å., Gustafson, I., Ohlsson, P.-Å., Olofsson, G. and Puu, G. (1992) On the deposition of phospholipids onto planar supports with the Langmuir-Blodgett technique using factorial experimental design. 2. Optimising lipid composition for maximum adhesion to platinum substrates, Colloid Surface 64, 289-298.
20. Sellström, Å., Gustafson, I., Ohlsson, P.-Å., Olofsson, G. and Puu, G. (1992) On the deposition of phospholipids onto planar supports with the Langmuir-Blodgett technique using factorial experimental design. 1. Screening of various factors and supports, Colloid Surface 64, 275-287.
21. Kalb, E., Frey, S. and Tamm, L.K. (1992) Formation of supported planar bilayers by fusion of vesicles to supported phospholipid monolayers, Biochim. Biophys. Acta 1103, 307-316.
22. Seifert, K., Fendler, K. and Bamberg, E. (1993) Charge Transport By Ion Translocating Membrane-Proteins On Solid Supported Membranes, Biophys. J. 64, 384-391.
23. Tjärnhage, T. and Puu, G. (1996) Liposome and phospholipid adsorption on a platinum surface studied in a flow cell designed for simultaneous quartz crystal microbalance and ellipsometry measurements, Colloid Surface B 8, 39-50.
24. Lang, H., Duschl, C. and Vogel, H. (1994) A new class of thiolipids for the attachment of lipid bilayers on gold surfaces, Langmuir 10, 197-210.
25. Lang, H., Duschl, C., Grätzel, M. and Vogel, H. (1992) Self-assembly of thiolipid molecular layers on gold surfaces: optical and electrochemical characterisation, Thin Solid Films 210/211, 818-821.
26. Lu, X., Leitmannova-Ottova, A. and Tien, H.T. (1996) Biophysical aspects of agar-gel supported bilayer lipid membranes: a new method for forming and studying planar bilayer lipid membranes, Bioelectrochem. Bioenerg. 39, 285-289.
27. Yuan, H., Leitmannova-Ottova, A. and Tien, H.T. (1996) An agarose-stabilised BLM: a new method for forming bilayer lipid membranes, Mater. Sci. Eng. 4, 33-38.
28. Beyer, D., Elender, G., Knoll, W., Kuhner, M., Maus, S., Ringsdorf, H. and Sackmann, E. (1996) Influence of anchor lipids on the homogeneity and mobility of lipid bilayers on thin polymer films, Angew. Chem. Int. Ed. Engl. 35, 1682-1685.
29. Kalb, E. and Tamm, L.K. (1992) Incorporation of cytochrome b 5 into supported phospholipid bilayers by vesicle fusion to supported monolayers, Thin Solid Films 210/211, 763-765.
30. Naumann, R., Schmidt, E.K., Jonczyk, A., Fendler, K., Kadenbach, B., Liebermann, T., Offenhausser, A. and Knoll, W. (1999) The peptide-tethered lipid membrane as a biomimetic system to incorporate cytochrome c oxidase in a functionally active form, Biosens. Bioelectron. 14, 651-662.
31. Gritsch, S., Nollert, P., Jahnig, F. and Sackmann, E. (1998) Impedance spectroscopy of porin and gramicidin pores reconstituted into supported lipid bilayers on indium-tin-oxide electrodes, Langmuir 14, 3118-3125.
32' McAuley, K.E., Fyfe, P.K., Ridge, J.P., Isaacs, N.W., Cogdell, R.J. and Jones, M.R. (1999) Structural details of an interaction between cardiolipin and an integral membrane protein, Proc. Natl. Acad. Sci. U. S. A. 96, 14706-14711.
33' Franz, H., Dante, S., Wappmannsberger, T., Petry, W., de Rosa, M. and Rustichelli, F. (1998) An X-ray reflectivity study of monolayers and bilayers of archae lipids on a solid substrate, Thin Solid Films 329, 52-55.
34' Byron, O. and Gilbert, R.J.C. (2000) Neutron scattering: good news for biotechnology, Curr. Opin. Biotechnol. 11, 72-80.
35' Lšsche, M., Erdelen, C., Rump, E., Ringsdorf, H., Kjaer, K. and Vaknin, D. (1994) On the Lipid Head Group Hydration of Floating Surface Monolayers Bound to Self-Assembled Molecular Protein Layers, Thin Solid Films 242, 112-117.
36' Lukes, P.J., Petty, M.C. and Yarwood, J. (1992) An infrared study of the incorporation of ion channel peptides into Langmuir-Blodgett films of phosphatidic acid, Langmuir 8, 3043-3050.
37' Kšnig, S., Sackmann, E., Richter, D., Zorn, R., Carlile, C. and Bayerl, T.M. (1994) Molecular-Dynamics at Water in Oriented DPPC Multilayers Studied by Quasi-Elastic Neutron-Scattering and Deuterium- Nuclear Magnetic-Resonance Relaxation, J. Chem. Phys. 100, 3307-3316.
38' Kuhl, T.L., Leckband, D.E., Lasic, D.D. and Israelachvili, J.N. (1994) Modulation of Interaction Forces Between Bilayers Exposing Short-Chained Ethylene-Oxide Headgroups, Biophys. J. 66, 1479-1488.
39' Fujiwara, I., Ohnishi, M. and Seto, J. (1992) Atomic force microscopy study of protein-incorporating Langmuir-Blodgett film, Langmuir 8, 2219-2222.

40' Löfås, S., Malmqvist, M., Rönnberg, I., Stenberg, E., Liedberg, B. and Lundström, I. (1991) Bioanalysis with Surface-Plasmon Resonance, Sens. Actuator B-Chem. 5, 79-84.
41. Gizeli, E., Lowe, C.R., Liley, M. and Vogel, H. (1996) Detection of supported lipid layers with the acoustic Love waveguide device: Application to biosensors, Sens. Actuator B-Chem. 34, 295-300.
42. Puu, G., Gustafsson, I., Artursson, E. and Ohlsson, P.-Å. (1995) Retained activities of some membrane proteins in stable lipid bilayers on a solid support, Biosens. Bioelectron. 10, 463-476.
43' Cornell, B.A., Braach-Maksvytis, V.L.B., King, L.G., Osman, P.D.J., Raguse, B., Wieczorek, L. and Pace, R.J. (1997) A biosensor that uses ion-channel switches, Nature 387, 580-583.
44' Rueda, M., Navarro, I., Ramirez, G., Prieto, F., Prado, C. and Nelson, A. (1999) Electrochemical impedance study of Tl^+ reduction through gramicidin channels in self-assembled gramicidin-modified dioleoylphosphatidylcholine monolayers on mercury electrodes, Langmuir 15, 3672-3678.
45. Macdonald, J.R., Ed. (1987) Impedance Spectroscopy, John Wiley & Sons, New York.
46. Blodgett, K. (1935) Films built by depositing successive monomolecular layers on a solid surface, J. Am. Chem. Soc. 57, 1007-1022.
47. Blodgett, K. and Langmuir, I. (1937) Built-up films of barium stearate and their optical properties, Phys. Rev. 51, 964-982.
48. Petty, M.C. (1992) Possible application for Langmuir-Blodgett films, Thin Solid Films 210/211, 417-426.
49. Procarione, W.L. and Kauffman, J.W. (1974) The electrical properties of phospholipid bilayer Langmuir films, Chem. Phys. Lipids 12, 251-260.
50. Hollars, C.W. and Dunn, R.C. (1998) Submicron structure in L-alpha-dipalmitoylphosphatidylcholine monolayers and bilayers probed with confocal, atomic force, and near-field microscopy, Biophys. J. 75, 342-353.
51. Steinem, C., Janshoff, A., Ulrich, W.-P., Sieber, M.-P. and Galla, H.-J. (1996) Impedance analysis of supported lipid bilayer membranes: a scrutiny of different preparation techniques, Biochim. Biophys. Acta 1279, 169-180.
52. Ohlsson, P.-Å., Puu, G. and Sellström, Å. (1994) On the deposition of phospholipids onto planar supports with the Langmuir-Blodgett technique using factorial experimental design. 3. Optimising lipid composition for maximal adhesion to chromium substrates, Colloid Surface B 3, 39-48.
53. Stelzle, M. and Sackmann, E. (1989)Sensitive detection of protein adsorption to supported lipid bilayers by frequency-dependent capacitance measurements and microelectrophoresis, Biochim. Biophys. Acta 981, 135-142.
54. Fare, T.L., Rusin, K.M. and P.P. Bey, J. (1991) Langmuir-Blodgett deposited valinomycin-phospolipid films on platinum: an a.c. impedance response to potassium, Sensors and Actuators B 3, 51-62.
55. Miller, C., Cuendet, P. and Grätzel, M. (1990) K^+ sensitive bilayer supporting electrodes, J. Electroanal. Chem. 278, 175-192.
56. Terrettaz, S., Vogel, H. and Grätzel, M. (1992) Ca^{2+}-sensitive monolayer electrodes, J. Electroanal. Chem. 326, 161-176.
57. Zaba, B.N., Wilkinson, M.C., Taylor, D.M., Lewis, T.J. and Laidman, D.L. (1987) Electrochemical characteristics of platinum electrodes coated with cytochrome b_5-phospholipid monolayers, Febs letters 213, 49-54.
58. Brian, A.A. and McConnell, H.M. (1984) Allogeneic stimulation of cytotoxic T cells by supported planar membranes, Proc. Natl. Acad. Sci. USA 81, 6159-6163.
59. Fringeli, U.P. (1989) Structure-activity relationship in biomembranes investigated by infrared-ATR spectroscopy, Schlunegger, U.P., Ed., Springer-Verlag, Berlin - Heidelberg.
60. Wenzl, P., Fringeli, M., Goette, J. and Fringeli, U.P. (1994) Supported phospholipid bilayers prepared by the "LB/vesicle method": A Fourier transform infrared attenuated total reflection spectroscopic study on structure and stability, Langmuir 10, 4253-4264.
61. Johnson, S.J., Bayerl, T.M., McDermott, D.C., Adam, G.W., Rennie, A.R., Thomas, R.K. and Sackmann, E. (1991) Structure of an Adsorbed Dimyristoylphosphatidylcholine Bilayer Measured with Specular Reflection of Neutrons, Biophys. J. 59, 289-294.
62. Hetzer, M., Heinz, S., Grage, S. and Bayerl, T.M. (1998) Asymmetric molecular friction in supported phospholipid bilayers revealed by NMR measurements of lipid diffusion, Langmuir 14, 982-984.
63. Helm, C.A., Israelachvili, J.N. and McGuiggan, P.M. (1992) Role of Hydrophobic Forces in Bilayer Adhesion and Fusion, Biochemistry 31, 1794-1805.

64. Puu, G. and Gustafson, I. (1997) Planar lipid bilayers on solid supports from liposomes - factors of importance for kinetics and stability, BBA-Biomembranes 1327, 149-161.
65. Stelzle, M., Miehlich, R. and Sackmann, E. (1992) 2-Dimensional Microelectrophoresis in Supported Lipid Bilayers, Biophys. J. 63, 1346-1354.
66. Groves, J.T., Boxer, S.G. and McConnell, H.M. (1997) Electric field-induced reorganisation of two-component supported bilayer membranes, Proc. Natl. Acad. Sci. U. S. A. 94, 13390-13395.
67. Ottenbacher, D., Kindervater, R., Gimmel, P., Klee, B., Jähnig, F. and Göpel, W. (1992) Developing biosensors with pH-ISFET transducers utilising lipid bilayer membranes with transport proteins, Sensor. Actuat. B 6, 192-196.
68. Puu, G., Artursson, E., Gustafsson, I., Lundström, M. and Jass, J. (2000) Distribution and stability of membrane proteins in lipid membranes on solid supports, Biosens. Bioelectron. 15, 31-41.
69. Salafsky, J., Groves, J.T. and Boxer, S.G. (1996) Architecture and function of membrane proteins in planar supported bilayers: A study with photosynthetic reaction centers, Biochemistry, 35, 14773-14781.
70. Groves, J.T., Wulfing, C. and Boxer, S.G. (1996) Electrical manipulation of glycan phosphatidyl inositol tethered proteins in planar supported bilayers, Biophys. J. 71, 2716-2723.
71. Cezanne, L., Lopez, A., Loste, F., Parnaud, G., Saurel, O., Demange, P. and Tocanne, J.F. (1999) Organisation and dynamics of the proteolipid complexes formed by annexin V and lipids in planar supported lipid bilayers, Biochemistry 38, 2779-2786.
72. Stelzle, M., Weismüller, G. and Sackmann, E. (1993) On the application of supported bilayers as receptive layers for biosensors with electrical detection, J. Phys. Chem. 97, 2974 - 2981.
73. Plant, A.L., Brighamburke, M., Petrella, E.C. and Oshannessy, D.J. (1995) Phospholipid Alkanethiol Bilayers For Cell-Surface Receptor Studies By Surface-Plasmon Resonance, Anal. Biochem. 226, 342-348.
74. Gizeli, E., Liley, M., Lowe, C.R. and Vogel, H. (1997) Antibody binding to a functionalised supported lipid layer: A direct acoustic immunosensor, Anal. Chem. 69, 4808-4813.
75. Cooper, M.A., Try, A.C., Carroll, J., Ellar, D.J. and Williams, D.H. (1998) Surface plasmon resonance analysis at a supported lipid monolayer, BBA-Biomembranes 1373, 101-111.
76. Boncheva, M., Duschl, C., Beck, W., Jung, G. and Vogel, H. (1996) Formation and characterisation of lipopeptide layers at interfaces for the molecular recognition of antibodies, Langmuir 12, 5636-5642.
77. Plant, A.L., Gueguetcherkeri, M. and Yap, W. (1994) Supported Phospholipid/Alkanethiol biomimetic membranes: Insulating properties, Biophys. J. 67, 1126-1133.
78. Steinem, C., Janshoff, A., Hohn, F., Sieber, M. and Galla, H.J. (1997) Proton translocation across bacteriorhodopsin containing solid supported lipid bilayers, Chem. Phys. Lipids 89, 141-152.
79. Rao, N.M., Plant, A.L., Silin, V., Wight, S. and Hui, S.W. (1997) Characterisation of biomimetic surfaces formed from cell membranes, Biophys. J. 73, 3066-3077.
80. Pierrat, O., Lechat, N., Bourdillon, C. and Laval, J.M. (1997) Electrochemical and surface plasmon resonance characterisation of the step-by-step self-assembly of a biomimetic structure onto an electrode surface, Langmuir 13, 4112-4118.
81. Pierrat, O., Bourdillon, C., Moiroux, J. and Laval, J.M. (1998) Enzymatic electrocatalysis studies of Escherichia coli pyruvate oxidase, incorporated into a biomimetic supported bilayer, Langmuir 14, 1692-1696.
82. Cullison, J.K., Hawkridge, F.M., Nakashima, N. and Yoshikawa, S. (1994) A study of Cytochrome c oxidase in lipid bilayer membranes on electrode surfaces, Langmuir 10, 877-882.
83. Burgess, J.D., Rhoten, M.C. and Hawkridge, F.M. (1998) Cytochrome c oxidase immobilised in stable supported lipid bilayer membranes, Langmuir 14, 2467-2475.
84. Yonetani, T. and Ray, G.S. (1965) Studies on Cytochrome Oxidase. VI: Kinetics of the aerobic oxidation of ferrocytochrome c by cytochrome oxidase, J. Biol. Chem. 240, 3392-3398.
85. Burgess, J.D., Rhoten, M.C. and Hawkridge, F.H. (1998) Observation of the resting and pulsed states of cytochrome c oxidase in electrode-supported lipid bilayer membranes, J. Am. Chem. Soc. 120, 4488-4491.
86. Tsukihara, T., Aoyama, H., Yamashita, E., Tomizaki, T., Yamaguchi, H., Shinzawaitoh, K., Nakashima, R., Yaono, R. and Yoshikawa, S. (1995) Structures of Metal Sites of Oxidised Bovine Heart Cytochrome- C-Oxidase At 2.8 Angstrom, Science 269, 1069-1074.
87. Hongyo, K.-i., Joseph, J., Huber, R.J. and Janata, J. (1987) Experimental observation of chemically modulated admittance of supported phospholipid membranes, Langmuir 3, 827-830.

88. Florin, E.-L. and Gaub, H.E. (1993) Painted supported lipid membranes, Biophys. J. 64, 375-383.
89. Salamon, Z. and Tollin, G. (1997) Interaction of horse heart cytochrome c with lipid bilayer membranes: Effects on redox potentials, J. Bioenerg. Biomembr. 29, 211-221.
90. Terrettaz, S., Vogel, H. and Gratzel, M. (1998)Determination of the surface concentration of crown ethers in supported lipid membranes by capacitance measurements, Langmuir 14, 2573-2576.
91. Tien, H.T. and Salamon, Z. (1989) Formation of self-assembled lipid bilayers on solid substrates, Bioelectrochem. Bioenerg. 22, 211-218.
92. Tien, H.T. and Ottova, A.L.(1999) From self-assembled bilayer lipid membranes (BLMs) to supported BLMs on metal and gel substrates to practical applications, Colloid Surface A 149, 217-233.
93. Ottova, A., Tvarozek, V., Racek, J., Sabo, J., Ziegler, W., Hianik, T. and Tien, H.T. (1997) Self-assembled BLMs: biomembrane models and biosensor applications, Supramol. Sci. 4, 101-112.
94. Tien, H.T. and Ottova, A.L. (1998) Supported planar lipid bilayers (s-BLMs) as electrochemical biosensors, Electrochim. Acta 43, 3587-3610.
95. Tien, H.T., Barish, R.H., Gu, L.Q. and Ottova, A.L. (1998) Supported bilayer lipid membranes as ion and molecular probes, Analyt. Sci. 14, 3-18.
96. Tien, H.T., Wurster, S.H. and Ottova, A.L. (1997) Electrochemistry of supported bilayer lipid membranes: Background and techniques for biosensor development, Bioelectrochem. Bioenerg. 42, 77-94.
97. Siontorou, C.G., Nikolelis, D.P. and Krull, U.J. (1997) A carbon dioxide biosensor based on haemoglobin incorporated in metal supported bilayer lipid membranes (BLMs): Investigations for enhancement of response characteristics by using platelet- activating factor, Electroanalysis 9, 1043-1048.
98. Rehak, M., Snejdarkova, M. and Hianik, T. (1997) Acetylcholine minisensor based on metal-supported lipid bilayers for determination of environmental pollutants, Electroanalysis 9, 1072-1077.
99. Hianik, T., Kaatze, U., Sargent, D.F., Krivanek, R., Halstenberg, S., Pieper, W., Gaburjakova, J., Gaburjakova, M., Pooga, M. and Langel, U. (1997) A study of the interaction of some neuropeptides and their analogs with bilayer lipid membranes and liposomes, Bioelectrochem. Bioenerg. 42, 123-132.
100. Duschl, C., Liley, M., Corradin, G. and Vogel, H. (1994) Biologically addressable monolayer structures formed by templates of sulphur-bearing molecules., Biophys. J. 67, 1229-1237.
101. Duschl, C., Liley, M. and Vogel, H. (1994) Micrometer-Scale lateral structuring of organic thiolate layers through self-organisation, Angew. Chem. Int. Ed. Engl. 33, 1274-1276.
102. Jenkins, A.T.A., Boden, N., Bushby, R.J., Evans, S.D., Knowles, P.F., Miles, R.E., Ogier, S.D., Schonherr, H. and Vancso, G.J. (1999) Microcontact printing of lipophilic self-assembled monolayers for the attachment of biomimetic lipid bilayers to surfaces, J. Am. Chem. Soc. 121, 5274-5280.
103. Jenkins, A.T.A., Bushby, R.J., Boden, N., Evans, S.D., Knowles, P.F., Liu, Q.Y., Miles, R.E. and Ogier, S.D. (1998) Ion-selective lipid bilayers tethered to microcontact printed self-assembled monolayers containing cholesterol derivatives, Langmuir 14, 4675-4678.
104. Steinem, C., Janshoff, A., von dem Bruch, K., Reihs, K., Goossens, J. and Galla, H.J. (1998) Valinomycin-mediated transport of alkali cations through solid supported membranes, Bioelectrochem. Bioenerg. 45, 17-26.
105 Naumann, R., Jonczyk, A., Kopp, R., Esch, J.v., Ringsdorf, H., Knoll, W. and Gräber, P. (1995) Incorporation of membrane proteins in solid supported lipid layers, Angew. Chem. Int. Ed. Engl. 34, 2056-2058.
106. Bunjes, N., Schmidt, E.K., Jonczyk, A., Rippmann, F., Beyer, D., Ringsdorf, H., Graber, P., Knoll, W. and Naumann, R. (1997) Thiopeptide-supported lipid layers on solid substrates, Langmuir 13, 6188-6194.
107. Naumann, R., Jonczyk, A., Hampel, C., Ringsdorf, H., Knoll, W., Bunjes, N. and Graber, P. (1997) Coupling of proton translocation through ATPase incorporated into supported lipid bilayers to an electrochemical process, Bioelectrochem. Bioenerg. 42, 241-247.
108. Schmidt, E.K., Liebermann, T., Kreiter, M., Jonczyk, A., Naumann, R., Offenhausser, A., Neumann, E., Kukol, A., Maelicke, A. and Knoll, W. (1998) Incorporation of the acetylcholine receptor dimer from *Torpedo californica* in a peptide supported lipid membrane investigated by surface plasmon and fluorescence spectroscopy, Biosens. and Bioelectron. 13, 585-591.

109. Raguse, B., Braach-Maksvytis, V., Cornell, B.A., King, L.G., Osman, P.D.J., Pace, R.J. and Wieczorek, L. (1998) Tethered lipid bilayer membranes: Formation and ionic reservoir characterisation, Langmuir 14, 648-659.
110. Kuhner, M., Tampe, R. and Sackmann, E. (1994) Lipid Monolayer and Bilayer Supported On Polymer-Films - Composite Polymer-Lipid Films On Solid Substrates, Biophys. J. 67, 217-226.
111. Arya, A., Krull, U.J., Thompson, M. and Wong, A.E. (1985) Langmuir-Blodgett deposition of lipid films on hydrogel as a basis for biosensor development, Anal. Chim. Acta 173, 331-336.
112. Bruckner-Lea, C., Petelenez, D. and Janata, J. (1990) Use of poly(octadec-1-ene-maleic anhydride) for interfacing bilayer membranes to solid supports in sensor applications, Microchim. Acta 1, 169-185.
113. Spinke, J., Yang, J., Wolf, H., Liley, M., Ringsdorf, H. and Knoll, W. (1992) Polymer-supported bilayer on a solid substrate., Biophys. J. 63, 1667-1671.
114. Erdelen, C., Häussling, L., Naumann, R., Ringsdorf, H., Wolf, H., Yang, J., Liley, M., Spinke, J. and Knoll, W. (1994) Self-assembled disulfide-functionalised amphiphile copolymers on gold, Langmuir 10, 1246-1250.
115. Decher, G. (1997) Fuzzy Nanoassemblies: Toward Layered Polymeric Multicomposites, Science 277, 1232-1237.
116. Lindholm-Sethson, B. (1996) Electrochemistry at ultrathin organic films at planar gold electrodes, Langmuir 12, 3305-3314.
117. Lindholm-Sethson, B., Gonzalez, J.C. and Puu, G. (1998) Electron transfer to a gold electrode from cytochrome oxidase in a lipid bilayer via a polyelectrolyte film, Langmuir 14, 6705-6708.
118. Majewski, J., Wong, J.Y., Park, C.K., Seitz, M., Israelachvili, J.N. and Smith, G.S. Structural studies of polymer-cushioned lipid bilayers, Biophys. J. 75, 2363-2367, 1998.
119. Wong, J.Y., Majewski, J., Seitz, M., Park, C.K., Israelachvili, J.N. and Smith, G.S. (1999) Polymer-cushioned bilayers. I. A structural study of various preparation methods using neutron reflectometry, Biophys. J. 77, 1445-1457.
120. Seitz, M., Wong, J.Y., Park, C.K., Alcantar, N.A. and Israelachvili, J. (1998) Formation of tethered supported bilayers via membrane-inserting reactive lipids, Thin Solid Films 329, 767-771.
121. Gyorvary, E., Wetzer, B., Sleytr, U.B., Sinner, A., Offenhausser, A. and Knoll, W. (1999) Lateral diffusion of lipids in silane-, dextran-, and S-layer- supported mono- and bilayers, Langmuir 15, 1337-1347.
122. Hillebrandt, H., Wiegand, G., Tanaka, M. and Sackmann, E. (1999) High electric resistance polymer/lipid composite films on indium-tin-oxide electrodes, Langmuir, 15, 8451-8459.
123. Pagano, R.E. and Miller, I.R. (1973) Transport of Ions Across Lipid Monolayers. IV. Reduction of Polarographic Currents by Spread monolayers, J. Colloid Interface Sci., 45, 126-137.
124. Miller, I.R. (1981) Structural and energetic aspects of charge transport in lipid layers and in biological membranes, in Milazzo, G. (ed), Topics in Bioelectrochemistry and Bioenergitics, Vol. 4, John Wiley & Sons, Ltd., pp. 161-224.
125. Miller, I.R., Doll, L. and Lester, D.S. (1992) Interaction of Alamethicin, Melittin and Protein-Kinase-C With Pure and Phospholipid Monolayer Covered Mercury-Electrode Surfaces, Bioelectrochem. Bioenerg. 28, 85-103.
126. Nelson, A. and Benton, A. (1986) Phospholipid Monolayers At the Mercury Water Interface, J. Electroanal. Chem. 202, 253-270.
127. Nelson, A. and Auffret, N. (1988) Phospholipid Monolayers of Di-Oleoyl Lecithin At the Mercury Water Interface, J. Electroanal. Chem. 244, 99-113.
128. Bizzotto, D. and Nelson, A. (1998) Continuing electrochemical studies of phospholipid monolayers of dioleyl phosphatidylcholine at the mercury - electrolyte interface, Langmuir 14, 6269-6273.
129. Nelson, A. (1991) Electrochemical Studies of Thallium(I) Transport Across Gramicidin Modified Electrode-Adsorbed Phospholipid Monolayers, J. Electroanal. Chem. 303, 221-236.
130. Nelson, A. (1996) Influence of biologically active compounds on the monomolecular gramicidin channel function in phospholipid monolayers, Langmuir 12, 2058-2067.
131. Nelson, A. (1997) Influence of fixed charge and polyunsaturated compounds on the monomolecular gramicidin channel function in phospholipid monolayers: Further studies, Langmuir 13, 5644-5651.
132. Nelson, A. and Bizzotto, D. (1999) Chronoamperometric study of Tl(I) reduction at gramicidin-modified phospholipid-coated mercury electrodes, Langmuir 15, 7031-7039.
133. Rueda, M., Navarro, I., Ramirez, G., Prieto, F. and Nelson, A. (1998) Impedance measurements with phospholipid-coated mercury electrodes, J. Electroanal. Chem. 454, 155-160.

134. Nelson, A. (1991) Electrochemical Studies of Antibiotic-23187 (A23187) Mediated Permeability to Divalent Heavy-Metal Ions in Phospholipid Monolayers Adsorbed On Mercury-Electrodes, J. Chem. Soc. Faraday T. 87, 1851-1856.
135. Nelson, A. (1992) Voltammetry of Retinal in Phospholipid Monolayers Adsorbed On Mercury, J. Electroanal. Chem. 335, 327-343.
136. Moncelli, M.R., Becucci, L., Nelson, A. and Guidelli, R. (1996) Electrochemical modelling of electron and proton transfer to ubiquinone-10 in a self-assembled phospholipid monolayer, Biophys. J. 70, 2716-2726.
137. Zawisza, I., Bilewicz, R., Luboch, E. and Biernat, J.F. (2000) Complexation of metal ions by azocrown ethers in Langmuir- Blodgett monolayers, J. Chem. Soc.-Dalton T. 4, 499-503.
138. Moncelli, M.R. and Becucci, L. (1995) The intrinsic pKa values for phosphatidic acid in monolayers deposited on mercury electrodes. J. Electroanal. Chem. 385, 183-189.
139. Moncelli, M.R., Becucci, L. and Guidelli, R. (1994)The Intrinsic Pka, Values For Phosphatidylcholine, Phosphatidylethanolamine, and Phosphatidylserine in Monolayers Deposited On Mercury-Electrodes, Biophys. J. 66, 1969-1980.
140. Moncelli, M.R. and Becucci, L. (1996) Effect of the molecular structure of more common amphiphilic molecules on their interactions with dioleoyl- phosphatidylcholine monolayers, Bioelectrochem. Bioenerg. 39, 227-234.
141. Dolfi, A., Tadini-Buoninsegni, F., Aloisi, G. and Moncelli, M.R. (1999) Bacteriorhodopsin-containing membrane fragments adsorbed on mercury-supported biomimetic membranes, Electrochem. Comm. 1, 131-134.
142. Bartlett, P.N., Brace, K., Calvo, E.J. and Etchenique, R. (2000) In situ characterisation of phospholipid coated electrodes, J. Mater. Chem. 10, 149-156.
143. Csucs, G. and Ramsden, J.J. (1998) Interaction of phospholipid vesicles with smooth metal-oxide surfaces, BBA-Biomembranes 1369, 61-70.

FUNCTIONAL STRUCTURE OF THE SECRETIN RECEPTOR

P. ROBBERECHT[1], M. WAELBROECK[1], AND N. MOGUILEVSKY[2].
[1] Department of Biochemistry and Nutrition, Faculty of Medicine
[2] Department of Applied Genetics, IBMM, Faculty of Sciences, Université Libre de Bruxelles, Belgium.

Abstract

Secretin is a 27 amino acid peptide secreted by specialised endocrine cells of the gut that regulates the pancreatic exocrine - and the liver bile secretions. Its role as a neuropeptide is likely but not yet proven. Secretin acts through interaction with a membrane receptor coupled to the $G_{\alpha s}$ protein and thus stimulates cyclic AMP production. It belongs to a new subgroup of receptors named the GPCRB that includes - among others - the VIP, the PACAP, the Glucagon, the Glucagon like peptide 1, the Calcitonin and the Parathormone receptors. The expression in CHO cells of the wild type and of mutated receptors permits the identification of domains involved in ligand recognition, in signal transduction, in desensitisation and in receptor internalisation.

1. Introduction

Bayliss and Starling, in 1902, discovered that acid extracts of the mucosa of the upper intestine contained a substance that, if injected into the blood stream caused the pancreas to secrete. They named the substance Secretin. They suggested a new term "hormone" to describe such chemical messengers between different organs. The purification of Secretin was undertaken at the Karolinska Institute by Jorpes and Mutt and the amino acid sequence definitively established in 1970. At the same time, Bodansky synthesised the 27 amino acid peptide that had a biological activity indistinguishable from that of the natural hormone [1]. Secretin had marked similarities to Glucagon that was sequenced before. Later on, sequences similarities were found with newly discovered peptides VIP, GIP, GRF, PACAP, PHI, GLP-1, and GLP-2. The existence of a large family of biologically active peptides with sequence similarities but different properties was established [2].

Using isolated pancreatic acini or acinar cells, it was shown that Secretin acted through interaction with a specific membrane receptor coupled to adenylate cyclase [3].

M. De Cuyper and J.W.M. Bulte (eds.), Physics and Chemistry Basis of Biotechnology, 167–176.

In 1991, the rat secretin receptor was cloned by Ishiara et al. [4]. The receptor sequence was related to that of the calcitonin and the parathormone receptors [5,6]. The knowledge of the sequence of the secretin receptor initiated the cloning by analogy of the VIP, PACAP, glucagon, GRF, GIP, and GLP-1 receptors, revealing the existence of a large new family of G protein coupled receptors known as the GPCR B family [7].

2. The secretin receptor

2.1. GENERAL ARCHITECTURE

The mRNA of the rat [4], rabbit [8] and human [9,10] secretin receptors, coded for proteins of 449, 445, and 440 amino acid residues respectively. Sequence analysis and modelling predicted 23 amino acid residues for a signal peptide, 120 residues for the amino terminal extra-cellular domain (N-EC), the existence of seven transmembrane domains (TM1 to TM7) connected by three extracellular (EC1 to EC3) and three intracellular loops (IC1 to IC3) and an intracellular carboxyl terminus of 50 residues (Fig. 1).

RAT SECRETIN RECEPTOR

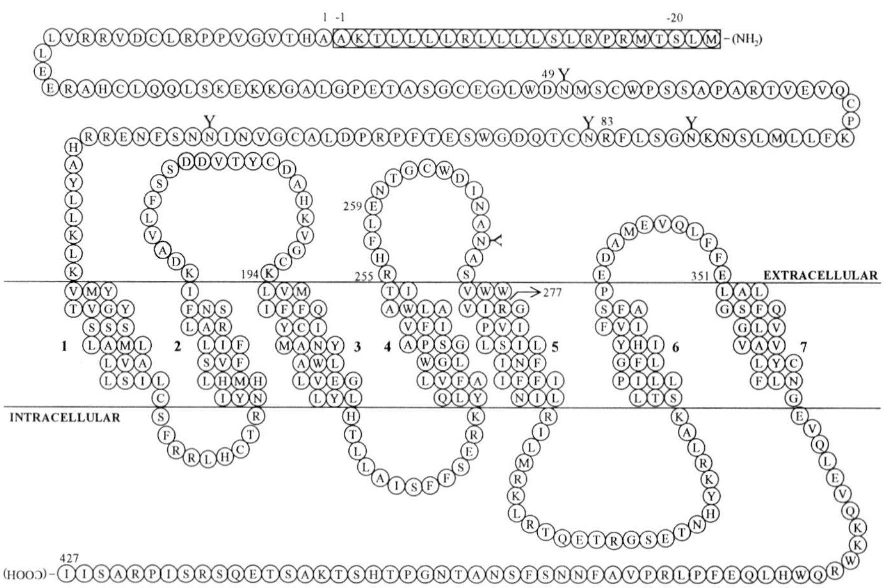

Figure 1. A secondary structure model representation of the rat secretin receptor. The putative signal peptide is bordered by a rectangle and Y indicates the glycosylable asparagine residues.

Four glycosylable asparagine residues were found in the N-EC; a fifth residue was present in the EC2 domain of the rat and human sequences, but not in the rabbit one. A potential site for O-glycosylation is also present in the N-EC. Ten highly conserved Cys residues were noticed in the extracellular part of the receptor : 7 in the N-EC, 2 in EC1 and 1 in EC2 domains. In a tentative to establish the disulfide bridges, we mutated each Cys residue in serine. Six of the point-mutant receptors, C25 S, C44 S, C53 S, C67 S, C85 S, and C101 S, in the N-EC domain could not be detected by binding studies or by functional assay of adenylate cyclase activation : the mutations resulted in functional inactivation of the receptor, but it was not established if this lack of activity was due to a lack of expression at the cell surface or to a severe misfolding of an otherwise normally expressed receptor. In contrast, the four other point-mutant receptors C11 S, C186 S, C193 S and C263 S, that means the first Cys residue that followed the signal peptide and the three Cys residues in the EC domains, were able to bind secretin and to be activated by secretin. However the affinity for the ligand was lower than that of the wild-type receptor. These results suggested that Cys residues 24, 44, 53, 67, 85, and 101 were necessary for receptor function and that two putative disulfide bridges formed by Cys residues 11, 186, 193, and 263 were functionally relevant but not essential for receptor expression. Secretin activated adenylate cyclase with a very high EC_{50} value (concentration required for 50 % activation of the enzyme) through the quadruple mutant C11,186,193,263 S, the four triple mutants and through double mutants C186,193 S and C186,263 S, suggesting that, in the wild-type receptor, disulfide bridges are formed between the N-EC Cys residue 11 and the first residue 186 of EC1 on one hand and between the second residue of EC1 193 and the residue 263 of EC2. Several results indicated that in the mutant receptors alternative disulfide bridges can be formed between Cys 186 and 193 or 263, suggesting that these residues are in close spatial proximity in the wild-type receptor [11].

Mutations, deletions, or exchange of parts of the secretin receptor by the corresponding domain of the parent receptors VIP, PACAP, or glucagon have been performed to delineate the areas necessary for an appropriate structure. It is however difficult to make the difference between receptors modifications leading to an impaired structure from those that modulate ligand binding without major change in the general structure. Mutations in the N-EC domain of Asp residue 49 into Arg (but not by an uncharged polar or hydrophobic residue), of the Arg residue 83 by an Asp or by a Leu residue led to almost inactive receptors. Similarly, mutations in the EC1 domain of Asp 174 into Ala but not Asn, or mutation in the EC2 domain of Arg 255 by an Asp or a Glu residue also severely impaired ligand recognition [12]. No functional receptor could be obtained after replacement of N-EC or EC3 by the corresponding sequences of the glucagon receptor [13]. At the opposite, deletion of the highly charged sequence 30-33 Lys-Glu-Lys-Lys did not affect ligand recognition and receptor expression (unpublished data). Taken as a whole, these results suggest that several parts of the receptor contributed to the functional structure of the secretin receptor.

2.2. FUNCTIONAL DOMAINS

2.2.1. Ligand binding domain

At variance with what is observed for the GPCR A group of receptors that included the bioactive amines, the small peptides, the nucleotides and the prostaglandin receptors, the N-EC domain is of paramount importance for ligand binding and discrimination among the ligands. Chow reported [14] that the exodomain alone of the secretin receptor binds secretin with a high affinity. Replacement of the secretin receptor N-EC by the VIP receptor counterpart resulted in biological responsiveness rather typical of VIP receptor. The replacement of the VIP receptor N-EC domain by the secretin counterpart led to a secretin-preferring receptor [15,16]. However, from this last chimera it appears that the EC1 domain plays a complementary role for the secretin receptor [17,18]. Refining the chimeric approach, we obtained unexpected results: replacement of the rat secretin receptor sequences 1-35, 36-81, 35-121, by the rat VPAC1 receptor counterpart led to constructions that could not be studied due to a lack of expression or to a low affinity. The replacement of the sequences 13-18 and 23-29 of the secretin receptor by the VPAC1 receptor sequence did not change the receptor's characteristics. Introduction of the VPAC1 receptor sequences 103-110 or 116-120 in the secretin receptor led to constructions with a lower affinity for secretin or a higher affinity for VIP respectively [19]. Holtman et al. [15] reported the importance of the first 10 residues of the N-EC1 sequence for secretin recognition, a finding confirmed by Olde et al. [17]. It appears thus that several parts of the amino-terminal extracellular domain of the receptor contribute to the ligand recognition. By testing the functional properties of modified ligands on chimeric VPAC/secretin receptors, we obtained indirect arguments for an interaction between the carboxyl-terminal sequence 23-27, the sequence 8-15 and residue 16 of secretin with the N-EC domain of the receptor [20-22]. It was also suggested that the Lys residue 15 of secretin contacts with the Asp 98 of the receptor [23]. By photo-affinity labelling of the secretin receptor with a secretin probe that had incorporated a photo-labile p-benzoyl-L-phenylalanine into position 22 and 6, it was shown by Miller's group that Leu 22 interacted with a receptor sequence comprised between amino acid 1 to 30 of the receptor and Phe 6 with Val 4 of the receptor [24,25]. The N-EC domain is not the sole domain interacting with the ligand: the His 189-Lys 190 sequence of the C-terminus of EC1 and Phe 257-Leu 258, Asn 260-Thr 261 in the amino-terminal half of EC2 also provide critical determinants, but the ligand counterpart was not established [18]. Several arguments suggest that the amino terminus of secretin interacted directly with the secretin receptor : a) deletion of His 1, Ser 2 and Asp 3 decreased 1000 to 10,000-fold peptide affinity and decreased the peptide efficacy : secretin (2-27) and (3-27) are partial agonists whereas secretin (4-27) is an antagonist; b) replacement of Asp 3 by Glu and Asn reduced secretin affinity 10 and 300-fold respectively [26]. We first observed [13] that the Lys residue 173 boarding TM2 and EC1 was in the vicinity of the binding region of Asp 3 as its mutation into Ileu decreased peptide affinity and makes Asn 3 and Glu 3 secretin indistinguishable. We then explored the amino acid residues surrounding that Lys residue : mutation of Asp 174 into Asn not only reduced secretin affinity but also changed receptor's selectivity as judged by a decreased secretin / Asn 3 secretin ratio [27]. When Arg 166 (in TM2) was mutated to Gln, Asp or Leu, the receptor had a very low affinity for secretin but an up to 10-fold higher affinity for Asn 3 secretin. Mutation of Asn 170 was of no consequence on receptor affinity and selectivity. Arg 166 is an excellent candidate for a precise contact between the ligand and the receptor : a) the positively charged residue is conserved in the whole family of receptors except the calcitonin receptor; b) receptor modelling suggests it to be rather accessible as its side chain is facing an internal pocket limited by the other TM [28]; c) according to this model, the residue is buried relatively deeply in the transmembrane domain suggesting that the ligand may enter into the receptor to change its conformation and promote the G protein activation. Analysis of the sequence of the receptor indicated that two Tyr residues located in TM1 at the same level as Arg 166 in TM2 might be involved also in the positioning of secretin amino-terminus sequence. We mutated Tyr 124 and Tyr 128 into Ala and His. Mutations of Tyr 124 reduced 10-fold the affinity of secretin and all the analogs tested. Mutation of Tyr 128 into Ala reduced 50-fold secretin affinity but also modified the capability of the receptor to discriminate the analogs modified in position 3. Ala 128 secretin receptor was unable to discriminate secretin from Glu 3 secretin. His 128 secretin receptor was almost comparable to the wild-type receptor [29].

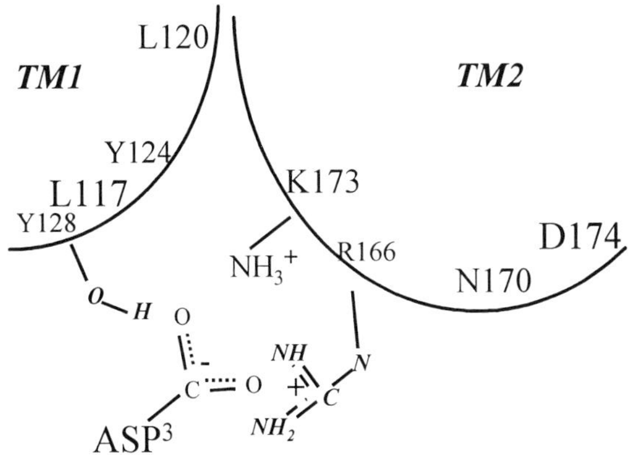

Figure 2.*Schematic representation of the possible interactions between the Asp 3 residue of secretin and the lateral chains of the amino acids of TM1 and TM2.*

Taken together, the analysis of the receptors mutated in TM1 and TM2 supported the hypothesis that secretin Asp 3 (or Glu) residues are anchored by ionic interactions with the Lys 173 and Arg 166 residues while the Asp 3 (or Asn) residues may form hydrogen bonds with Tyr 128. Replacement of the two basic residues by uncharged residues decreased markedly the affinity for secretin and Glu 3 secretin but not for the uncharged Asn 3 secretin analog. In contrast, replacement of Tyr 128 by an Ala residue decreased the receptor affinity for secretin and the isosteric Asn 3 secretin analog, but not for the Glu 3 secretin. Replacement of Tyr 128 by His which may also form hydrogen bonds had little consequence on the recognition profile. A model for the positioning of the Asp 3 residue in the receptor is presented in Fig. 2.

2.2.2. Coupling of the receptor to the G protein.

It was shown that secretin- but not secretin analogs in which a Tyr residue was introduced for a Leu residue in position 10 and 13, stimulated adenylate cyclase activity at low concentrations and phosphatidylinositol bisphosphate hydrolysis at higher concentrations [30]. The dual activation of the cyclic AMP and the inositol trisphosphate / Ca++ pathways, that was reported for other parents receptors [31,32] was not described for the wild-type recombinant secretin receptor expressed in CHO cells (may be it was not yet tested). So far, the preferential, if not the sole, coupling mechanism of the secretin receptor is $G_{\alpha s}$ and adenylate cyclase activation. There is to our knowledge no study investigating the precise domain of the receptor responsible for that coupling. Mutation of His 156 into Arg (in TM2 near the inner face of the membrane) or Thr 322 into Pro (in TM6 near the inner face of the membrane) resulted in a constitutively active receptor, that means that the unstimulated receptor may couple to adenylate cyclase. Interestingly, the double mutant Arg 156 / Pro 322 had a greater activity than each individual mutant and secretin behaves like an inverse agonist, that means it reduced the basal adenylate cyclase activity [33]. These data suggested that the coupling to adenylate cyclase through interaction with $G_{\alpha s}$ involved a large area at the inner face of the membrane.

2.2.3. Desensitisation of the receptor.

As most- if not all- the G protein coupled receptors, the secretin receptor desensitises rapidly and is internalised through a not yet identified mechanism, insensitive to sucrose but sensitive to Con A [34]. After secretin exposure, the secretin receptor expressed in Cos or CHO cells is rapidly phosphorylated, mainly on Thr residues located essentially in the carboxyl-terminal intracellular tail [35]. Truncation of this part of the receptor does not alter the ligand recognition, nor the coupling to adenylate cyclase but prevents its phosphorylation [36]. However, the truncated receptor remains still internalisable. Thus a dual mechanism limits secretin action : a rapid phosphorylation decreases the cyclic AMP production and an internalisation independent of phosphorylation occurs. In transfected HEK 293 cells, it was demonstrated that protein kinases A and C do not markedly affect receptor phosphorylation that is mainly induced by the GRK2 and GRK5 kinases [37].

3. Conclusions and perspectives

Secretin is mainly a hormone controlling the pancreatic fluid secretion through interaction with acinar and duct pancreatic cells [3] and the bile flow through interaction with cholangiocytes [38]. Intravenous injections of secretin were used for years to evaluate the exocrine pancreatic function in human. The natural porcine secretin used first was contaminated by the neurohormone cholecystokinin-pancreozymin (CCK-PZ) that contracts vigorously the gallbladder; since the availability of essentially pure porcine secretin obtained from the Karolinska Institute, and later on from Kabi-Vitrum, the effect of secretin and CCK-PZ have been dissociated. The functional secretin-cholecystokinin test was performed by measuring fluid and enzyme secretion in duodenal aspirates. This was a good clinical test when performed by trained physicians. Its interest dropped however when retrograde cholangio-wirsungography was introduced as a routine endoscopic procedure : the visualisation of the pancreatic and the biliary tract was more helpful for the clinician than the functional tests. Furthermore, echography became the best non-invasive technique for gallbladder volume and contractility evaluation. Catheterisation of the main pancreatic duct with a balloon catheter allowed the analysis of pure pancreatic juice in response to secretin. This technique was however difficult to perform and was considered as an invasive one limited to experimental studies. More recently, the imaging of the pancreatic tree was possible by magnetic resonance cholangiopancreatography after secretin injection. This non-invasive technique will probably be in a near future the reference for pancreatic exploration. The systematic injection of secretin to patients suffering of gastrointestinal tract diseases will hopefully led to the discovery of secretin-linked pathologies (hypersecretory responses that could be due to constitutively active secretin receptors, absence of secretin response ...).

Secretin, or preferably stable analogues - peptidic or non peptidic - could be of a great therapeutic value : repeated administration of secretin by inducing a high pancreatic flow rate might reduce the formation of pancreatic stones; secretin administered after fragmentation of pancreatic calculi by ultrasound bombardment will efficiently eliminate the stone fragments.

Secretin is also a neuropeptide regulating tyrosine hydroxylase activity in PC 12 cells and in sympathetic ganglia [39]. It activates also adenylate cyclase in striatal neurons from embryonic mouse brain grown in cultures [40]. Its effect on adult rat brain occurs however at high concentrations only [41]. Recently, case reports of dramatic improvements of children suffering from autism were reported after secretin administration [42]. However, a recent double-blind study concluded that a single dose of human secretin was not an effective treatment for autism [43]. The use of more stable analogs with a facilitated penetration into the brain should be evaluated.

How may the knowledge of the interactions between ligand and receptor contribute to the conception of peptidominetic ligands of potential therapeutic use ? Actually, most - if not all - peptidominetics have been discovered by chance (for instance the antibiotic erythromycin mimics the effect of the hormone Motilin by interacting with the motilin receptor) or by a systematic screening of large chemical libraries. This approach requires only an appropriate cellular model expressing the receptor of interest coupled to an easily detected effector system and the access to a large number of molecules. Another approach, based on the knowledge of the binding domain of the receptor is intellectually more rewarding but difficult : it requires a feed-back interaction between receptor's modellers and biologists, but to start modelling, at least some firm structural data must be available. This is not the case for the G protein coupled receptors : the only available x-ray structure is that of bacteriorhodopsin that possesses seven transmembrane helices related to that of the receptors. The arrangement of the helices can be predicted in the receptor, but their fiability must still be validated. There is not data on the structure, or the folding of the aminoterminal extracellular domain of the receptor and as detailed in this review this domain is particularly important for ligand binding. Several laboratories try to produce sufficient amounts of the isolated aminoterminal domain to try a crystallisation of this large part of the receptor, but there is still no reported success in the literature.

Acknowledgements

This work was supported by an "Action de Recherche Concertée" from the Communauté Française de Belgique.

References

1. Mutt, V. Secretin (1980) Isolation, Structure and Functions, in G.B. Jerzy Glass (eds.), *Gastrointestinal Hormones,* Raven Press, New York. pp 85-126.
2. Christophe, J., Svoboda, M., Dehaye, J.P., Winand, J., Vandermeers Piret, M.C., Vandermeers, A., Cauvin, A., Gourlet, P., and Robberecht, P. (1989) The VIP/PHI/secretin/helodermin/helospectin/GRF family : structure-function relationship of the natural peptides, their precursors and synthetic analogues as tested in vitro on receptors and adenylate cyclase in a panel of tissue membranes, in J. Martinez (ed.), *Peptide hormones as prohormones: Processing, Biological Activity, Pharmacology,* Ellis Horwood Limited, Chichester pp. 211-243.
3. Zhou, Z.C., Gardner, J.D., and Jensen, R.T. (1987) Receptors for vasoactive intestinal peptide and secretin on guinea pig pancreatic acini, *Peptides 8*, 633-637.
4. Ishihara, T., Nakamura, S., Kaziro, Y., Takahashi, T., Takahashi, K., and Nagata, S. (1991) Molecular cloning and expression of a cDNA encoding the secretin receptor, *EMBO. J. 10*, 1635-1641.
5. Lin, H.Y., Harris, T.L., Flannery, M.S., Aruffo, A., Kaji, E.H., Gorn, A., Kolakowski, L.F. Jr, Lodish, H.F., and Goldring, S.R. (1991) Expression cloning of an adenylate cyclase-coupled calcitonin receptor, *Science 254,* 1022-1024.
6. Juppner, H., Abou Samra, A.B., Freeman, M., Kong, X.F., Schipani, E., Richards, J., Kolakowski, L.F. Jr, Hock, J., Potts, J.T. Jr, Kronenberg, H.M., et a.l. (1991) A G protein-linked receptor for parathyroid hormone and parathyroid hormone-related peptide, *Science 254,* 1024-1026.
7. Horn, F., Weare, J., Beukers, M.W., Horsch, S., Bairoch, A., Chen, W., Edvardsen, O., Campagne, F., and Vriend, G. (1998)GPCRDB: an information system for G protein-coupled receptors, *Nucleic Acids Res. 26,* 275-279.
8. Svoboda, M., Tastenoy, M., De Neef, P., Delporte, C., Waelbroeck, M., and Robberecht, P. (1998) Molecular cloning and in vitro properties of the recombinant rabbit secretin receptor, *Peptides 19,* 1055-1062.
9. Paolo, E., De Neef, P., Moguilevsky, N., Petry, H., Cnudde, J., Bollen, A., Waelbroeck, M., and Robberecht, P. (1999) Properties of a recombinant human secretin receptor: a comparison with the rat and rabbit receptors, *Pancreas 19,* 51-55.
10. Patel, D.R., Kong, Y., and Sreedharan, S.P. (1995) Molecular cloning and expression of a human secretin receptor, *Mol. Pharmacol. 47,* 467-473.

11. Vilardaga, J.P., Di Paolo, E., Bialek, C., De Neef, P., Waelbroeck, M., Bollen, A., and Robberecht, P. (1997) Mutational analysis of extracellular cysteine residues of rat secretin receptor shows that disulfide bridges are essential for receptor function, *Eur. J. Biochem. 246,* 173-180.
12. Di Paolo, E., Vilardaga, J.P., Petry, H., Moguilevsky, N., Bollen, A., Robberecht, P., and Waelbroeck, M. (1999) Role of charged amino acids conserved in the vasoactive intestinal polypeptide/secretin family of receptors on the secretin receptor functionality, *Peptides 20,* 1187-1193.
13. Vilardaga, J.P., di Paolo, E., de Neef, P., Waelbroeck, M., Bollen, A., and Robberecht, P. (1996) Lysine 173 residue within the first exoloop of rat secretin receptor is involved in carboxylate moiety recognition of Asp 3 in secretin, *Biochem. Biophys. Res. Commun. 218,* 842-846.
14. Chow, B.K. (1997) Functional antagonism of the human secretin receptor by a recombinant protein encoding the N-terminal ectodomain of the receptor, *Recept. Signal. Transduct. 7,* 143-150.
15. Holtmann, M.H., Hadac, E.M., and Miller, L.J. (1995) Critical contributions of amino-terminal extracellular domains in agonist binding and activation of secretin and vasoactive intestinal polypeptide receptors. Studies of chimeric receptors, *J. Biol. Chem. 270,* 14394-14398.
16. Vilardaga, J.P., De Neef, P., Di Paolo, E., Bollen, A., Waelbroeck, M., and Robberecht, P. (1995) Properties of chimeric secretin and VIP receptor proteins indicate the importance of the N-terminal domain for ligand discrimination, *Biochem. Biophys. Res. Commun. 211,* 885-891.
17. Olde, B., Sabirsh, A., and Owman, C. (1998) Molecular mapping of epitopes involved in ligand activation of the human receptor for the neuropeptide, VIP, based on hybrids with the human secretin receptor, *J. Mol. Neurosci. 11,* 127-134.
18. Holtmann, M.H., Ganguli, S., Hadac, E.M., Dolu, V., and Miller, L.J. (1996) Multiple extracellular loop domains contribute critical determinants for agonist binding and activation of the secretin receptor, *J. Biol. Chem. 271,* 14944-14949.
19. Robberecht, P., and Waelbroeck, M. (1998) A critical view of the methods for characterisation of the VIP/PACAP receptor subclasses, *Ann. N. Y. Acad. Sci. 865,* 157-163.
20. Gourlet, P., Vilardaga, J.P., De Neef, P., Vandermeers, A., Waelbroeck, M., Bollen, A., and Robberecht, P. (1996) Interaction of amino acid residues at positions 8-15 of secretin with the N-terminal domain of the secretin receptor, *Eur. J. Biochem. 239,* 349-355.
21. Gourlet, P., Vilardaga, J.P., De Neef, P., Waelbroeck, M., Vandermeers, A., and Robberecht, P. (1996) The C-terminus ends of secretin and VIP interact with the N-terminal domains of their receptors, *Peptides 17,* 825-829.
22. Gourlet, P., Vandermeers, A., Vandermeers Piret, M.C., De Neef, P., Waelbroeck, M., and Robberecht, P. (1996) Effect of introduction of an arginine16 in VIP, PACAP and secretin on ligand affinity for the receptors, *Biochim. Biophys. Acta 1314,* 267-273.
23. Holtmann, M.H., Hadac, E.M., Ulrich, C.D., and Miller, L.J. (1996) Molecular basis and species specificity of high affinity binding of vasoactive intestinal polypeptide by the rat secretin receptor, *J. Pharmacol. Exp. Ther. 279,* 555-560.
24. Dong, M., Wang, Y., Pinon, D.I., Hadac, E.M., and Miller, L.J. (1999) Demonstration of a direct interaction between residue 22 in the carboxyl-terminal half of secretin and the amino-terminal tail of the secretin receptor using photoaffinity labelling, *J. Biol. Chem. 274,* 903-909.
25. Dong, M., Wang, Y., Hadac, E.M., Pinon, D.I., Holicky, E., and Miller, L.J. (1999) Identification of an interaction between residue 6 of the natural peptide ligand and a distinct residue within the amino-terminal tail of the secretin receptor, *J. Biol. Chem. 274,* 19161-19167.
26. Vilardaga, J.P., Ciccarelli, E., Dubeaux, C., De Neef, P., Bollen, A., and Robberecht, P. (1994) Properties and regulation of the coupling to adenylate cyclase of secretin receptors stably transfected in Chinese hamster ovary cells, *Mol. Pharmacol. 45,* 1022-1028.
27. Di Paolo, E., De Neef, P., Moguilevsky, N., Petry, H., Bollen, A., Waelbroeck, M., and Robberecht, P. (1998) Contribution of the second transmembrane helix of the secretin receptor to the positioning of secretin, *FEBS Lett. 424,* 207-210.
28. Donnelly, D. (1997) The arrangement of the transmembrane helices in the secretin receptor family of G-protein-coupled receptors, *FEBS Lett. 409,* 431-436.
29. Di Paolo, E., Petry, H., Moguilevsky, N., Bollen, A., De Neef, P., Waelbroeck, M., and Robberecht, P. (1999) Mutations of aromatic residues in the first transmembrane helix impair signalling by the secretin receptor, *Receptors. Channels 6,* 309-315.

30. Trimble, E.R., Bruzzone, R., Biden, T.J., Meehan, C.J., Andreu, D., and Merrifield, R.B. (1987) Secretin stimulates cyclic AMP and inositol trisphosphate production in rat pancreatic acinar tissue by two fully independent mechanisms, *Proc. Natl. Acad. Sci. U. S. A. 84,* 3146-3150.
31. Van Rampelbergh, J., Poloczek, P., Francoys, I., Delporte, C., Winand, J., Robberecht, P., and Waelbroeck, M. (1997) The pituitary adenylate cyclase activating polypeptide (PACAP I) and VIP (PACAP II VIP1) receptors stimulate inositol phosphate synthesis in transfected CHO cells through interaction with different G proteins, *Biochim. Biophys. Acta 1357,* 249-255.
32. Delporte, C., Poloczek, P., de Neef, P., Vertongen, P., Ciccarelli, E., Svoboda, M., Herchuelz, A., Winand, J., and Robberecht, P. (1995) Pituitary adenylate cyclase activating polypeptide (PACAP) and vasoactive intestinal peptide stimulate two signalling pathways in CHO cells stably transfected with the selective type I PACAP receptor, *Mol. Cell. Endocrinol. 107,* 71-76.
33. Ganguli, S.C., Park, C.G., Holtmann, M.H., Hadac, E.M., Kenakin, T.P., and Miller, L.J. (1998) Protean effects of a natural peptide agonist of the G protein-coupled secretin receptor demonstrated by receptor mutagenesis, *J. Pharmacol. Exp. Ther. 286,* 593-598.
34. Mundell, S.J., and Kelly, E. (1998) The effect of inhibitors of receptor internalisation on the desensitisation and resensitisation of three Gs-coupled receptor responses, *Br. J. Pharmacol. 125,* 1594-1600.
35. Ozcelebi, F., Holtmann, M.H., Rentsch, R.U., Rao, R., and Miller, L.J. (1995) Agonist-stimulated phosphorylation of the carboxyl-terminal tail of the secretin receptor, *Mol. Pharmacol. 48,* 818-824.
36. Holtmann, M.H., Roettger, B.F., Pinon, D.I., and Miller, L.J. (1996) Role of receptor phosphorylation in desensitisation and internalisation of the secretin receptor, *J. Biol. Chem. 271,* 23566-23571.
37. Shetzline, M.A., Premont, R.T., Walker, J.K., Vigna, S.R., and Caron, M.G. (1998) A role for receptor kinases in the regulation of class II G protein-coupled receptors. Phosphorylation and desensitisation of the secretin receptor, *J. Biol. Chem. 273,* 6756-6762.
38. Glaser, S.S., Rodgers, R.E., Phinizy, J.L., Robertson, W.E., Lasater, J., Caligiuri, A., Tretjak, Z., LeSage, G.D., and Alpini, G. (1997) Gastrin inhibits secretin-induced ductal secretion by interaction with specific receptors on rat cholangiocytes, *Am. J. Physiol. 273,* G1061-G1070.
39. Roskoski, R. Jr, White, L., Knowlton, R., and Roskoski, L.M. (1989) Regulation of tyrosine hydroxylase activity in rat PC12 cells by neuropeptides of the secretin family, *Mol. Pharmacol. 36,* 925-931.
40. Chneiweiss, H., Glowinski, J., and Premont, J. (1986) Do secretin and vasoactive intestinal peptide have independent receptors on striatal neurons and glial cells in primary cultures, *J. Neurochem. 47,* 608-613.
41. Fremeau, R.T. Jr, Korman, L.Y., and Moody, T.W. (1986) Secretin stimulates cyclic AMP formation in the rat brain, *J. Neurochem. 46,* 1947-1955.
42. Horvath, K., Stefanatos, G., Sokolski, K.N., Wachtel, R., Nabors, L., and Tildon, J.T. (1998) Improved social and language skills after secretin administration in patients with autistic spectrum disorders, *J. Assoc. Acad. Minor. Phys. 9,* 9-15.
43. Sandler, A.D., Sutton, K.A., DeWeese, J., Girardi, M.A., Sheppard, V., and Bodfish, J.W. (1999) Lack of benefit of a single dose of synthetic human secretin in the treatment of autism and pervasive developmental disorder, *N. Engl. J. Med. 341,* 1801-1806.

COLD-ADAPTED ENZYMES

D. GEORLETTE, M. BENTAHIR, P. CLAVERIE, T. COLLINS, S. D'AMICO, D. DELILLE[‡], G. FELLER, E. GRATIA, A. HOYOUX, T. LONHIENNE, M-A. MEUWIS, L. ZECCHINON AND CH. GERDAY
Laboratory of Biochemistry, Institute of Chemistry, University of Liège, B6, 4000 Sart-Tilman, Belgium
Laboratoire Océanologique de Banyuls, Université P. and M. Curie (Paris 6), CNRS UA117, F66650 Banyuls-Sur-Mer, France.

In the last few years, increased attention has been focused on a class of organisms called psychrophiles. These organisms, hosts of permanently cold habitats, display metabolic fluxes more or less comparable to those exhibited by mesophilic organisms at moderate temperatures. Psychrophiles have evolved by producing, among others, "cold-evolved" enzymes which have to cope with the reduction of chemical reaction rates induced by low temperatures. Thermal compensation in these enzymes is reached, in most cases, through a high catalytic efficiency as compared to their mesophilic counterparts at temperatures ranging from 0 to 30°C. Optimisation of the catalytic parameters can originate from an increased flexibility of either a selected area of the molecular edifice or of the overall protein structure. The increased resilience probably enhances abilities to undergo conformational changes during catalysis at low temperatures. In return, this flexibility would be responsible for the weak thermal stability of the psychrophilic enzymes. Structure modelling and recent crystallographic data have allowed the identification of the structural parameters that could be involved in a higher plasticity. The current consensus is that only subtle modifications of the conformation of cold-adapted enzymes can be related to the structural flexibility and that each enzyme adopts its own strategy. Moreover, it appears that there is a continuum in the strategy of protein adaptation to temperature, since known structural factors involved in protein stability of thermophiles are either reduced in number or modified, in order to increase flexibility in psychrophilic enzymes. The possible biotechnological purposes of cold-adapted enzymes are numerous. Due to their attractive properties, i.e. high specific activity and low thermal stability, they are very promising in various fields such as the detergent and food industries, molecular biology, bioremediation, etc.

M. De Cuyper and J.W.M. Bulte (eds.), Physics and Chemistry Basis of Biotechnology, 177–196.

1. Introduction

It has been well established that most enzymes should be stored at low temperatures due to the fact that at temperatures around freezing, enzyme activity is usually minimised and protein stability maximised. However, a class of organisms called psychrophiles, living in permanently cold environments, display metabolic fluxes more or less comparable to those exhibited by mesophilic organisms at moderate temperatures. These organisms produce "cold-evolving enzymes" that have to cope with the reduction of chemical reaction rates induced by low temperatures. One can therefore address the following questions: "What is the mechanism of adaptation? Can these enzymes be of any help in understanding protein folding and the relationship between stability and activity? Do they offer a significant advantage for biotechnological applications?".

The temperature dependence of biological reaction rates is commonly reflected in an equation proposed by S. Arrhenius at the end of the last century:

$$k = Ae^{-Ea/RT} \quad (1)$$

in which Ea is the activation energy, R the gas constant (8.31 kJ mol^{-1}) and T the temperature in Kelvin. A decrease in temperature leads to a modification in the rate of a chemical reaction; indeed for most biological systems, a decrease of 10°C divides the rate of a chemical reaction occurring in the system by a factor ranging between 2-3, corresponding to Q_{10}, which expresses the ratio of reaction rates measured at an interval temperature of 10°C. For instance, the activity of a mesophilic enzyme is about 80 times lower when the reaction temperature is shifted from 37 to 0°C. Nevertheless, metabolic rates of Antarctic fishes are only slightly lower than those of temperate water species [1] and the generation times of psychrophilic bacteria near 0°C are of the same order as those of mesophilic microorganisms at 37°C [2-4]. This clearly indicates that mechanisms of temperature compensation are involved.

2. Enzymes and low temperatures

The effect of temperature on "temperate" enzymes from mesophilic organisms has been extensively studied. When the reaction rates mediated by such enzymes are measured at various temperatures and extrapolated to low temperatures, one finds that at about 0°C, the reaction rate is close to zero. It is likely that this drop is generated by the reduction in catalytic efficiency (k_{cat}/K_m), caused by a decrease in the ability of the enzyme to modify its conformation or 'breathe' during catalysis. When temperature decreases, water viscosity increases and gives rise to decreased k_{cat}/K_m, due to a reduced motion of both solutes and water. Moreover, a decrease in temperature also induces a reduction in salt solubility and an increase in gas solubility. Decreases in the pH of biological buffers are also observed with decreasing temperature, affecting both the protein solubility and the charge of amino acids, particularly histidine residues [5, 6]. Temperate enzymes are also susceptible to cold-denaturation leading to the loss of enzyme activity at low temperatures [7]. This phenomenon, which is a very general property of globular proteins, affects, in particular, multimeric enzymes such as ATPases and fatty-acid synthetases as well as proteins showing a high degree of hydrophobicity [8]. Cold-denaturation is thought to result from interactions between polar and non polar groups in protein and water [9], a process thermodynamically favoured at low temperatures. The breakdown of hydrophobic forces that are important for protein folding and stability causes protein unfolding. Enzymes from cold-adapted organisms must be resistant to this low temperature induced-denaturation. Therefore hydrophobic forces might be less important in psychrophilic enzymes.

3. Cold-adaptation: generality and strategies

Proteins and especially enzymes from organisms adapted to high temperature environments have been the subject of considerable interest in both basic and applied research for a number of years; the main objective has been to understand how structural and functional properties enable thermophiles to live in conditions where most normal proteins and nucleic acids would be denatured. Such information could contribute to a better understanding of the structural basis of thermostability and adaptive strategies, and could be applied to the biotechnology industry [10-13]. Because of their attractive biotechnological applications, cold-adapted bacteria are now receiving increased attention (for review, see Gerday *et al.* [14] and Margesin *et al.* [15]). Cold-adapted organisms live at the lowest temperatures allowing life development; either prokaryotic or eukaryotic; they have successfully colonised permanently cold habitats such as polar and alpine regions or deep-sea waters [16-19]. To survive under such stress conditions, they have developed various adaptive strategies from the level of single molecules to that of the whole organism. Indeed, cold adaptation has led to the modification of enzyme kinetics, the synthesis of specialised molecules known as antifreeze molecules, cryoprotectors and cold-shock proteins, the regulation of membrane fluidity, ion channels, microtubules polymerisation, frost hardening and seasonal dormancy [20-27]. In permanently cold habitats, low temperatures have constrained psychrophiles to develop enzymatic tools allowing metabolic rates compatible to life which are close to those of temperate organisms. Such an objective could be reached by increasing the enzyme concentrations compensating for low k_{cat} values. However, this strategy is improbable due to its energy cost. Therefore it has been reported only in a very few cases during acclimation of ectotherms [28, 29]. Another strategy would be the expression of specific isotypes kinetically adapted to a given environment [20, 30]. This kind of adaptation is only feasible when multicopies of genes are available. This process is particularly useful for seasonal reversible adaptation mechanisms, as observed in fish [31] and nematode enzymes [32]. Another alternative in cold-adaptation would be the design of enzymes characterised by temperature-independent reaction rates. In these perfectly evolved enzymes, the reaction rate would be only diffusion controlled. The analysis of the Arrhenius equation (1) shows that in this case, E_a tends to zero and therefore the exponential term E_a/RT tends to 1, leading to fast and mostly temperature-independent reactions. Such enzymes are relatively rare; some known examples are carbonic anhydrase, acetylcholinesterase or triosephosphate isomerase. More generally however, most organisms living in cold habitats have adapted their metabolic rates by increasing the catalytic potential of their enzymes. Due to their different metabolic pathways, extracellular and intracellular enzymes have possibly adopted different adaptation strategies. Indeed, extracellular enzymes often encounter high substrate concentration, allowing them to perform catalysis at maximum speed. In this case, the adaptation will be oriented towards the catalytic constant k_{cat}. However, for exoenzymes secreted in a liquid medium poor in substrate, one can argue that they should also have highly optimised K_m values. On the other hand, intracellular enzymes usually perform catalysis at low substrate concentration, leading to the adaptation of k_{cat} or K_m or both and the constant to be taken into account is k_{cat}/K_m. Thus, it is often difficult to decide which parameter is to be optimised in the context of cold-adaptation and there are no general rules.

4. Kinetic evolved-parameters

The catalytic cycle of an enzyme is made up of three main phases: recognition and binding of the substrate, conformational changes induced by the substrate, leading to product formation and finally, release of the product. Each of these phases involves weak interactions sensitive to temperature changes. Depending on the type of these weak interactions and on the substrate concentration, the temperature dependence of the enzymatic reaction can be close to that of an uncatalysed reaction, or quite different. Indeed, in the case of Michaelis-Menten kinetics, at high substrate concentrations, the concentration of K_m, which roughly represents a measure of the affinity of the enzyme for the substrate, becomes negligible. Therefore, at a fixed enzyme concentration, the thermodependence of the reaction rate will be dependent only on k_{cat}, the rate constant. However, at subsaturating substrate concentrations, the situation will be quite different, since K_m will also have an influence on reaction rate. Thus the term to be considered is catalytic efficiency, k_{cat}/K_m.

Efficient binding of substrate by the enzyme is mediated by the nature and strength of weak interactions which are of two types: (i) interactions formed with a negative modification of enthalpy and hence exothermic (Van Der Waals, Hydrogen, Electrostatic) and (ii) interactions formed with a positive modification of enthalpy, and thus endothermic (Hydrophobic). The formers are destabilised by an increase of temperature, whereas the latter will tend to be stabilised by moderately high temperatures. K_m will be differentially affected by temperature according to the contribution of each type of bond to the enzyme-substrate interaction. If hydrophobic interactions are dominant, moderately elevated temperatures will have little effect on K_m and the Q_{10} value will be kept constant. On the contrary, at low temperatures, K_m will increase and Q_{10} values will reach values well over 3.

Many scientists have investigated the adaptation process of psychrophilic enzymes and these works have been summarised in recent publications [14, 33-36]. These studies have revealed that the dominating character of these cold-adapted enzymes is their high specific activity compared to their mesophilic homologues at temperatures ranging from 0 to 30°C. At higher temperatures, heat-denaturation occurs. The specific activity of many extracellular enzymes has been extensively studied: α-amylase [37], β-lactamase [38], chymotrypsin [39], elastase [40], lipase [41, 42], Ca^{2+}-Zn^{2+} protease [43], subtilisin [44-46], trypsin [47] and xylanase [48]. It emerged from these studies that all these enzymes have a general tendency to optimise their catalytic efficiency k_{cat}/K_m at the temperature of the habitat and that they all show higher thermosensitivity. The only particular case is β-lactamase (cephalosporinase) from *Psychrobacter immobilis* A8 [38]. This enzyme displays a low level of thermal stability and a low optimal temperature of activity. In contrast to other cold-adapted enzymes, however, it does not show an increased specific activity when compared to the most active mesophilic counterparts. In this particular case, a high level of specific activity is probably not required as the concentration of β-lactam antibiotics in the Antarctic environment is low [38]. Nevertheless, another element has to be taken into consideration. Indeed, class C β-lactamases have almost diffusion-controlled reaction rates [49, 50]. Consequently, activity of both psychrophilic and mesophilic cephalosporinases is weakly influenced by temperature with Q_{10} as low as 1.1. Such perfectly evolved enzymes therefore have very few possibilities to further improve k_{cat} values in the context of cold adaptation. The fact that the psychrophilic enzyme is thermolabile does, however, raise the following question: if the psychrophilic β-lactamase is almost perfectly evolved, why should it be thermolabile? Is it the result of the lack of selective pressure for a stable protein, or is heat-lability a common property required for efficient catalysis at low temperatures? As mentioned below, the latter proposal seems to apply here.

Different mechanisms can be proposed to increase the catalytic efficiency of psychrophilic extracellular enzymes. An unusual one is a decrease in substrate binding-strength (increased K_m value) which could contribute to lowering the activation energy and consequently increasing the reaction rate. This first possibility has only been reported in a few cases [37, 47, 51]. One has to emphasise here that care should be taken when using small chromogenic substrates for determining the thermal dependence of K_m, rather than the natural macromolecular one. Small chromogenic substrates are very useful, but may have quite distinct binding modes, as seen in the case of psychrophilic α-amylase, where the temperature dependence of K_m strongly varies as a function of the substrate used [37].

Regarding intracellular or periplasmic enzymes such as multimeric alcohol dehydrogenase [52], L-lactate dehydrogenase [53], triosephosphate isomerase [54, 55], glucose-6-phosphate dehydrogenase [56], citrate synthase [57], PspPI methyltransferase [58], β-galactosidase (Hoyoux *et al*, personal communication), monomeric DNA polymerase [59], phosphoglycerate kinase [60] and DNA ligase [61], kinetic adaptations are unclear. Indeed even if these enzymes reveal a higher thermolability when compared to mesophilic enzymes, the higher specific activity is not a general characteristic: k_{cat}/K_m values are not optimised, except in the case of psychrophilic β-galactosidase, phosphoglycerate kinase and DNA ligase. Here again, artefactual measurements, due to partial denaturation for example, have to be taken into account along with the use of inappropriate substrates. Specific cofactors can also be involved. In the case of multimeric proteins, preserving appropriate intermolecular interactions could prevent or limit the molecular adjustments required to achieve high catalytic efficiency at low temperatures

5. Activity/thermolability/flexibility

Enzyme catalysis generally involves the "breathing" of all or of a particular region of the enzyme, enabling the accommodation of the substrate. The ease with which such movement can occur may be one of the determinants of catalytic efficiency. Therefore optimising a function of an enzyme at a given temperature requires a proper balance between two often opposing factors: structural rigidity, allowing the retention of a specific 3D conformation at the physiological temperature and flexibility, allowing the protein to perform its catalytic function [62, 63]. At room temperature, a thermophilic enzyme would be therefore stable, rigid and poorly active. This is certainly due to an increase in molecular edifice rigidity dictated by the low thermal energy in the surroundings, thus preventing essential movement of residues. In order to secure the appropriate stability at high temperatures, thermophilic enzymes appear to have a very rigid and compact structure at moderate temperatures, which, in most cases, is characterised by a tightly packed hydrophobic core and maximal exposure of charged residues at the surface [64, 65].

Hydrophobic interactions have been initially proposed as the major stabilising force in proteins [66, 67]. However, in 1995, Ragone and Colonna [68] suggested that hydrophobic interactions would not stabilise proteins having melting temperatures of about 87°C or above. So, other forces, such as salt bridges and hydrogen bonds, would be expected to play a major role in the extra thermostabilisation of such proteins. Their hypothesis is corroborated by studies suggesting that hydrogen bonding and the hydrophobic effect make comparable contributions to the stability of globular proteins [69, 70]. Since thermophily is correlated with the rigidity of a protein, psychrophily, at the opposite end of the temperature scale, should be characterised by a more flexible structure to compensate for the lower thermal energy provided by the low temperature habitat [20]. This plasticity would enable a good complementarity with the substrate at a low energy cost, thus explaining the high specific activity of psychrophilic enzymes. In return, this flexibility would be responsible for the weak thermal stability of psychrophilic enzymes. The weak stability of cold-adapted enzymes has been demonstrated by the drastic shift of their apparent optimal temperature of activity, the low resistance of the protein to denaturing agents and the high susceptibility of the structure to unfold at moderate temperatures. Until now, attempts to correlate this weak stability to an increased conformational flexibility have failed. Indeed, there is no direct experimental demonstration of such relationship, contrary to what was found in the case of thermophily [71]. In order to assess the flexibility of a protein structure, care has to be taken to avoid the use of a technique which gives an average measure of flexibility that does not correctly reflect the local flexibility required for catalysis. Some techniques have been extensively used to demonstrate the putative increased flexibility of psychrophilic enzymes. Hydrogen-Deuterium (H/D) exchange offers a good quantitative parameter for measuring the average conformational flexibility of a macromolecule. For instance, it has been shown that the thermostable D-glyceraldehyde-3-phosphate dehydrogenase is more rigid than its mesophilic counterpart at 25°C, thus suggesting that conformational flexibility is similar at corresponding physiological temperatures [72]. This fact has been confirmed by H-D exchange studies performed on thermophilic and mesophilic 3-isopropylmalate dehydrogenases [71]. Indeed, the thermostable enzyme is more rigid at room temperature than its mesophilic homologue, whereas the enzymes display nearly identical flexibility under their respective optimal working conditions. The authors argued that evolutionary adaptation tends to maintain a "corresponding state" regarding conformational flexibility: conformational fluctuations necessary for catalytic function are restricted at room temperature in the thermophilic enzyme, suggesting a close relationship between conformational flexibility and enzyme function [73]. The fact that more rigid thermostable proteins reach the flexibility of thermolabile proteins at higher temperatures was already mentioned by Vihinen and colleagues [74]. In addition, T4-lysosyme site-directed mutagenesis revealed that enzyme residues involved in function are not optimised for stability, supporting the 'stability-function' relationship [75]. Nevertheless H/D exchange rate measurements recorded by Fourier transformed infrared spectroscopy (FTIR) carried out with psychrophilic and mesophilic α-amylases failed to reveal any significant difference in the rate and magnitude of amide proton replacement, indicating either that there is no real difference or that the experiment temperature was not appropriate. Moreover, loops are well known to be the most mobile parts of enzyme structures. External loops in the psychrophilic α-amylase are shorter and the lower bulk of exchangeable protons can possibly compensate for an improved H/D exchange due to flexibility. NMR experiments on small psychrophilic enzymes should prove to be an attractive alternative in further investigation of the expected stability-flexibility relationship. So far, probably the only case in which the higher flexibility of cold enzymes has been demonstrated is in a fluorescence spectroscopy experiment on a Ca^{2+}-Zn^{2+} protease isolated from psychrophilic and mesophilic *Pseudomonas* sp. It was clearly shown that the quenching effect of acrylamide was much more important in the case of the psychrophilic enzyme [43]. However, it has yet to be demonstrated that increased flexibility gives rise to higher specific activity at low and moderate temperatures.

In fact, the relationship between stability, specific activity and flexibility is much more complex than was expected. Diverse directed-mutagenesis experiments on psychrophilic enzymes have led to a better understanding of the activity-flexibility-stability relationship. In such experiments, weak interactions mediated by residues occurring in mesophilic or thermophilic enzymes were introduced in psychrophilic enzymes specifically devoid of these interactions. The main goal of these experiments was to stabilise the psychrophilic mutants and to record the changes of the specific activity associated with the mutation. Attention was particularly focused on the α-amylase from the psychrophile *Pseudoalteromonas haloplanktis*. Among the different mutations restoring a putative mesophilic character, the double-mutation N150D-

V196F, engineering a salt bridge (N150D) and an aromatic interaction (V196F), caused decreased activity and significantly increased melting temperature (44°C to 46.4°C), establishing a correlation between increased stability and lowered activity (D'Amico *et al.*, personal communication). Some other site-directed mutagenesis experiments have been carried on subtilisin [45] and a moderately stable thermolysin like protease [76]. Both studies revealed that the stability of the mutated enzymes was drastically improved due to a small number of mutations, while increasing or retaining the original catalytic properties of the enzymes, which does not correspond with the stability-specific activity-flexibility assumption. Moreover, a recent *in vitro* evolution experiment has proved that stability and catalytic activity are not systematically inversely related, since random mutagenesis on the *Bacillus subtilis* p-nitrobenzyl esterase led to an increase in stability (>14°C increase in Tm) without compromising catalytic activity [77]. In fact, reduced stability may not necessarily arise from a general reduction in strength of intramolecular forces, but from weakened interactions in one or a few important regions of the structure [78].

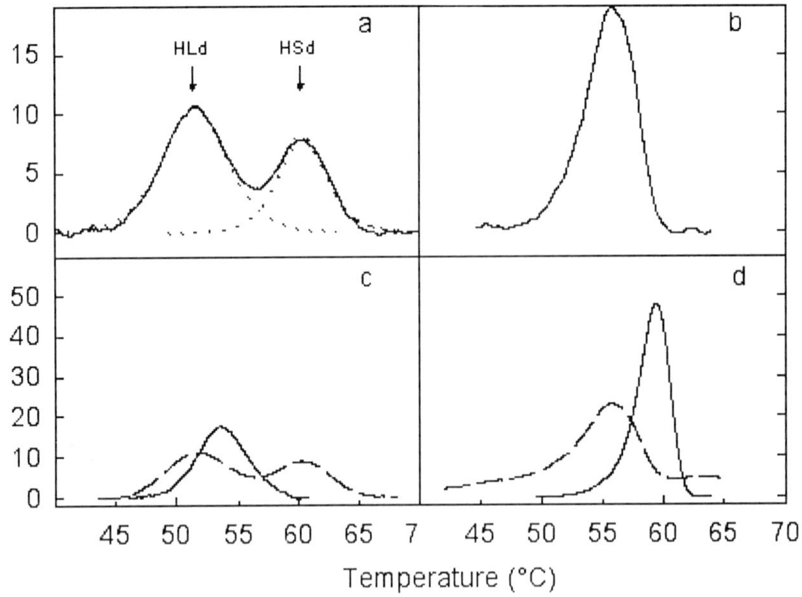

Figure 1. *Thermal unfolding of the psychrophilic Pseudomonas sp. TACII 18 (left) and yeast (right) phosphoglycerate kinase (PGK). a to d: baseline-subtracted thermograms recorded by DSC. a and b: free enzymes (the deconvolution in two transitions is dotted). c and d: PGK in the presence of 5 mM 3-phosphoglyceric acid and 5 mM Mg-ADP (solid lines). In c and d, the thermogram for free enzyme is given as a dashed line for comparison. HLd, heat-labile domain; HSd, heat-stable domain.*

Some recent works have corroborated this statement. In the case of citrate synthase [79], a calculation of crystallographic temperature (*B*) factors, which, to a certain extent, reflect the flexibility or disorder of the crystal structure in a given region, revealed an unexpected higher flexibility for the thermophilic enzyme when considering the overall protein structure. However, in the psychrophilic homologue, the small domain showed a higher degree of flexibility than did the large domain. This inequality could promote activity at low temperatures, since a precise positioning of the small domain following substrate binding is necessary for efficient catalysis. As mentioned by Fields and Somero [80], notothenioid A4-lactate dehydrogenase has adapted to cold by increasing flexibility of small areas of the molecule which affect the mobility of adjacent active-site structures. The increased flexibility may reduce energy barriers to rate-governing shifts in conformation and thereby, increase k_{cat}. Moreover, crystallisation of malate dehydrogenase from the Arctic *Aquaspirillium arcticum* revealed that the active site of this enzyme is more flexible than that of the thermophilic one, facilitating efficient catalysis at low temperatures [81]. Finally, differential scanning calorimetry performed on the phosphoglycerate kinase from the Antarctic *Pseudomonas* sp. TACII 18 [60] revealed unusual variations of its conformational stability in free (Figure 1a) and liganted forms (Figure 1c). As shown in Figure 1a, the psychrophilic PGK is characterised by a heat-labile (HLd) and a heat-stable (HSd) domain, whereas its mesophilic homologue has two domains unfolding simultaneously. In the substrate-free form of psychrophilic enzyme (Figure 1a), the heat-labile and the thermostable domain unfold independently. The existence of a stabilising domain has previously been mentioned in xylanases [82], in which the increased stability of one domain promotes the stability of the whole molecule. Similarly, Bentahir and his colleagues have proposed that the PGK heat labile domain acts as a destabilising domain, providing the required flexibility around the active site for catalysis at low temperatures. Thus cold adapted proteins have evolved, either by displaying a reduced stability of all calorimetric units giving rise to native states of the lowest stability [83] or by being constituted of different elements, some controlling protein stability and others conferring the required flexibility for efficient catalysis at the habitat temperature.

6. Structural comparisons

The thermostability of thermophilic enzymes has been extensively investigated and several possible determinants of this stability have been proposed [84]. In contrast, structural comparisons of cold active enzymes with their mesophilic and thermophilic counterparts have been limited, until recently, to the analysis of homology models and/or sequence alignments for a limited number of enzymes from bacteria, yeast and fish, such as α-amylase [37, 85], triosephosphate isomerase [54], subtilisin [44], trypsin (fish) [86], β-lactamase [38], elastase (fish) [87], 3-isopropylmalate dehydrogenase [88], lipase [41], elongation factor (EF) 2 [89] and EF-G [90], xylanase (yeast) [48] and phosphoglycerate kinase [60]. These studies have revealed that only subtle modifications of the enzyme conformation account for low stability and that the

adaptation strategy is unique to each enzyme. Moreover neither the amino-acid residues involved in the catalytic process nor the topology of the active site of the enzymes are affected to any extent in cold-adapted proteins. This suggests that cold-adaptation does not involve new catalytic mechanisms, but rather that conventional mechanisms are modified to operate better at low temperatures. In fact, these works have revealed that the main features possibly implicated in cold-adaptation consisted of fewer salt links (especially with a reduced number of arginine residues), hydrogen bonds, isoleucine clusters and proline residues in loops, extended and highly charged surface loops, an increased number of glycine and serine residues close to functional motifs, a reduction in the hydrophobicity of the enzyme and an increase in the number of interactions between the enzyme surface and the solvent. All these parameters have been reviewed by Feller et al. [33, 91] and Gerday et al. [34]. Interestingly, the same structural factors have been implicated in the stability of thermophilic proteins [62, 92-96] suggesting that there is a continuum in the strategy of protein adaptation to temperature.
The putative parameters involved in cold adaptation have been confirmed by recent resolutions of crystallographic 3D structures of some cold enzymes: salmon trypsin [78], α-amylase from *Alteromonas haloplanktis* [97-99], Ca^{2+}-Zn^{2+} protease from *Pseudomonas aeruginosa* [100], triosephosphate isomerase from *Vibrio marinus* [55], citrate synthase from the Antarctic bacterial strain DS2-R [79] and malate dehydrogenase from *Aquaspirillium arcticum* [81]. For instance, these enzymes frequently display a weakening of the intramolecular interactions. However, an increase in intramolecular ion pairs has been observed in the heat-labile citrate synthase, demonstrating that, as previously mentioned, all psychrophilic enzymes do not use the same strategy to increase structural resilience. The authors explained that this increase may serve to prevent cold denaturation by counteracting the reduced thermodynamic stability originating from an increase in solvent-hydrophobic interactions. Exposure of hydrophobic residues to solvent is indeed destabilising, due to the ordering of water molecules. Therefore, the elimination of destabilising hydrophobic interactions with solvent may be necessary for hyperthermostability, whereas their presence in thermolabile enzymes may be an important factor in preserving the protein structure at low temperatures. Increased solvent-hydrophobic residues interaction has been reported for salmon trypsin and a cold-active α-amylase. Decreased stability of cold-enzymes is also reached by a general weakening of interdomain or intersubunit interactions, which emerges as a critical force in stabilising hyperthermophilic enzymes [95, 101, 102]. As observed for salmon trypsin, citrate synthase and malate dehydrogenase, a precise distribution of surface charges favours a better electrostatic attraction of substrates, contributing to an increased catalytic efficiency. Finally a better accessibility of the catalytic cavity can improve substrate accommodation; this has been described for salmon trypsin [78], α-amylase (Figure 2) [103] and citrate synthase [79] and could lead to a higher specific activity at low temperatures. All the above-mentioned factors may possibly improve the overall or local flexibility of the molecular edifice.

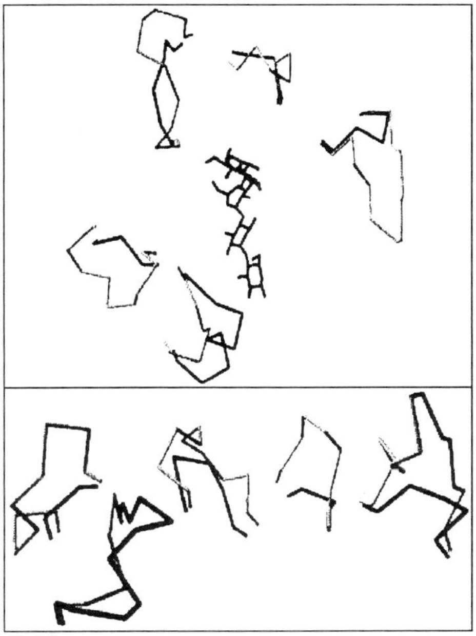

Figure 2. *Active site accessibility.* Upper panel*: superimposition of the variable loops bordering the active site of the psychrophilic α-amylase from* Pseudoalteromonas haloplanktis *(backbone in black) and of pig pancreatic α-amylase (backbone in grey). The carbohydrate inhibitor acarbose bound to the active site is also shown (centre of the figure, in black). These variable loops are markedly shorter in the cold-active enzyme.* Lower panel*: tangential view of the molecular surfaces facing the external medium (upper side). Acarbose is removed for clarity. The loops around the catalytic cleft of the cold-active enzyme are less protruding and favour active site accessibility. Picture generated by Swiss-PDBViewer [127] using data from references [98, 99].*

7. Fundamental and biotechnological applications

It has already been demonstrated that cold-evolved enzymes constitute useful tools for fundamental studies in protein folding. Moreover, psychrophilic organisms and their products offer a high potential for biotechnological applications [15, 104, 105]. For example, a well known and important application of cold enzymes, such as proteases, lipases, α-amylases and cellulases, is their use in the detergent industry. Indeed, cold washing allows energy savings and a reduction in wear and tear. However, the thermal instability and storage of these enzymes can constitute a considerable drawback. It is, though, always possible to engineer recombinant enzymes in which reasonable stability is coupled with high catalytic efficiency [45]. In the textile industry, the use of a cold-adapted cellulase would offer an excellent alternative to stone-washing of jeans and biopolishing of cotton fibres [106]. Pre-treatment of tissues with cellulases, under the appropriate conditions, would reduce pill-formation (due to cotton fibre ends protruding from the main tissue fibres) and increase both the durability and the softness of the tissue. The current treatment, however, leads to an alteration of the main fibre, as a result of the resistance of the mesophilic enzyme to heat-inactivation. The use of psychrophilic cellulase, which is easily inactivated, would allow circumventing this problem. One can also take advantage of the high thermosensitivity of psychrophilic enzymes in the food industry. The possible applications of cold-evolved enzymes are numerous. The main goals include the improvement of taste and product quality as well as the minimisation of mesophilic microorganism contamination. In order to prevent early degradation due to microbial growth, low temperature is often mandatory in the food industry. Thus, properties of psychrophilic enzymes, i.e. high specific activity at low and moderate temperatures and low thermal stability, would allow food to retain their freshness and flavour during the enzymatic reaction and following enzyme inactivation. Hence enzymes are promising in various fields such as lipid extraction , processing of fruit juices (pectinases) and cheese, modification of food texture, improvement in the quality of milk (β-galactosidases), bread (amylases, proteases, xylanases) and alcoholic drinks.

Some cold-evolved enzymes, such as DNA ligase, alkaline phosphatase and uracil-DNA glycosylase, also constitute new and performing tools in molecular biology. DNA ligases are essential in molecular biology. In fact, the ligation reaction requires a low temperature, in order to ensure sufficient temporal base pairing (through hydrogen bonding) to allow the formation of a phosphodiester bond. Commercially mesophilic DNA ligases have, however, relatively poor activity at temperatures below 15°C and require long incubation times. Under such conditions, the action of residual nucleases is favoured, which can interfere with the ligation reaction. *Pseudoalteromonas haloplanktis* DNA ligase displays a high catalytic efficiency at low and moderate temperatures, compared to its mesophilic counterpart [61]. In addition, a relatively low inactivation temperature can be used, which does not denature DNA, and therefore, this enzyme represents a novel tool in biotechnology. Radioactive end-labelling of nucleic acids by T4 polynucleotide kinase requires the removal of the existing phosphates at the 5' ends of DNA by alkaline phosphatase (APase). This enzyme must be inactivated after the reaction, to prevent degradation of labelled ATP and the loss of label from the substrates. However, known mesophilic APases display great thermal stability, leading to incomplete enzyme inactivation and interference with subsequent kinase and ligase reaction. Hence the remarkable temperature sensitivity of psychrophilic HK47 APase appears to be a useful feature for 5' end-labelling [107]. Uracil-DNA glycosylase (UNG), produced by a psychrophilic marine bacterium, is also a potential tool for the biotechnology market [108]. UNG belong to a specific class of DNA repair enzymes [109]. These enzymes are mainly used to prevent carryover contamination in polymerase chain reaction (PCR), and hence avoiding false positive results, due to the contamination of PCR samples by products from previous amplifications. Prior to the actual PCR, the enzyme must be heat-inactivated. However, *E. coli* UNG, usually used, is not completely inactivated during heat treatment, leading to degradation of the newly synthesised PCR product. This problem can be overcome by using the heat-labile UNG [108]. We have also to emphasise that the development of a 'cold' gene expression system is under investigation [110-112]. In fact, since cold-active enzymes are usually thermolabile, ordinary expression systems such as the *E. coli* expression system, may not be the most suitable for the expression of cold-active enzymes. So far, some cold-shock proteins have been well characterised [26]. CspA is the major cold-shock protein in *E. coli* and recently, an expression system with a *cspA* promoter, which is controlled by cold-shock treatment, has been developed [113]. Furthermore, other cold-inducible promoters in *E. coli* have also been reported [114]. Therefore, the availability of such a cold-shock inducible expression system could enhance the expression yield by minimising proteolytic degradation as well as the accumulation of the recombinant product in insoluble precipitates. In addition, Remaut *et al.* [112] have developed an expression system in which *E. coli*-derived expression controlling elements are introduced into psychrophilic hosts, using as vector system a broad-host-range plasmid. It was shown that *E. coli* $lacI^{q}$-*Ptrc* repressor-promoter system is operative at temperatures as low as 4°C in two different psychrotrophic species.

Another interesting aspect of some Antarctic microorganisms is their production of polyunsaturated fatty acids (PUFAs), including eicosapentaenoic acid (EPA) and docosahexaenoic acid (DHA) (for review, see [115, 116]). In fact, they attract considerable attention as sources of pharmaceutical agents, functional foods, and supplement nutrition. Among other, PUFAs are essential for normal growth and development of the larvae of many aquaculture species. While bacteria have previously been considered for use in aquaculture feeds, their lack of essential PUFA was seen as a major drawback [117]. However, it now appears that some strains of Antarctic bacteria also produce high levels of PUFA [118, 119]. Use of such PUFA-producing microheterotrophs in aquaculture diets, livestock and human diets is now an expanding area of interest.

Over the past decade numerous environments and especially low temperature habitats have been contaminated by oils, which has led to the investigation of hydrocarbon degradation by Antarctic micro-organisms [120-122]. Degradation of xenobiotic compounds including diesel oil and polychlorinated biphenyls by psychrophilic bacteria could offer a possible alternative to physicochemical methods. Moreover, the addition of such bacteria to contaminated cold area should help to enhance the biodegradation of hydrocarbons [18, 123-126].

8. Conclusions

In the last few years, increased attention has been focused on enzymes produced by cold-adapted micro-organisms. It has emerged that psychrophilic enzymes represent an extremely powerful tool in both protein folding investigations and for biotechnological purposes. Such enzymes are characterised by an increased thermosensitivity and, most of them, by a higher catalytic efficiency at low and moderate temperatures, when compared to their mesophilic counterparts. The high thermosensitivity probably originates from an increased flexibility of either a selected area of the molecular edifice or the overall protein structure, providing enhanced abilities to undergo conformational changes during catalysis at low temperatures. Structure modelling and recent crystallographic data have allowed to elucidate the structural parameters that could be involved in this higher resilience. It was demonstrated that each psychrophilic enzyme adopts its own adaptive strategy. It appears, moreover, that there is a continuum in the strategy of protein adaptation to temperature, as the previously mentioned structural parameters are implicated in the stability of thermophilic proteins. Additional 3D crystal structures, site-directed and random mutagenesis experiments should now be undertaken to further investigate the stability-flexibility-activity relationship.

Acknowledgements

We are grateful to N. Gerardin and R. Marchand for advises technical assistance. We acknowledge the "Institut Français de Recherche et de Technologie polaire" for generously accommodating our research fellows at the French Antarctic Station J.S. Dumont d'Urville in Terre Adélie.

References

1. Clarke, A. (1983) Life in cold water: the physiological ecology of polar marine ectotherms, *Oceanogr. Mar. Biol. Ann. Rev.* 21, 341-453.
2. Morita, R. Y. (1975) Psychrophilic bacteria, *Bacteriol. Rev.* 39, 144-167.
3. Mohr, P. W. and Krawiec, S. (1980) Temperature characteristics and Arrhenius plots for nominal psychrophiles, mesophiles and thermophiles, *J. Gen. Microbiol.* 121, 311-317.
4. Feller, G., Narinx, E., Arpigny, J.-L., Zekhnini, Z., Swings, J. and Gerday, C. (1994) Temperature dependence of growth, enzyme secretion and activity of psychrophilic Antarctic bacteria, *Appl. Microbiol. Biotechnol.* 41, 477-479.
5. Yancey, P. H. and Somero, G. N. (1978) Temperature dependence of intracellular pH: its role in the conservation of pyruvate apparent Km values of vertebrate lactate dehydrogenases, *J. Comp. Physiol. B.* 125, 129-134.
6. Somero, G. N. (1981) pH-temperature interactions on proteins: principles of optimal pH and buffer system design, *Mar. Biol. Lett.* 2, 163-178.
7. Privalov, P. L. (1990) Cold denaturation of proteins, *Crit. Rev. Biochem. Mol. Biol.* 25, 281-305.
8. Creighton, T. E. (1991) Stability of folded conformations, *Curr. Opin. Struct. Biol.* 1, 5-16.
9. Makhatadze, G. I. and Privalov, P. L. (1995) Energetics of protein structure, *Adv. Protein. Chem.* 47, 307-425.
10. Adams, M. W. (1993) Enzymes and proteins from organisms that grow near and above 100 degrees C, *Annu. Rev. Microbiol.* 47, 627-658.
11. Pennisi, E. (1997) In industry, extremophiles begin to make their mark, *Science.* 276, 705-706.
12. Stetter, K. O., Fiala, G., Huber, G., Huber, R. and Segerer, G. (1990) Hyperthermophilic microorganisms, *FEMS Microbiol. Rev.* 75, 117-124.
13. Stetter, K. O. (1999) Extremophiles and their adaptation to hot environments, *FEBS Lett.* 452, 22-25.
14. Gerday, C., Aittaleb, M., Bentahier, M., Chessa, J. P., Claverie, P., Collins, T., D'Amico, S., Dumont, J., Garsoux, G., Georlette, D., Hoyoux, A., Lonhienne, T., Meuwis, M.-A. and Feller, G. (2000) Cold-adapted enzymes: from fundamentals to biotechnology, *Trends Biotechnol.* 18, 103-107.
15. Margesin, R. and Schinner, F. (1999) *Biotechnological applications of cold-adapted organisms*, Springer-Verlag, Heidelberg.

16. Gilichinsky, D. and Wagener, S. (1995) Microbial life in permafrost. A historical review, *Permafrost Periglacial Pro.* 6, 243-250.
17. Bowman, J. P., McCammon, S. A., Brown, M. V., Nichols, D. S. and McMeekin, T. A. (1997) Diversity and association of psychrophilic bacteria in Antarctic sea ice, *Appl. Environ. Microbiol.* 63, 3068-3078.
18. Margesin, R. and Schinner, F. (1997) Efficiency of indigenous and inoculated cold-adapted soil microorganisms for biodegradation of diesel oil in alpine soils, *Appl. Environ. Microbiol.* 63, 2660-2664.
19. Morita, Y., Nakamura, T., Hasan, Q., Murakami, Y., Yokoyama, K. and Tamiya, E. (1997) Cold-active enzymes from cold-adapted bacteria, *J. Am. Oil. Chem. Soc.* 74, 441-444.
20. Hochachka, P. W. and Somero, G. N. (1984) Temperature adaptation. In: *Biochemical adaptations* (Hochacka, P. W. and Somero, G. N., eds) pp. 355-449, Princeton University Press, Princeton.
21. Storey, K. B. and Storey, J. M. (1988) Freeze tolerance in animals, *Physiol. Rev.* 68, 27-84.
22. Jaenicke, R. (1990) Proteins at low temperature, *Phil. Trans. R. Soc. Lond. B.* 326, 535-553.
23. Franks, F. (1985) *Biophysics and biochemistry at low temperatures*, Cambridge University Press, Cambridge.
24. Mayr, B., Kaplan, T., Lechner, S. and Scherer, S. (1996) Identification and purification of a family of dimeric major cold shock protein homologs from the psychrotrophic *Bacillus cereus* WSBC 10201, *J. Bacteriol.* 178, 2916-2925.
25. Berger, F., Morellet, N., Menu, F. and Potier, P. (1996) Cold shock and cold acclimation proteins in the psychrotrophic bacterium *Arthrobacter globiformis* SI55, *J. Bacteriol.* 178, 2999-3007.
26. Phadtare, S., Alsina, J. and Inouye, M. (1999) Cold-shock response and cold-shock proteins, *Curr. Opin. Microbiol.* 2, 175-180.
27. Ewart, K. V., Lin, Q. and Hew, C. L. (1999) Structure, function and evolution of antifreeze proteins, *Cell Mol. Life Sci.* 55, 271-283.
28. Crawford, D. L. and Powers, D. A. (1989) Molecular basis of evolutionary adaptation at the lactate dehydrogenase- B locus in the fish *Fundulus heteroclitus*, *Proc. Natl. Acad. Sci. U. S. A.* 86, 9365-9369.
29. Crawford, D. L. and Powers, D. A. (1992) Evolutionary adaptation to different thermal environments via transcriptional regulation, *Mol. Biol. Evol.* 9, 806-813.
30. Somero, G. N. (1995) Proteins and temperature, *Annu. Rev. Physiol.* 57, 43-68.
31. Baldwin, J. and Hochachka, P. W. (1970) Functional significance of isoenzymes in thermal acclimatisation. Acetylcholinesterase from trout brain, *Biochem. J.* 116, 883-887.
32. Jagdale, G. B. and Gordon, R. (1997) Effect of temperature on the activities of glucose-6-phosphate dehydrogenase and hexokinase in entomopathogenic nematodes (*Nematoda steinernematidae*), *Comp Biochem. Physiol. A Physiol.* 118, 1151-1156.
33. Feller, G., Arpigny, J. L., Narinx, E. and Gerday, C. (1997) Molecular adaptations of enzymes from psychrophilic organisms, *Comp. Biochem. Physiol.* 118, 495-499.
34. Gerday, C., Aittaleb, M., Arpigny, J. L., Baise, E., Chessa, J. P., Garsoux, G., Petrescu, I. and Feller, G. (1997) Psychrophilic enzymes: a thermodynamic challenge, *Biochim. Biophys. Acta.* 1342, 119-131.
35. Gerday, C., Aittaleb, M., Arpigny, J. L., Baise, E., Chessa, J. P., François, J. M., Petrescu, I. and Feller, G. (1999) Cold enzymes: a hot topic. In: *Cold adapted organisms: Ecology, Physiology, Enzymology and Molecular Biology* (Margessi, R. and Schinner, F., eds) pp. 257-275, Springer-Verlag, Heidelberg.
36. Russell, N. J. (2000) Toward a molecular understanding of cold activity of enzymes from psychrophiles, *Extremophiles.* 4, 83-90.
37. Feller, G., Payan, F., Theys, F., Qian, M., Haser, R. and Gerday, C. (1994) Stability and structural analysis of α-amylase from the antarctic psychrophile *Alteromonas haloplanctis* A23, *Eur. J. Biochem.* 222, 441-447.
38. Feller, G., Zekhnini, Z., Lamotte-Brasseur, J. and Gerday, C. (1997) Enzymes from cold-adapted microorganisms. The class C β-lactamase from the antarctic psychrophile *Psychrobacter immobilis* A5, *Eur. J. Biochem.* 244, 186-191.

39. Asgeirsson, B. and Bjarnason, J. B. (1991) Structural and kinetic properties of chymotrypsin from Atlantic cod (*Gadus morhua*). Comparison with bovine chymotrypsin, *Comp. Biochem. Physiol. B.* 99, 327-335.
40. Asgeirsson, B. and Bjarnason, J. B. (1993) Properties of elastase from Atlantic cod, a cold-adapted proteinase, *Biochim. Biophys. Acta.* 1164, 91-100.
41. Arpigny, J. L., Lamotte, J. and Gerday, C. (1997) Molecular adaptation to cold of an Antarctic bacterial lipase, *J. Mol. Catal. B.* 3, 29-35.
42. Choo, D. W., Kurihara, T., Suzuki, T., Soda, K. and Esaki, N. (1998) A cold-adapted lipase of an Alaskan psychrotroph, *Pseudomonas* sp. strain B11-1: gene cloning and enzyme purification and characterisation, *Appl. Environ. Microbiol.* 64, 486-491.
43. Chessa, J.-P., Petrescu, I., Bentahir, M., Van Beeumen, J. and Gerday, C. (2000) Purification, physico-chemical characterisation and sequence of a heat-labile alkaline metaloprotease isolated from a psychrophilic *Pseudomonas* species, *Biochim. Biophys. Acta.* In press.
44. Davail, S., Feller, G., Narinx, E. and Gerday, C. (1994) Cold adaptation of proteins. Purification, characterisation, and sequence of the heat-labile subtilisin from the antarctic psychrophile *Bacillus* TA41, *J. Biol. Chem.* 269, 17448-17453.
45. Narinx, E., Baise, E. and Gerday, C. (1997) Subtilisin from psychrophilic antarctic bacteria: characterisation and site-directed mutagenesis of residues possibly involved in the adaptation to cold, *Protein Eng.* 10, 1271-1279.
46. Kristjansson, M. M., Magnusson, O. T., Gudmundsson, H. M., Alfredsson, G. A. and Matsuzawa, H. (1999) Properties of a subtilisin-like proteinase from a psychrotrophic *Vibrio* species. Comparison with proteinase K and aqualysin I, *Eur. J. Biochem.* 260, 752-760.
47. Simpson, B. K. and Haard, N. F. (1984) Purification and characterisation of trypsin from the Greenland cod (*Gadus ogac*). 1. Kinetic and thermodynamic characteristics, *Can. J. Biochem. Cell. Biol.* 62, 894-900.
48. Petrescu, I., Lamotte-Brasseur , J., Chessa, J.-P., Ntarima, P., Claeyssens, M., Devreese, M., Marino, G. and Gerday, C. (2000) The xylanase from the psychrophilic yeast *Cryptococcus adeliae*, *Extremophiles* In press.
49. Fersht, A. (1985) *Enzyme structure and mechanism*, W. H. Freeman and Company, New York.
50. Fisher, J., Belasco, J. G., Khosla, S. and Knowles, J. R. (1980) β-Lactamase proceeds via an acyl-enzyme intermediate. Interaction of the *Escherichia coli* RTEM enzyme with cefoxitin, *Biochemistry.* 19, 2895-2901.
51. Feller, G., Narinx, E., Arpigny, J.-L., Aittaleb, M., Baise, E., Genicot, S. and Gerday, C. (1996) Enzymes from psychrophilic organisms, *FEMS Microbiol. Rev.* 18, 189-202.
52. Tsigos, I., Velonia, K., Smonou, I. and Bouriotis, V. (1998) Purification and characterisation of an alcohol dehydrogenase from the Antarctic psychrophile *Moraxella* sp. TAE123, *Eur. J. Biochem.* 254, 356-362.
53. Vckovski, V., Schlatter, D. and Zuber, H. (1990) Structure and function of L-lactate dehydrogenases from thermophilic, mesophilic and psychrophilic bacteria, IX. Identification, isolation and nucleotide sequence of two L-lactate dehydrogenase genes of the psychrophilic bacterium *Bacillus psychrosaccharolyticus*, *Biol. Chem. Hoppe Seyler.* 371, 103-110.
54. Rentier-Delrue, F., Mande, S. C., Moyens, S., Terpstra, P., Mainfroid, V., Goraj, K., Lion, M., Hol, W. G. and Martial, J. A. (1993) Cloning and overexpression of the triosephosphate isomerase genes from psychrophilic and thermophilic bacteria. Structural comparison of the predicted protein sequences, *J. Mol. Biol.* 229, 85-93.
55. Alvarez, M., Zeelen, J. P., Mainfroid, V., Rentier-Delrue, F., Martial, J. A., Wyns, L., Wierenga, R. K. and Maes, D. (1998) Triose-phosphate isomerase (TIM) of the psychrophilic bacterium *Vibrio marinus*. Kinetic and structural properties, *J. Biol. Chem.* 273, 2199-2206.
56. Ciardiello, M. A., Camardella, L. and di Prisco, G. (1995) Glucose-6-phosphate dehydrogenase from the blood cells of two antarctic teleosts: correlation with cold adaptation, *Biochim. Biophys. Acta.* 1250, 76-82.
57. Gerike, U., Danson, M. J., Russell, N. J. and Hough, D. W. (1997) Sequencing and expression of the gene encoding a cold-active citrate synthase from an Antarctic bacterium, strain DS2-3R, *Eur. J. Biochem.* 248, 49-57.

58. Rina, M., Caufrier, F., Markaki, M., Mavromatis, K., Kokkinidis, M. and Bouriotis, V. (1997) Cloning and characterisation of the gene encoding PspPI methyltransferase from the Antarctic psychrotroph *Psychrobacter* sp. strain TA137. Predicted interactions with DNA and organisation of the variable region, *Gene.* 197, 353-360.
59. Schleper, C., Swanson, R. V., Mathur, E. J. and DeLong, E. F. (1997) Characterisation of a DNA polymerase from the uncultivated psychrophilic archaeon *Cenarchaeum symbiosum*, *J. Bacteriol.* 179, 7803-7811.
60. Bentahir, M., Feller, G., Aittaleb, M., Lamotte-Brasseur, J., Himri, T., Chessa, J. P. and Gerday, C. (2000) Structural, kinetic, and calorimetric characterisation of the cold-active phosphoglycerate kinase from the antarctic *Pseudomonas* sp. TACII18, *J. Biol. Chem.* 275, 11147-11153.
61. Georlette, D., Jónsson, Z. O., Van Petegem, F., Chessa, J.-P., Van Beeumen, J., Hübscher, U. and Gerday, C. (2000) A DNA ligase from the psychrophile *Pseudoalteromonas haloplanktis* gives insights into the adaptation of proteins to low temperatures, *Eur. J. Biochem.*, In press.
62. Jaenicke, R. (1991) Protein stability and molecular adaptation to extreme conditions, *Eur. J. Biochem.* 202, 715-728.
63. Jaenicke, R. (1996) Protein folding and association: in vitro studies for self-organisation and targeting in the cell, *Curr. Top. Cell. Regul.* 34, 209-314.
64. Vieille, C., Burdette, D. S. and Zeikus, J. G. (1996) Thermozymes, *Biotechnol. Annu. Rev.* 2, 1-83.
65. Scandurra, R., Consalvi, V., Chiaraluce, R., Politi, L. and Engel, P. C. (1998) Protein thermostability in extremophiles, *Biochimie.* 80, 933-941.
66. Dill, K. A. (1990) Dominant forces in protein folding, *Biochemistry.* 29, 7133-7155.
67. Doig, A. J. and Williams, D. H. (1991) Is the hydrophobic effect stabilising or destabilising in proteins? The contribution of disulphide bonds to protein stability, *J. Mol. Biol.* 217, 389-398.
68. Ragone, R. and Colonna, G. (1995) Do globular proteins require some structural peculiarity to best function at high temperatures? *J. Am. Chem. Soc.* 117, 16-20.
69. Makhatadze, G. I. and Privalov, P. L. (1994) Hydration effects in protein unfolding, *Biophys. Chem.* 51, 291-309.
70. Pace, C. N., Shirley, B. A., McNutt, M. and Gajiwala, K. (1996) Forces contributing to the conformational stability of proteins, *Faseb J.* 10, 75-83.
71. Zavodszky, P., Kardos, J., Svingor and Petsko, G. A. (1998) Adjustment of conformational flexibility is a key event in the thermal adaptation of proteins, *Proc. Natl. Acad. Sci. U. S. A.* 95, 7406-7411.
72. Wrba, A., Schweiger, A., Schultes, V., Jaenicke, R. and Zavodsky, P. (1990) Extremely thermostable D-glyceraldehyde-3-phosphate dehydrogenase from the eubacterium *Thermotoga maritima*, *Biochemistry.* 29, 7584-7592.
73. Jaenicke, R. and Zavodszky, P. (1990) Proteins under extreme physical conditions, *FEBS Lett.* 268, 344-349.
74. Vihinen, M. (1987) Relationship of protein flexibility to thermostability, *Protein Eng.* 1, 477-480.
75. Shoichet, B. K., Baase, W. A., Kuroki, R. and Matthews, B. W. (1995) A relationship between protein stability and protein function, *Proc. Natl. Acad. Sci. U. S. A.* 92, 452-456.
76. Van den Burg, B., Vriend, G., Veltman, O. R., Venema, G. and Eijsink, V. G. (1998) Engineering an enzyme to resist boiling, *Proc. Natl. Acad. Sci. U. S. A.* 95, 2056-2060.
77. Giver, L., Gershenson, A., Freskgard, P. O. and Arnold, F. H. (1998) Directed evolution of a thermostable esterase, *Proc. Natl. Acad. Sci. U. S. A.* 95, 12809-12813.
78. Smalas, A. O., Heimstad, E. S., Hordvik, A., Willassen, N. P. and Male, R. (1994) Cold adaptation of enzymes: structural comparison between salmon and bovine trypsin, *Proteins.* 20, 149-166.
79. Russell, R. J., Gerike, U., Danson, M. J., Hough, D. W. and Taylor, G. L. (1998) Structural adaptations of the cold-active citrate synthase from an Antarctic bacterium, *Structure.* 6, 351-361.
80. Fields, P. A. and Somero, G. N. (1998) Hot spots in cold adaptation: localised increases in conformational flexibility in lactate dehydrogenase A(4) orthologs of Antarctic notothenioid fishes, *Proc. Natl. Acad. Sci. U. S. A.* 95, 11476-11481.
81. Kim, S. Y., Hwang, K. Y., Kim, S. H., Sung, H. C., Han, Y. S. and Cho, Y. J. (1999) Structural basis for cold adaptation. Sequence, biochemical properties, and crystal structure of malate dehydrogenase from a psychrophile *Aquaspirillium arcticum*, *J. Biol. Chem.* 274, 11761-11767.

82. Fontes, C. M., Hazlewood, G. P., Morag, E., Hall, J., Hirst, B. H. and Gilbert, H. J. (1995) Evidence for a general role for non-catalytic thermostabilising domains in xylanases from thermophilic bacteria, *Biochem. J.* 307, 151-158.
83. Feller, G., d'Amico, D. and Gerday, C. (1999) Thermodynamic stability of a cold-active α-amylase from the Antarctic bacterium *Alteromonas haloplanctis*, *Biochemistry.* 38, 4613-4619.
84. Goldman, A. (1995) How to make my blood boil, *Structure.* 3, 1277-1279.
85. Feller, G., Lonhienne, T., Deroanne, C., Libioulle, C., Van Beeumen, J. and Gerday, C. (1992) Purification, characterisation, and nucleotide sequence of the thermolabile α-amylase from the antarctic psychrotroph *Alteromonas haloplanctis* A23, *J. Biol. Chem.* 267, 5217-5221.
86. Genicot, S., Rentier-Delrue, F., Edwards, D., Van Beeumen, J. and Gerday, C. (1996) Trypsin and trypsinogen from an Antarctic fish: molecular basis of cold adaptation, *Biochim. Biophys. Acta.* 1298, 45-57.
87. Aittaleb, M., Hubner, R., Lamotte-Brasseur, J. and Gerday, C. (1997) Cold adaptation parameters derived from cDNA sequencing and molecular modelling of elastase from Antarctic fish *Notothenia neglecta*, *Protein Eng.* 10, 475-477.
88. Wallon, G., Lovett, S. T., Magyar, C., Svingor, A., Szilagyi, A., Zavodszky, P., Ringe, D. and Petsko, G. A. (1997) Sequence and homology model of 3-isopropylmalate dehydrogenase from the psychrotrophic bacterium *Vibrio* sp. I5 suggest reasons for thermal instability, *Protein Eng.* 10, 665-672.
89. Thomas, T. and Cavicchioli, R. (1998) Archaeal cold-adapted proteins: structural and evolutionary analysis of the elongation factor 2 proteins from psychrophilic, mesophilic and thermophilic methanogens, *FEBS Lett.* 439, 281-286.
90. Berchet, V., Thomas, T., Cavicchioli, R., Russell, N. J. and Gounot, A. M. (2000) Structural analysis of the elongation factor G protein from the low-temperature-adapted bacterium *Arthrobacter globiformis* SI55, *Extremophiles.* 4, 123-130.
91. Feller, G. and Gerday, C. (1997) Psychrophilic enzymes: molecular basis of cold adaptation, *Cell. Mol. Life Sci.* 53, 830-841.
92. Matthews, B. W. (1993) Structural and genetic analysis of protein stability, *Annu. Rev. Biochem.* 62, 139-160.
93. Arpigny, J. L., Feller, G., Davail, S., Génicot, S., Narinx, E., Zekhnini, Z. and Gerday, C. (1994) Molecular adaptations of enzymes from thermophilic and psychrophilic organisms. In: *Adv. Compa. Env. Physi.* (Gilles, R., ed.) pp. 269-295, Springer-Verlag, Berlin-Heidelberg.
94. Chan, M. K., Mukund, S., Kletzin, A., Adams, M. W. and Rees, D. C. (1995) Structure of a hyperthermophilic tungstopterin enzyme, aldehyde ferredoxin oxidoreductase, *Science.* 267, 1463-1469.
95. Korndorfer, I., Steipe, B., Huber, R., Tomschy, A. and Jaenicke, R. (1995) The crystal structure of holo-glyceraldehyde-3-phosphate dehydrogenase from the hyperthermophilic bacterium *Thermotoga maritima* at 2.5 Å resolution, *J. Mol. Biol.* 246, 511-521.
96. Jaenicke, R., Schurig, H., Beaucamp, N. and Ostendorp, R. (1996) Structure and stability of hyperstable proteins: glycolytic enzymes from hyperthermophilic bacterium *Thermotoga maritima*, *Adv. Protein Chem.* 48, 181-269.
97. Aghajari, N., Feller, G., Gerday, C. and Haser, R. (1996) Crystallisation and preliminary X-ray diffraction studies of α-amylase from the antarctic psychrophile *Alteromonas haloplanctis* A23, *Protein Sci.* 5, 2128-2129.
98. Aghajari, N., Feller, G., Gerday, C. and Haser, R. (1998) Structures of the psychrophilic *Alteromonas haloplanctis* α-amylase give insights into cold adaptation at a molecular level, *Structure.* 6, 1503-1516.
99. Aghajari, N., Feller, G., Gerday, C. and Haser, R. (1998) Crystal structures of the psychrophilic α-amylase from *Alteromonas haloplanctis* in its native form and complexed with an inhibitor, *Protein Sci.* 7, 564-572.
100. Villeret, V., Chessa, J. P., Gerday, C. and Van Beeumen, J. (1997) Preliminary crystal structure determination of the alkaline protease from the Antarctic psychrophile *Pseudomonas aeruginosa*, *Protein Sci.* 6, 2462-2464.

101. Yip, K. S., Stillman, T. J., Britton, K. L., Artymiuk, P. J., Baker, P. J., Sedelnikova, S. E., Engel, P. C., Pasquo, A., Chiaraluce, R. and Consalvi, V. (1995) The structure of *Pyrococcus furiosus* glutamate dehydrogenase reveals a key role for ion-pair networks in maintaining enzyme stability at extreme temperatures, *Structure.* 3, 1147-1158.
102. Lim, J. H., Yu, Y. G., Han, Y. S., Cho, S., Ahn, B. Y., Kim, S. H. and Cho, Y. (1997) The crystal structure of an Fe-superoxide dismutase from the hyperthermophile *Aquifex pyrophilus* at 1.9 Å resolution: structural basis for thermostability, *J. Mol. Biol.* 270, 259-274.
103. D'Amico, S., Gerday, C. and Feller, G. (2000) Structural similarities and evolutionary relationships in chloride-dependent α-amylases, *Gene*, In press.
104. Marshall, C. J. (1997) Cold-adapted enzymes, *Trends Biotechnol.* 15, 359-364.
105. Russell, N. J. (1998) Molecular adaptations in psychrophilic bacteria: potential for biotechnological applications, *Adv. Biochem. Eng. Biotechnol.* 61, 1-21.
106. Cummings, S. P. and Black, G. W. (1999) Polymer hydrolysis in a cold climate, *Extremophiles.* 3, 81-87.
107. Kobori, H., Sullivan, C. W. and Shizuya, H. (1984) Heat-labile alkaline phosphatase from Antarctic bacteria: rapid 5' end labelling of nucleic acids, *Proc. Natl. Acad. Sci. U. S. A.* 81, 6691-6695.
108. Sobek, H., Schmidt, M., Frey, B. and Kaluza, K. (1996) Heat-labile uracil-DNA glycosylase: purification and characterisation, *FEBS Lett.* 388, 1-4.
109. Savva, R., McAuley-Hecht, K., Brown, T. and Pearl, L. (1995) The structural basis of specific base-excision repair by uracil-DNA glycosylase, *Nature.* 373, 487-493.
110. Tutino, L. M., Fontanella, B., Moretti, M. A., Duilio, A., Sannia, G. and Marino, G. (1999) Plasmids from antarctic bacteria. In: *Cold adapted organisms: Ecology, Physiology, Enzymology and Molecular Biology* (Margessi, R. and Schinner, F., eds) pp. 335-347, Springer-Verlag, Berlin-Heidelberg.
111. Ohgiya, S., Hoshino, T., Okuyama, H., Tanaka, S. and Ishizaki, K. (1999) Biotechnology of enzymes from cold-adapted organisms. In: *Biotechnological applications of cold adapted organisms* (Margessi, R. and Schinner, F., eds) pp. 17-34, Springer-Verlag, Berlin-Heidelberg.
112. Remaut, E., Bliki, C., Iturriza-Gomara, M. and Keymeulen, K. (1999) Development of regulatable expression systems for cloned genes in cold-adapted bacteria. In: *Biotechnological applications of cold adapted organisms* (Margessi, R. and Schinner, F., eds) pp. 1-16, Springer -Verlag, Berlin-Heidelberg.
113. Vasina, J. A. and Baneyx, F. (1996) Recombinant protein expression at low temperatures under the transcriptional control of the major *Escherichia coli* cold shock promoter *cspA*, *Appl. Environ. Microbiol.* 62, 1444-1447.
114. Qoronfleh, M. W., Debouck, C. and Keller, J. (1992) Identification and characterisation of novel low-temperature-inducible promoters of *Escherichia coli*, *J. Bacteriol.* 174, 7902-7909.
115. Nichols, D., Bowman, J., Sanderson, K., Nichols, C. M., Lewis, T., McMeekin, T. and Nichols, P. D. (1999) Developments with Antarctic microorganisms: culture collections, bioactivity screening, taxonomy, PUFA production and cold-adapted enzymes, *Curr. Opin. Biotech.* 10, 240-246.
116. Okuyama, H., Morita, N. and Yumoto, I. (1999) Cold-adapted microorganisms for use in food biotechnology. In: *Biotechnological applications of cold adapted organisms* (Margessi, R. and Schinner, F., eds) pp. 101-115, Springer-Verlag, Berlin-Heidelberg.
117. Brown, M. R., Barrett, S. M., Volkman, J. K., Nearhos, S. P., Nell, J. A. and Allan, G. L. (1996) Biochemical composition of new yeasts and bacteria evaluated as food for bivalve aquaculture, *Aquaculture.* 143, 341-360.
118. Bowman, J. P., McCammon, S. A., Nichols, D. S., Skerratt, J. H., Rea, S. M., Nichols, P. D. and McMeekin, T. A. (1997) *Shewanella gelidimarina* sp. nov. and *Shewanella frigidimarina* sp. nov., novel Antarctic species with the ability to produce eicosapentaenoic acid (20:5 omega 3) and grow anaerobically by dissimilatory Fe(III) reduction, *Int. J. Syst. Bacteriol.* 47, 1040-1047.
119. Bowman, J. P., Gosink, J. J., McCammon, S. A., Lewis, T. E., Nichols, D. S., Nichols, P. D., Skerratt, J. H., Staley, J. T. and McMeekin, T. A. (1998) *Colwellia demingiae* sp. nov., *Colwellia hornerae* sp. nov., *Colwellia rossensis* sp. nov. and *Colwellia psychrotropica* sp. nov.: psychrophilic Antarctic species with the ability to synthesise docosahexaenoic acid (22:6 omega 3), *Int. J. Syst. Bacteriol.* 48, 1171-1180.
120. Karl, D. M. (1992) The grounding of Bahia Paraiso: microbiology of the 1989 Antarctic oil spill,. *Microbial. Ecol.* 24, 170-189.

121. Delille, D., Bassères, A. and Dessommes, A. (1997) Seasonal variation of bacteria in sea ice contaminated by diesel fuel and dispersed crude oil, *Microb. Ecol.* 33, 97-105.
122. Cavanagh, J. E., Nichols, P. D., Franzmann, P. D. and McMeekin, T. A. (1998) Hydrocarbon degradation by Antarctic coastal bacteria, *Antarct. Sci.* 10, 386-397.
123. Margesin, R. and Schinner, F. (1997) Laboratory bioremediation experiments with soil from a diesel-oil contaminated site. Significant role of cold-adapted microorganisms and fertilisers, *J. Chem. Technol. Biotechnol.* 70, 92-98.
124. Margesin, R. and Schinner, F. (1998) Oil biodegradation potential in alpine habitats, *Arctic Alpine Res.* 30, 262-265.
125. Master, E. R. and Mohn, W. W. (1998) Psychrotolerant bacteria isolated from arctic soil that degrade polychlorinated biphenyls at low temperatures, *Appl. Environ. Microbiol.* 64, 4823-4829.
126. Timmis, K. N. and Pieper, D. H. (1999) Bacteria designed for bioremediation, *Trends Biotechnol.* 17, 200-204.
127. Guex, N. and Peitsch, M. C. (1997) SWISS-MODEL and the Swiss-PdbViewer: an environment for comparative protein modelling, *Electrophoresis.* 18, 2714-2723.

MOLECULAR AND CELLULAR MAGNETIC RESONANCE CONTRAST AGENTS

J.W.M. BULTE AND L.H. BRYANT JR.
Laboratory of Diagnostic Radiology Research, Clinical Center, National Institutes of Health, Bethesda, MD 20892

Summary

Certain chemical structures have magnetic properties that enable faster proton relaxation, allowing their use as MR contrast agents or magnetopharmaceuticals when found biocompatible. Larger complexes such as macromolecular or particulate contrast agents have recently emerged as a distinct subgroup of magnetic materials, suitable for contrast enhancement of the blood pool and specific tissues. This chapter describes the latest advances in chemical engineering and molecular/cellular biology, which are producing an entirely new class of (targeted) MR contrast agents that can be used for high-resolution imaging of biologic processes at the molecular and cellular level.

1. Introduction

Image contrast in MR imaging is largely determined by the magnetic relaxation times of tissues. Following a radiofrequency (RF) pulse at the resonance frequency (42.57 MHz/Tesla), protons absorb the electromagnetic radiation and return or "relax" to the lowest energy state of alignment with the applied magnetic field. This magnetic relaxation is described by the time constant T1, which represents the time for 63% of the relaxation to take place. The RF pulse also creates a transverse magnetisation that precesses about the applied magnetic field. The time constant T2 refers to the time it takes for 63% of the transverse magnetisation to decay following multiple spin echoes. MR contrast agents alter the magnetic relaxation times and may affect both T1 and T2 by dipole-dipole interactions. Their efficiency is usually expressed as relaxivity (R), which represents the reciprocal of the relaxation time per unit concentration of metal, with units $mM^{-1}s^{-1}$. Because of dephasing effects, one can classify contrast material into "T1 agents" and "T2 agents". This classification is according to their predominant effect on relaxation, but, because of dephasing effects, the T2 relaxivity (R2) is usually higher than R1, even for T1 agents.

M. De Cuyper and J.W.M. Bulte (eds.), Physics and Chemistry Basis of Biotechnology, 197–211.

As compared to single ionic chelates, larger magnetic complexes have higher (molecular) relaxivities and may have a longer blood half-life. They are currently being explored as potential blood pool and tissue-specific MR contrast agents. Magnetic nanostructures that primarily affect T1 are all based on paramagnetic chelates containing Gd(III), Mn(II) or Fe(III), bound to a larger molecule or polymer backbone. Examples of such "carrier" molecules include albumin [1-3], poly-L-lysine [4,5], dextran [6,7], liposomes [8-10], and Starburst® dendrimers [11,12]. Magnetic nanostructures that predominantly affect T2 include Dy(III) containing polymers [13,14] and superparamagnetic iron oxides [15-17], where the magnetically active core is Fe_3O_4, γFe_2O_3, or FeO_xOH_y.

Depending on the applied field strength, chemical environment, and tissue biodistribution, contrast agents may enhance the relaxation of water protons by different mechanisms. Much of our understanding has come from variable-field relaxometry, and analysing the resulting nuclear magnetic relaxation dispersion (NMRD) profiles. Until recently, the technique has been limited to the study of T1 relaxation mechanisms, but a unique instrument is now available in our laboratory that can obtain both 1/T1 and 1/T2 NMRD profiles. This variable-field T1-T2 relaxometer is considered to be a key instrument for the development of new magnetopharmaceuticals. It also serves as a routine instrument to determine the specific uptake of cellular and molecular contrast agents *in vitro*, analogous to the use of a liquid scintillation counter in nuclear medicine.

The non-invasive nature of MR imaging, its high spatial resolution (up to 20 μm isotropic, i.e. near cellular resolution), and the ability of 3D volume acquisition continuously or repeatedly at different time points, all make contrast-enhanced MR imaging a unique and powerful tool to study biologic processes at the molecular and cellular level. Indeed, the interest in the development of molecular and cellular MR contrast agents appears to have increased significantly over the last few years, and the present chapter attempts to give an overview of the latest developments in the field.

2. Magnetically labelled antibodies

With the introduction of the monoclonal antibody (mab) technology by Köhler and Milstein in 1975 [18], the opportunity was created to produce antibodies in large quantities with a high degree of purity and specific for a single antigenic epitope. With the ability to conjugate radioisotopes to mab's it was shown a few years later that tumour nodules could be detected specifically and non-invasively, initially using ^{131}I labelled anti-CEA mabs [19]. This was followed by reports that improved radioimmunodetection could be achieved when Fab or $F(ab')_2$ fragments were used instead of the intact immunoglobulin [20,21].

When MR imaging was introduced in the early 1980s along with gadolinium chelates as MR contrast agents, the preparation and use of magnetically labelled antibodies seemed a natural extension of the earlier work carried out with radiolabelled antibodies. In this way the detailed anatomic information on the MR images could then potentially be specifically altered or marked in order to detect (neoplastic) disease in its earliest stages. The first reports on the use of Gd-DTPA-labelled mab appeared around 1985 and described the attachment of a few to a maximum of 15 chelates per antibody [22-25]. The outcome of these early studies was disappointing, in that no specific contrast enhancement could be observed. Using the exact same tumour-animal model system but ^{153}Gd instead of ^{157}Gd, Anderson-Berg et al. [24] showed that this is an issue of sensitivity: radioisotopes can be detected as tracer molecules in nanomolar concentrations, whereas paramagnetic chelates require micro- to millimolar doses in order to be detectable. Since the amount of linkable chelates per molecule is limited to approximately 5-10 (too heavily loaded mabs lose their immunoreactivity), other strategies had to be pursued.

One strategy is the use of large carrier molecules such as poly-L-lysine [26-28] or crosslinked albumin-gelatin complexes [29], which are loaded with chelates, and covalently linked to mabs. In this approach, about 50-100 chelates can be bound per mab. A detailed review about the achievable "molecular" relaxivity (i.e. total relaxation enhancement per mab or molecule taking into account all attached metal ions) using different strategies has been published elsewhere [30]. However, it should be kept in mind that in addition to (not always that simple) dose-signal enhancement requirements there are physiological biodistribution barriers which eventually impose their own limitations on the feasibility of immunospecific imaging using paramagnetically labelled antibodies. For instance, larger molecules or complexes may have a reduced blood half-life and/or vascular permeability, w both of which can reduce the specific uptake in the target tissue. Another factor is the total dose needed for specific (tumour) detection; an administered dose of 2 mg antibody per mouse (100 mg/kg) translates to 7 grams of protein for an equivalent human study.

Instead of linear polymers, such as poly-L-lysine, spherical molecules or complexes that allow multiple attachment of chelates may be used. Dendrimers are ideal molecules for this purpose: a molecular relaxivity of 60,000-80,000 $mM^{-1}s^{-1}$ can be achieved for a generation 10 (G10) dendrimer [12]. These molecules allow the attachment of mabs [31] or other receptor-specific molecules [32], and are currently being further explored for that purpose. Alternatively, particulate emulsions or liposomes may be employed, but these large complexes may exhibit limited tissue penetration. Their use has been most successful in targeting antigens that are expressed on the surface of endothelial cells and are exposed to the blood pool, such as the neovascular marker integrin $\alpha_v\beta_3$ [33,34] and fibrin [35].

The second approach for preparing magnetically labelled antibodies is to use superparamagnetic iron oxide nanoparticles. In general these particles have, on a (milli)molar metal basis, a significantly higher relaxivity than the paramagnetic (gadolinium) chelates. In addition, there are usually several thousand iron atoms per particle, thus amplifying their effectiveness as a contrast agent. For example, when a particle (size range in the order of 10-20 nm) containing 5000 iron atoms is linked to a few mabs and exhibits a measured relaxivity of 100 $mM^{-1}s^{-1}$, the actual molecular relaxivity will be in the order of 500,000 $mM^{-1}s^{-1}$. But perhaps even more important is the fact that these magnetic nanoparticles can be detected with scanning techniques that are very sensitive to local differences in magnetic susceptibility and microscopic field inhomogeneities, which causes a rapid dephasing of protons (T2 shortening), and can

be detected without refocusing of 180° pulses (T2* effect). Of paramount importance here is that water protons at distant sites can be affected, leading to a "blooming effect", i.e. an amplification of signal changes.
Magnetite particles need to be stabilised in order to prevent aggregation. Most commonly this is accomplished by a coating of dextran. Immunoglobulins can then be covalently linked to the polysaccharide coat using a periodate-oxidation/borohydride-reduction method, which, through the formation of Schiff bases as intermediates, covalently links the amine (lysine) groups of the mab to the alcohol groups of the dextran [36,37]. MION-(46L) iron oxide nanoparticles have been conjugated this way to polyclonal IgG for detection of induced inflammation [38), to mab fragments for the specific visualisation of cardiac infarct [39], and to intact mabs for immunospecific detection of intracranial small cell lung carcinoma [40], ICAM-1 gene expression [41], and oligodendrocyte progenitors [42]. Alternative ways of attaching mabs to magnetic nanoparticles include glutaraldehyde crosslinking [43], complexing through ultrasonic sonication [44,45], using the biotin-streptavidin system [46] and amine-sulfhydryl group linkage [47,48]. For *in vivo* applications, limited success (e.g. true specific immunodetection) has been achieved thus far but this is likely to improve with the development of smaller nanoparticles that facilitate endothelial penetration and exhibit longer blood half-lives.

3. Other magnetically labelled ligands

Either paramagnetic chelates or magnetic nanoparticles can be linked to molecules other than mabs in order to confer specificity for a targetable receptor. For the group of paramagnetic agents, it has been demonstrated that "folated" gadolinium-dendrimers can be targeted *in vitro* to folate-receptor bearing leukaemic cells [32], and induce significant specific changes in relaxation times that is inhibitable by free, non-conjugated folate. This may be used for *in vivo* imaging of folate-receptor overexpressing tumours [49], but further work including the use of non-targeted polymer controls is needed. Another approach of conferring specificity to a paramagnetic label is to link it to an antisense oligonucleotide; a specific proton relaxation enhancement has been achieved for 5S rRNA as a macromolecular target and its labelled complimentary 6mer antisense sequence [50].

For magnetic iron oxide particles, the first use of a targetable ligand employed the use of arabinogalactan [51,52] in lieu of bacterial dextran as the polysaccharide coating: in this way, specific uptake in hepatocytes is achieved through uptake of the asialoglycoprotein receptor. Similar results were obtained when asialofetuin was used as a coating [53], and may be useful for improved detection of hepatocellular carcinoma. (Synthetic) peptides can also be linked to MION-46 or other very small iron oxide particles. For instance, cholecystokinin- [54] and secretin-[55] linked particles have been employed for MR visualisation of their respective pancreatic receptor and may aid in the diagnosis of pancreatic cancer. Transferrin is another example of a targetable protein, since certain tumours are known to overexpress transferrin receptors. Transferrin-iron oxide particles have been used for specific detection of gliosarcoma

[56,57] and breast carcinoma [58], with and without transfection of the Tfr-encoding gene, respectively. The protein ferritin has been exploited to encapsulate superparamagnetic iron oxides [59], but targeting these compounds to ferritin receptors (with and without pre-saturation with apoferritin) have been unsuccessful so far [60], possibly as a result from structural changes of the protein during the synthesis process.

4. Magnetically labelled cells

Most of our understanding of the biological function of cells and their interaction with(in) tissues comes from a static viewpoint obtained by light or electron microscopy, techniques which are basically unaltered since their inception about 350 and 60 years ago, respectively. Cells can be labelled *ex vivo* using a vital dye (optical or electron-dense, rendering the cell viable) and given back to the organism, which is then sacrificed and processed following a certain amount of time in order to determine the history and fate of administered cells. MR imaging offers the "dye and let live" approach: if cells can be labelled *ex vivo* with MR contrast agents, then their fate could possibly be monitored *in vivo non-invasively and repeatedly*, so that unique dynamic information can be obtained about the cellular movements and interactions with tissues. Clearly, this technique holds enormous potential and could potentially revolutionise the field of cell biology.

Requirements for the "vital dye", obviously, is no alteration of the function or longevity of the labelled cell, so it needs to be internalised into the cellular cytoplasm or nucleus (membrane-bound contrast agents are likely to interfere with cell-tissue interactions and may detach easily from the cell membrane). The second requirement is that, while not overloading and potentially killing the cell, a sufficient degree of labelling needs to be achieved in order to be able to detect the cells by MR imaging (see also section 1 on the generally low sensitivity of MR contrast agents). Iron oxide particles are naturally a first choice as candidates for cellular contrast agents, given their high relaxivity, T2* signal amplification effects, limited toxicity (cells naturally need and contain iron), and biodegradability.

In the late 80s, clinical studies on the biodistribution of *ex vivo* ^{111}In-labeled tumour infiltrating lymphocytes (TILs) and peripheral blood lymphocytes (PBLs) were carried out on patients undergoing adoptive cellular immunotherapy [61,62], and this stimulated attempts to label lymphocytes with superparamagnetic iron oxides. Strategies to prepare magnetically labelled lymphocytes included incubation with liposome-encapsulated iron oxide particles [63], incubation with non-derivatised dextran-coated iron oxide particles [64-69], lectin-mediated uptake [70], and uptake mediated by the tat-peptide [71,72]. The HIV-1 tat-peptide contains a membrane translocating signal that efficiently shuttles MION nanoparticles into cells. In addition, granulocytes (neutrophils) have been labelled with iron oxides to image their localisation in areas of infection and inflammation [73], analogous to clinical nuclear medicine studies using ^{111}In-labeled leukocytes. In the above studies, magnetically labelled white blood cells were administered systemically (intravenous injection). A different approach is to inject tagged cells *in situ* (locally) in the tissue of interest, i.e. to transplant the cells.

The central nervous system has been a primary area of interest for neurotransplantation studies of iron oxide-labelled cells. Several groups have demonstrated that it is possible to depict magnetically labelled cells at the injection site [74-76]. Our group demonstrated, for the first time, that it is possible to visualise *cell migration*, at least up to 10 mm away from the site of transplantation [42].

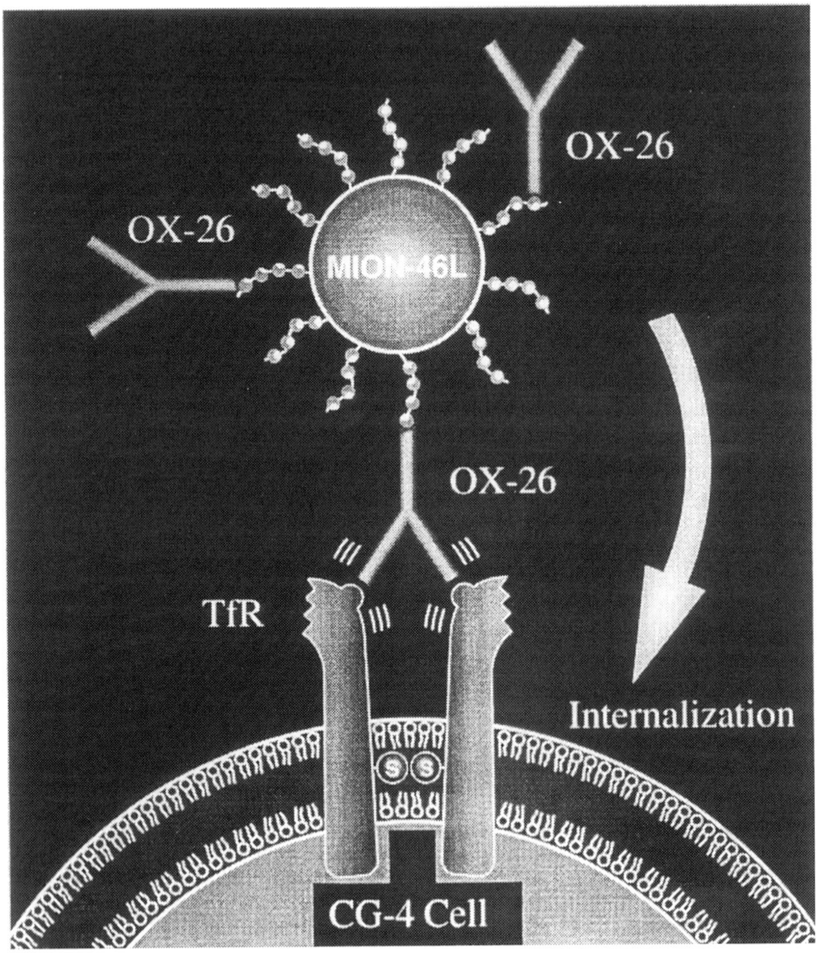

Figure 1. *Magnetically labelled cells: binding of an anti-transferrin receptor mab-iron oxide particle construct. Oligodendrocyte progenitor (CG-4) cells express high numbers of the transferrin receptor (Tfr). Upon crosslinking of the TfR by the specific OX-26 mab, the cell receives an internalising signal to endocytose the Tfr-OX-26 complex plus the covalently linked MION-46L iron oxide nanoparticles used as magnetic label.*

In this study, oligodendrocyte progenitors were first incubated with iron oxide particles that were covalently linked to anti-transferrin receptor(Tfr) mabs (See Fig. 1). Upon binding of the construct, the Tfr is being crosslinked by the mab which results in an internalising signal. The receptor-mediated endocytosis then stashes away the iron oxide particles in endosomes. Following transplantation of only 5 x 10^4 magnetically tagged cells, the dark areas on the MR images corresponded to the cellular migration mainly within the dorsal column, and corresponded to the areas of new myelination (see Fig.2). While the imaging in that study was performed at high resolution *ex vivo*, we have since demonstrated that these and similar cells can be monitored serially *in vivo*, even using lower resolution clinical MRI systems [77]. In parallel with these new technologies, another breakthrough development occurred that will have profound implications for the use of cellular therapies, namely the isolation and successful propagation of human embryonic stem (ES) cells [78, 79]. Using mouse ES cells, a number of different groups have shown the near unlimited potential of these cells to become differentiated, specific cell types that can be used to repair defunct or damaged tissue. Magnetic particles that can label cells non-specifically, regardless of tissue origin or animal species, yet show high cellular uptake ratios are highly desirable to further develop this field of cellular MR imaging. In collaboration with chemists at Temple University, our group has developed dendrimer-coated iron oxide particles or "magnetodendrimers" as a new type of non-specific magnetic tag that shows excellent cellular uptake and relaxation enhancement [80].

While superparamagnetic iron oxides appear, for good reason, the primary magnetic label of choice for tagging cells, paramagnetic chelates may also be used. It was shown that by injecting Gd-DTPA-dextran in just one single cell of an early (16-cell) stage developing frog embryo, the embryonic cell lineages and movement of differentiating cells could be followed by MR microscopic imaging [81]. Transferrin conjugated to poly-L-lysine has been used to co-transfer Gd-chelates and DNA into cells [82], and using the HIV tat-peptide described earlier cells have been tagged with Gd- and Dy-chelates [83]. The advantage of using Gd-chelates is that the inner-sphere water co-ordination may be manipulated in order to create "on-off" cellular switches; our group is currently pursuing this approach using paramagnetically labelled dendrimers [84].

5. Axonal and neuronal tracing

Nerves are normally isointense with the surrounding tissue and are therefore difficult to detect individually by MR imaging. Following injection of wheat germ agglutinin (WGA)-conjugated dextran-coated iron oxide particles into the rabbit forearm muscle, the particles are taken up and transported by median nerves, which can then be easily detected as separate structures on the images [85]. Similarly, using a rat model of focal crush injury to the sciatic nerve, it was shown that these nerves can become traceable following injection of MION particles directly at the site of injury [86]. Binding and transport of the iron oxides were comparable to slow axonal transport with a speed of about 5 mm a day. Interestingly, although WGA is being reported to have a high specific affinity for neurons, similar results were obtained for WGA-coated and non-coated particles that were either negatively or positively charged [86,87]. Another approach to detect nerves and, in particular, to trace neuronal connections is to use intravitreal injection of paramagnetic manganese ions (Mn^{2+}), allowing MRI visualisation of the olfactory pathway [88]. The mechanism of the neuronal cellular uptake of Mn^{2+} is not clear and possibly results from binding to voltage and ligand-gated Ca^{2+}-transporters. This pioneering work has sprang forward from the group's earlier observations that systemic Mn^{2+} administration plus opening of the blood-brain barrier enabled MRI detection of neuronal activation, presumably through mimicking of the calcium influx necessary for release of neurotransmitters [89].

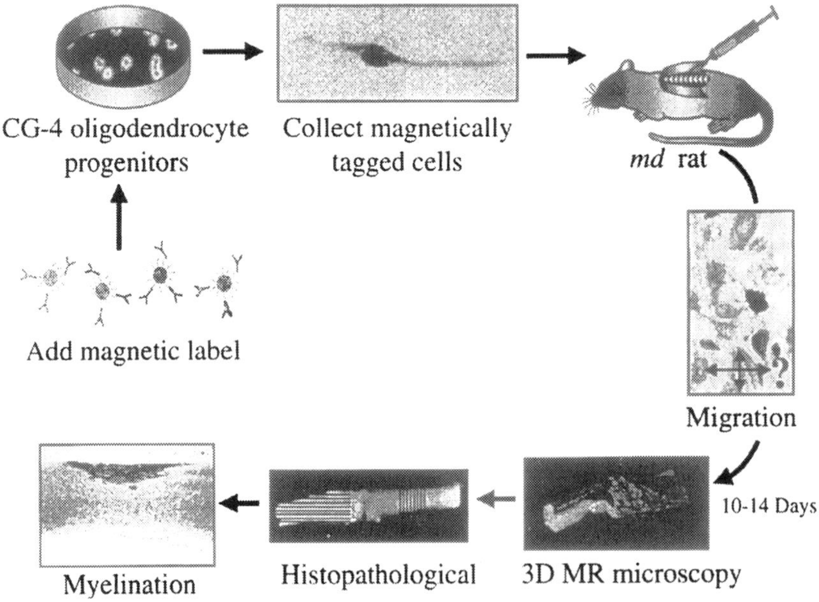

Figure 2. *Magnetically labelled cells: MR tracking of cell migration and myelination. Magnetically labelled oligodendrocyte progenitors are injected into the spinal cord of myelin deficient (md) rats. The cellular migration and distribution pattern at 10-14 days following transplantation can be visualised non-invasively and in three dimensions by MR microscopic (high-resolution) imaging. Histopathological correlation (performed afterwards) demonstrates that the dark areas of MR contrast correspond to the areas of cellular migration and the induced (new) myelination.*

6. Imaging of gene expression and enzyme activity

With the advent of gene therapy, that is, therapeutic cellular delivery of DNA encoding for missing or defective genes, there is an urgent need for a non-invasive technique that can monitor the cellular uptake, host DNA integration, and functional expression. The primary strategy to accomplish this would be co-transfection with a reporter gene, i.e. a gene of which the functional expression can be visualised directly (the assumption is that both genes will be co-expressed). While other imaging techniques have shown that this is indeed feasible, e.g. positron-emission tomography (PET) tracers linked to antiviral drugs that bind to the product of the reporter gene thymidine kinase (HSV-tk) [90-93] or bioluminescent and fluorescent proteins such as firefly luciferase and green fluorescent protein in the case of optical imaging [94], none of these techniques offer the microscopic resolution and deep tissue penetrating capabilities of MR imaging. However, as pointed out earlier, the sensitivity of MRI for detectable tracers is low.

Ideally the reporter gene would encode for the cellular expression and synthesis of a superparamagnetic iron oxide. We know that eukaryotic cells are capable of biological synthesis of magnetite: magnetoreceptor-bearing cells containing iron oxide particles have recently be identified in migrating trout fish [95], which use the earth's magnetic field for navigation. Similar particles can also be found in bees, salmon, pigeons and other birds. Studies on the genes involved with the preparation of magnetosomes (single domain magnetite particles coated with a lipid bilayer) have focussed on the magnetotactic bacterium *Magnetospirillum sp.* AMB-1 [96,97]. In order to achieve cellular expression of magnetite it is unlikely that just one single gene is needed; rather it would require a set of reporter genes of which the correct insertion, transcription and assembly may be extremely complex. Accomplishing this daunting task, however, would be of an immense value and would make MRI the primary imaging technique for the *in vivo* detection of gene expression. In the meantime, the group of Koretsky et al. who earlier pioneered a non-invasive detection of gene expression through use of ^{31}P NMR spectroscopy detectable tracers (creatine kinase, see [98], have inserted transferrin-receptor transfected tumour cells into the mouse, and were able to detect significant signal changes as a result from the increased uptake of iron [99]. The use of Tfr as a potential reporter gene, however, may have some complexities to it as the transferrin-bound iron is initially paramagnetic (predominant T1-effect), and the time course of subsequent metabolisation and transformation into antiferromagnetic ferritin (predominant T2-effect) may vary widely among different cell types. Another approach is to target iron oxide particles bound to transferrin to the Tfr-transfected cell lines [56,57], although this can be viewed as being merely a model of detecting an overexpressed receptor (e.g. other receptor-ligand systems may be used) rather than an attempt to use the intrinsic contrast-enhancing properties of a reporter gene. Another potential reporter gene for inducing "endogenous contrast", analogous to Koretsky's Tfr approach, is tyrosinase, an oxidoreductase that is an enzyme essential for the overproduction of melanin. Transfected cells incubated with high levels of iron show an increased uptake of iron and appear bright on T1-weighted images [100, 101].

Outside the imaging field, one of the most commonly used reporter genes is LacZ, that encodes for the enzyme β-galactosidase, and can be visualised (invasively) histochemically by conversion of the X-gal substrate resulting in a Prussian Blue stained endproduct. Meade et al. have created a paramagnetic Gd chelate in which the only accessible water-binding site (for inner sphere relaxation) is blocked by a galactose group [102]. In the presence of the β-galactosidase enzyme, the blocking sugar cap is removed, and the MR contrast agent "switched on". Indeed, the use of such a "functional" (as if there are many non-functional agents on the market these days) or "smart" or "intelligent" contrast agent has allowed tracking of the gene expression and cellular differentiation pattern of the *Xenopus laevis* embryo: by injecting a galactose-capped Gd-chelate in both cells of a two cell-stage embryo, but β-galactosidase DNA into only one cell, it is possible to follow the development of transfected progenitor cells (which all exhibit contrast enhancement) by MR microscopy [103].

Other examples of functional contrast agents include bioactivation of a prodrug and induction of relaxation enhancement by the enzyme alkaline phosphatase [104] and the development of a calcium-sensitive MR contrast agent [105]. Non-invasive, three-dimensional visualisation of such an intracellular secondary messenger concentration would be very valuable; however, for all these "smart" approaches, we should realise that two free parameters are being introduced: 1) the concentration of the enzyme or messenger and 2) the concentration of the administered contrast agent. Since both will be unknown and may vary in time, a correct interpretation of the obtained contrast may actually prove more difficult than expected (Robert Muller, personal communication). In addition, blocking of the contrast effect by a removable cap may not be complete but perhaps 80%, as a result of the remaining outer-sphere relaxation effect (Michael Tweedle, personal communication), complicating the interpretation possibly even further. Nevertheless, it is expected that these "smart" contrast agent technologies and variations thereof will mark the dawn of a new era and potentially revolutionise the field of molecular and cellular biology.

7. Conclusions

Several distinct species of molecular and cellular MR contrast agents have recently been developed. By exploiting their high relaxivity and tailoring their specificity, these agents appear to be promising for visualisation of selected biological events such as cell migration, transgene expression, and enzyme activity. Only an integral approach using the latest advances in biotechnology, molecular and cellular biology, biophysics, and chemistry will fully exploit the near boundless applications of molecular and cellular MR contrast agents.

References

1. Lauffer, R.B., Brady, T.J. (1985) Preparation and water relaxation properties of proteins labelled with paramagnetic metal chelates. *Magn. Reson. Imaging 3*, 11-16.
2. Schmiedl, U., Ogan, M.D., Moseley, M.R., Brasch, R.C. (1986) Comparison of the contrast-enhancing properties of albumin-(Gd-DTPA) and Gd-DTPA at 2.0 T: an experimental study in rats. *Amer. J. Roentgenol. 147*, 1263-1270.
3. Lauffer, R.B., Parmelee, D.J., Dunham, S.U., et al.(1998) MS-325: Albumin-targeted contrast agent for MR angiography. *Radiology 207*, 529-538.
4. Shreve, P., Aisen, A.M. (1986) Monoclonal antibodies labelled with polymeric paramagnetic ion chelates. *Magn. Reson. Med. 3*, 336-340.
5. Bogdanov, A.A., Weissleder, R.W., Frank, H.W., et al. (1993) A new macromolecule as a contrast agent for MR angiography - preparation, properties, and animal studies *Radiology 187*, 701-706.

6. Gibby, W.A., Bogdan, A., Ovitt, T.W. (1989) Cross-linked DTPA polysaccharides for magnetic-resonance imaging - synthesis and relaxation properties *Invest. Radiol. 24*, 302-309.
7. Wang, S.-C., Wikström, M.S., White, D.L., et al. (1990) Evaluation of Gd-DTPA-labelled dextran as an intravascular MR contrast agent: imaging characteristics in normal rat tissues. *Radiology 175*, 483-488.
8. Unger, E.C., Shen, D.-K., Fritz, T.A. (1993) Status of liposomes as MR contrast agents. *J. Magn. Reson. Imaging 3*, 195-198.
9. Trubetskoy, V.S., Canillo, J.A., Milshtein, A., Wolf, G.L., Torchilin, V.P. (1995) Controlled delivery of Gd-containing liposomes to lymph nodes - surface modification may enhance MRI contrast properties. *Magn. Reson. Imaging 13*, 31-37.
10. Storrs, R.W., Tropper, F.D., Li, H.Y., et al. (1995) Paramagnetic polymerised liposomes as new recirculating MR contrast agents. *J. Magn. Reson. Imaging 5*, 719-724.
11. Wiener, E.C., Brechbiel, M.W., Brothers, H. et al. (1994) Dendrimer-based metal chelates - a new class of magnetic resonance imaging contrast agents. *Magn. Reson. Med. 31*, 1-8.
12. Bryant, L.H., Brechbiel, M.W., Wu, C., Bulte, J.W.M., Herynek, V., Frank, J.A. (1999) Synthesis and relaxometry of high-generation (G = 5, 7, 9, and 10) PAMAM dendrimer-DOTA-gadolinium chelates. *J. Magn. Reson. Imaging 9*, 348-352.
13. Bulte, J.W.M., Wu, C., Brechbiel, M.W., Brooks, R.A., Vymazal, J., Holla, M., Frank, J.A. (1998) Dy-DOTA-PAMAM dendrimers as macromolecular T2 contrast agents: preparation and relaxometry. *Invest. Radiol. 33*, 841-845.
14. Zaharchuk, G., Bogdanov, A.A., Marota, J.J.A., et al. (1998) Continuous assessment of perfusion by tagging including volume and water extraction (CAPTIVE): A steady-state contrast agent technique for measuring blood flow, relative blood volume fraction, and the water extraction fraction. *Magn. Reson. Med. 40*, 666-678.
15. Stark, D.D., Weissleder, R., Elizondo, G, et al (1988). Superparamagnetic iron oxide: clinical application as a contrast agent for MR imaging of the liver. *Radiology 168*, 297-301.
16. Shen, T., Weissleder, R., Papisov, M., Bogdanov, A., Brady, T.J. (1993) Monocrystalline iron oxide nanocompounds (MION): Physicochemical properties. *Magn. Reson. Med. 29*, 599-604.
17. Bulte, J.W.M, and Brooks, R.A. (1997) Magnetic nanoparticles as contrast agents for MR imaging, in U. Häfeli, W. Schütt, J. Teller, M. Zborowksi (eds.), *Scientific and clinical applications of magnetic carriers*, Plenum Press, New York, pp. 527-543.
18. Köhler, G., Milstein, C. (1975). Continuous cultures of fused cells secreting antibody of predefined specificity. *Nature* 256, 495-497.
19. Mach, J.P., Buchegger, F., Forni, M. et al. (1981). Use of radiolabelled monoclonal anti-CEA antibodies for the detection of human carcinomas by external photoscanning and tomoscintigraphy. *Immunol. Today* 2, 239-249.
20. Larson, S.M., Brown, J.P., Wright, P.W., Carrasquillo, J.A., Hellstrom, I., Hellstrom, K.E. (1983). Imaging of melanoma with I-131-labeled monoclonal-antibodies *J. Nucl. Med.* 24, 123-129.
21. Moldofsky, P.J., Sears, H.F., Mulhern Jr., C.B., et al. (1984). Detection of metastatic tumour in normal-sized retroperitoneal lymph nodes by monoclonal antibody imaging. *N. Engl. J. Med.* 311, 106-107.
22. Unger, E.C., Totty, W.G., Neufeld, D.M., et al. (1985). Magnetic resonance imaging using gadolinium labelled monoclonal antibody. *Invest Radiol. 20*, 693-700.
23. Curtet, C, Tellier, C., Bohy, J., et al. (1986). Selective modification of NMR relaxation time in human colorectal carcinoma by using gadolinium-diethylenetriaminepentaacetic acid conjugated with monoclonal antibody 19-9. *Proc. Natl. Acad. Sci. USA 83*, 4277-4281.
24. Anderson-Berg, W.T., Strand, M., Lempert, T.E., Rosenbaum, A.E., Joseph, P.M. (1986). Nuclear magnetic resonance and gamma camera tumour imaging using gadolinium-labelled monoclonal antibodies. *J. Nucl. Med. 27,* 829-833.
25. Macrì, M.A., De Luca, F., Maraviglia, B., et al. (1989). Study of proton spin lattice relaxation variation induced by paramagnetic antibodies. *Magn. Reson. Med. 11*, 283-287.
26. Shreve, P., Aisen, A.M. (1986) Monoclonal antibodies labelled with polymeric paramagnetic ion chelates *Magn. Reson. Med* 3, 336-340.
27. Göhr-Rosenthal, S., Schmitt-Willich, H., Ebert, W., Conrad, J. (1993) The demonstration of human tumours on nude mice using gadolinium-labelled monoclonal antibodies for magnetic resonance imaging. *Invest Radiol. 28*, 789-795.

28. Curtet, C., Maton, F., Havet, T., et al. (1998) Polylysine-Gd-DTPA$_n$ and polylysine-Gd-DOTA$_n$ coupled to anti-CEA F(ab')$_2$ fragments as potential immunocontrast agents - relaxometry, biodistribution, and magnetic resonance imaging in nude mice grafted with human colorectal carcinoma. *Invest Radiol 33,* 752-761.
29. Kornguth, S.E., Turski, P.A., Perman, W.H., et al. (1987). Magnetic resonance imaging of gadolinium-labelled monoclonal antibody polymers directed at human T lymphocytes implanted in canine brain. *J. Neurosurg. 66,* 898-906.
30. Nunn, A.D., Linder, K.E., Tweedle, M.F. (1997) Can receptors be imaged with MRI agents? *Quart. J. Nucl. Med. 41,* 155-162.
31. Wu, C., Brechbiel, M.W., Kozak, R.W., Gansow, O.A. (1994) Metal-chelate-dendrimer-antibody constructs for use in radioimmunotherapy and imaging. *Biorg. Med. Chem. Lett.* 4, 449-454.
32. Wiener, E.C., Konda, S., Shadron, A., et al. (1997) Targeting dendrimer-chelates to tumours and tumour cells expressing the high-affinity folate receptor. *Invest Radiol 32,* 748-754.
33. Sipkins, D.A., Cheresh, D.A., Kazemi, M.R., Nevin, L.M., Bednarski, M.D., Li, K.C.P. (1998) Detection of tumour angiogenesis *in vivo* by $\alpha_v\beta_3$-targeted magnetic resonance imaging. *Nature Medicine 4,* 623-626.
34. Anderson, S.A., Rader, R.K., Westlin, W.F. et al. (2000) Magnetic resonance contrast enhancement of neovasculature with $\alpha_v\beta_3$-targeted nanoparticles. *Magn. Reson. Med.* 44, 433-439.
35. Lanza, G.M., Lorenz, C.H., Fischer, S.E., et al. (1998) Enhanced detection of thrombi with a novel fibrin-targeted magnetic resonance contrast agent. *Acad. Radiology 5,* S173-S176.
36. Sanderson, C.J., Wilson, D.V (1971) A simple method for coupling proteins to insoluble polysaccharides. *Immunology* 20, 1061-1065.
37. Dutton, A.H., Tokuyasu, K.T, Singer, S.J. (1979) Iron-dextran antibody conjugates: general method for simultaneous staining of two components in high-resolution immunoelectron microscopy. *Proc. Natl. Acad. Sci. USA* 76, 3392-3396.
38. Weissleder, R., Lee A.S., Fischman. A.J. et al. (1991) Polyclonal human immunoglobulin G labelled with polymeric iron oxide: antibody MR imaging. *Radiology 181,* 245-249.
39. Weissleder, R., Lee, A.S., Khaw, B.A., Shen, T., Brady, T.J. (1992) Antimyosin-labelled monocrystalline iron oxide allows detection of myocardial infarct: MR antibody imaging. *Radiology 182,* 381-385.
40. Remsen, L.G., McCormick, C.I., Roman-Goldstein, S. et. al. (1996) MR of carcinoma-specific monoclonal antibody conjugated to monocrystalline iron oxide nanoparticles: the potential for non-invasive diagnosis. *Amer. J. Neuroradiol. 17,* 411-418.
41. Bulte, J.W.M., Verkuyl, J.M., Herynek, V., et al. (1998) Magnetoimmunodetection of (transfected) ICAM-1 gene expression. Proceedings of the International Society for Magnetic Resonance in Medicine, Sixth Annual Meeting, Sydney, Australia: 307.
42. Bulte, J.W.M., Zhang, S.-C., van Gelderen, et al. (1999) Neurotransplantation of magnetically labelled oligodendrocyte progenitors: MR tracking of cell migration and myelination. *Proc. Natl. Acad. Sci. USA 96,* 15256-15261.
43. Renshaw PF, Owen CS, Evans AE, Leigh Jr JS (1986) Immunospecific NMR contrast agents. *Magn Reson Imaging 4,* 351-357.
44. Cerdan, S., Lötscher, H.R., Künnecke, B., Seelig, J. (1989) Monoclonal antibody-coated magnetite particles as contrast agents in magnetic resonance imaging of tumours. *Magn. Reson. Med. 12,* 151-163.
45. Suwa, T., Ozawa, S., Ueda, M., Ando, N., Kitajama, M (1998) magnetic resonance imaging of oesophageal squamous cell carcinoma using magnetite particles coated with ant-epidermal growth factor receptor antibody. *Int. J. Cancer* 75, 626-634.
46. Bulte, J.W.M., Hoekstra, Y., Kamman, R.L., et al. (1992) Specific MR imaging of human lymphocytes by monoclonal antibody guided dextran-magnetite particles. *Magn Reson Med 25,* 148-157.
47. Tiefenauer, L.X., Kühne, G., Andres, R.Y. (1993) Antibody-magnetite nanoparticles: in vitro characterisation of a potential tumour-specific contrast agent for magnetic resonance imaging. *Bioconj Chem 4,* 347-352.
48. Tiefenauer, L.X., Tschirky, A., Kühne, G., Andres, R.Y., (1996) In vivo evaluation of magnetite nanoparticles for use as a tumour contrast agent in MRI. *Magn Reson Imaging 14,* 391-402.
49. Wiener, E.C., Konda, S., Shadron, A., et al. (1997) Targeting dendrimer-chelates to tumours and tumour cells expressing the high-affinity folate receptor. *Invest Radiol 32,* 748-754.

50. Hines, J.V., Ammar, G.M., Buss, J., Schmalbrock, P. (1999) Paramagnetic oligonucleotides: contrast agents for magnetic resonance imaging with proton relaxation enhancement effects. *Bioconj. Chem. 10*, 155-158.
51. Josephson, L., Groman, E.V., Menz, E., Lewis, J.M., Bengele, H. (1990) A functionalised superparamagnetic iron oxide colloid as a receptor directed MR contrast agent. *Magn. Reson. Imaging 8*, 637-646.
52. Reimer, P., Weissleder, R., Lee, A.S., Wittenberg, J., Brady, T.J. (1990) Receptor imaging: application to MR imaging of liver cancer. *Radiology 177*, 729-734.
53. Schaffer, B.K., Linker, C., Papisov, M., et al. (1993) MION-ASF: biokinetics of an MR receptor agent. *Magn. Reson. Imaging 11*, 411-417.
54. Reimer, P., Weissleder, R., Shen, T., Knoefel, W.T., Brady, T.J. (1994) Pancreatic receptors: initial feasibility studies with a targeted contrast agent for MR imaging. *Radiology 193*, 527-531.
55. Shen, T.T., Bogdanov Jr., A., Bogdanova, A., Poss, K., Brady, T.J., Weissleder, R. (1996). Magnetically labelled secretin retains receptor affinity to pancreas acinar cells. *Bioconj. Chemistry 7*, 311-316.
56. Moore A, Basilion, J.P., Chiocca, E.A., Weissleder, R. (1998) Measuring transferrin receptor gene expression by NMR imaging. *Biochim. Biohys. Acta. 1402*, 239-249 (1998).
57. Weissleder, R., Moore, A., Mahmood, U. et al. (2000) *In vivo* magnetic resonance imaging of transgene expression. *Nature Medicine 6*, 351-354.
58. Kresse, M., Wagner, S., Pfefferer, D., Lawaczeck, R., Elste, V. Targeting of ultrasmall superparamagnetic iron oxides (SPIO) to tumour cells in vivo by using transferrin-receptor pathways. *Magn. Reson. Med. 40*, 236-242.
59. Bulte, J.W.M., Douglas, T., Mann, S., et al. (1994). Magnetoferritin: characterisation of a novel superparamagnetic MR contrast agent. *J. Magn. Reson. Imaging 4*, 497-505.
60. Bulte, J.W.M., Douglas, T., Mann, S., Vymazal, J., Laughlin, P.G., Frank, J.A. (1995) Initial assessment of magnetoferritin biokinetics and proton relaxation enhancement in rats. *Acad Radiol 2*, 871-878.
61. Griffith KD, Read EJ, Carrasquillo JA, Carter CS, Yang JC, Fisher B, Aebersold P, Packard BS, Yu MY, Rosenberg SA. (1989) In vivo distribution of adoptively transferred indium-111-labeled tumour infiltrating lymphocytes and peripheral-blood lymphocytes in patients with metastatic melanoma. *J. Natl. Cancer Inst.* 81, 1709-1717.
62. Fisher B, Packard BS, Read EJ, Carrasquillo JA, Carter CS, Topalian SL, Yang JC, Yolles P, Larson SM, Rosenberg SA (1989). Tumour-localisation of adoptively transferred in-111 labelled tumour infiltrating lymphocytes in patients with metastatic melanoma. *J. Clin. Oncol.* 7, 250-261 (1989).
63. Bulte, J.W.M., Ma, L.D., Magin, R.L., et al. (1993) Selective MR imaging of labelled human peripheral blood mononuclear cells by liposome mediated incorporation of dextran-magnetite particles. *Magn. Reson. Med. 29*, 32-37.
64. Yeh T-c., Zhang, W., Ildstad, S.T., Ho, C. (1993). Intracellular labelling of T-cells with superparamagnetic contrast agents. *Magn Reson Med 30*, 617-625.
65. Yeh T-c, Zhang, W., Ildstad, S.T., Ho, C. (1995). In vivo dynamic MRI tracking of rat T-cells labelled with superparamagnetic iron-oxide particles. *Magn Reson Med 33*, 200-208.
66. Weissleder, R., Cheng, H.-C., Bogdanova, A., Bogdanov Jr., A. (1997) Magnetically labelled cells can be detected by MR imaging. *J. Magn. Reson. Imaging 7*, 258-263.
67. Schoepf, U., Marecos, E., Melder, R., Jain, R., Weissleder, R. (1998) Intracellular magnetic labelling of lymphocytes for in vivo trafficking studies. *BioTechniques 24*, 642-651.
68. Sipe, J.C., Filippi, M., Martino, G., et al. (1999) Method for intracellular magnetic labelling of human mononuclear cells using approved iron contrast agents. *Magn. Reson. Imaging 17*, 1521-1523.
69. Dodd, S.J., Williams, M., Suhan, J.P., Williams, D.S., Koretsky, A.P., Ho, C. (1999) Detection of single mammalian cells by high-resolution magnetic resonance imaging. *Biophys. J. 76*, 103-109.
70. Bulte, J.W.M., Laughlin, P.G., Jordan, E.K., Tran, V.A., Vymazal, J., Frank, J.A. (1996) Tagging of T cells with superparamagnetic iron oxide: uptake kinetics and relaxometry. *Acad Radiol 3*, S301-303.
71. Josephson, L., Tung, C.-H., Moore, A., Weissleder, R. (1999) High-efficiency intracellular magnetic labelling with novel superparamagnetic-tat peptide conjugates. *Bioconj. Chem. 10*, 186-191.
72. Lewin, M., Carlesso, N., Tung, C-H., et al. (2000). Tat peptide-derivatised magnetic nanoparticles allow in vivo tracking and recovery of progenitor cells. *Nature Biotechnology 18*, 410-414.

73. Krieg, F.M., Andres, R.Y., Winterhalter, K.H. (1995) Superparamagnetically labelled neutrophils as potential abscess-specific contrast agent for MRI. *Magn. Reson. Imaging 13*, 393-400.
74. Norman, A.B., Thomas, S.R., Pratt, R.G., Lu, S.Y., Norgren, R.B. (1992) Magnetic resonance imaging of neural transplants in rat brain using a superparamagnetic contrast agent. *Brain Res 594*, 279-283.
75. Hawrylak, N., Ghosh, P., Broadus, J., Schlueter, C., Greenough, W.T., Lauterbur, P.C. (1993) Nuclear magnetic resonance (NMR) imaging of iron oxide-labelled neural transplants. *Exp. Neurology 121*, 181-192.
76. Franklin, R.J.M., Blaschuk, K.L., Bearchell, M.C., Prestoz, L.L.C, Setzu, A., Brindle, K.M., Ffrench-Constant, C. (1999). Magnetic resonance imaging of transplanted oligodendrocyte precursors in the rat brain. *Neuroreport* 10, 3961-3965.
77. Bulte, J.W.M., Zhang, S.-C., van Gelderen, P., Lewis, B.K., Duncan, I.D., Frank, J.A. (2000) 3D MR tracking of magnetically labelled oligosphere transplants: Initial *in vivo* experience in the LE (Shaker) rat brain. Proceedings of the International Society for Magnetic Resonance in Medicine, Eighth Annual Meeting, Denver, Colorado: 383.
78. Thomson, J.A., Itskovitz-Eldor, J., Shapiro, S.S., Waknitz, M.A., Swiergiel, J.J., Marshall, V.S., Jones, J.M. (1998) Embryonic stem cell lines derived from human blastocysts. *Science 282*, 1145-1147.
79. Shamblott, M.J., Axelman, J., Wang, S.P., Bugg, E.M., Littlefield, J.W., Donovan, P.J., Blumenthal, P.D., Huggins, G.R., Gearhart, J.D. (1998) Derivation of pluripotent stem cells horn cultured human primordial germ cells. *Proc. Natl. Acad. Sci. USA 95,* 13726-13731.
80. Bulte, J.W.M., Douglas, T., Strable, E., Moskowitz, B.M., Frank, J.A. (2000) Magnetodendrimers as a new class of cellular contrast agents. Proceedings of the International Society for Magnetic Resonance in Medicine, Eighth Annual Meeting, Denver, Colorado: 2061.
81. Jacobs, R.E., Fraser, S.E. (1994) Magnetic resonance microscopy of embryonic cell lineages and movements. *Science 263*, 681-684.
82. Kayyem, J.F., Kumar, R.M., Fraser, S.E., Meade, T.J. (1995) Receptor-targeted co-transport of DNA and magnetic resonance contrast agents. *Chem. Biol. 2*, 615-620.
83. Bhorade, R.M., Weissleder, R., Nakakoshi, T., Moore, A., Tung, C.H. (2000) Macrocyclic chelators with paramagnetic cations are internalised into mammalian cells via a HIV-Tat derived membrane translocation peptide. *Bioconj. Chem. 11*, 301-305.
84. Bryant Jr., L.H., Bulte, J.W.M., Combs, C.A., Lewis, B.K., Frank, J.A. (2000) Dendrimer-based cellular MR contrast agents: Development of a molecular "off-switch" for a macromolecular contrast agent. Proceedings of the International Society for Magnetic Resonance in Medicine, Eighth Annual Meeting, Denver, Colorado: 377.
85. Filler, A.G. (1994) Axonal transport and MR imaging: prospects for contrast agent development. *J. Magn Reson Imaging 4*, 259-267.
86. Enochs, W.S., Schaffer, B., Bhide, P.G., et al. (1993) MR imaging of slow axonal transport in vivo. *Exp. Neurology 123*, 235-242.
87. van Everdingen, K.J., Enochs, W.S., Bhide, P.G. et al. (1994) Determinants of in vivo MR imaging of slow axonal transport. *Radiology 193*, 485-491.
88. Pautler, R.G., Silva, A.C., Koretsky, A.P. (1998) In vivo neuronal tract tracing using manganese-enhanced magnetic resonance imaging. *Magn. Reson. Med. 40*, 740-748 (1998).
89. Lin, Y.J., Koretsky, A.P. (1997) Manganese ion enhances T-1-weighted MRI during brain activation: An approach to direct imaging of brain function. *Magn. Reson. Med. 38*, 378-388.
90. Tjuvajev, J.G., Stockhammer, G., Desai, R., Uehara, H., Watanabe, K., Gansbacher, B., Blasberg, R.G. (1995) Imaging the expression of transfected genes *in vivo*. *Cancer Res. 55*, 6126-6132.
91. Tjujavev, J.G., Finn, R., Watanabe, K., Joshi, R., Oku, T., Kennedy, J., Beattie, B., Koutcher, J., Larson, S., Blasberg, R.G. (1996) Noninvasive imaging of herpes virus thymidine kinase gene transfer and expression: a potential method for monitoring clinical gene therapy. *Cancer Res. 56*, 4087-4095.
92. Tjuvajev, J.G. Avril, N., Oku, T., Sasajima, T., Miyagawa, T., Joshi, R., Safer, M., Beattie, B., DiResta, G., Daghighian, F., Augensen, F., Koutcher, J., Zweit, J., Humm, J., Larson, S.M., Finn, R., Blasberg, R. (1998) Imaging herpes virus thymidine kinase gene transfer and expression by positron emission tomography. *Cancer Res. 58*, 4333-4341.
93. Gambhir, S.S., Barrio, J.R., Phelps, M.E., Iyer, M., Namavari, M., Satyamurthy, N., Wu, L., Green, L.A., Bauer, E., MacLaren, D.C., Nguyen, K., Berk A.J., Cherry, S.R., Herschman, H.R. (1999) Imaging adenoviral-directed reporter gene expression in living animals with positron emission tomography. *Proc. Natl. Acad. Sci. USA 96,* 2333-2338.

94. Contag, C.H., Jenkins, D., Contag, P.R., Negrin, R.S. (2000) Use of reporter genes for optical measurements of neoplastic disease in vivo. *Neoplasia 2*, 41-52.
95. Diebel, C.E., Proksch, R., Green, C.R., Neilson, P., Walker, M.M. (2000). Magnetite defines a vertebrate magnetoreceptor. *Nature 406*, 299-302.
96. Matsunaga, T. (1997) Genetic analysis of magnetic bacteria. *Mater. Sci. Eng. C 4*, 287-289.
97. Matsunaga, T., Kamiya, S., Tsujimura, N. (1997) Production of a protein (enzyme, antibody, protein A)-magnetite complex by genetically engineered magnetic bacteria *Magnetospirillum* Sp. AMB-1, in U. Häfeli, W. Schütt, J. Teller, M. Zborowski (eds.), *Scientific and clinical applications of magnetic carriers*, Plenum Press, New York, pp. 287-294.
98. Koretsky, A.P., Brosnan, M.J., Chen, L., Chen, J.D., VanDyke, T. (1990). NMR detection of creatine-kinase expressed in liver of transgenic mice - determination of free ADP levels. *Proc. Natl. Acad. Sci. USA 87*, 3112-3116.
99. Koretsky, A.P., Lin, Y.-J., Schorle, H., Jaenisch, R. (1996) Genetic control of MRI contrast by expression of the transferrin receptor. Proceedings of the International Society for Magnetic Resonance in Medicine, Fourth Annual Meeting, New York, New York: 69.
100. Enochs, W.S., Petherick, P., Bogdanova, A., Mohr, U., Weissleder, R. (1997 Paramagnetic metal scavenging by melanin: MR imaging. *Radiology 204*, 417-423.
101. Weissleder, R., Simonova, M., Bogdanova, A., Bredow, S., Enochs, W.S., Bogdanov, A. Jr. (1997) MR imaging and scintigraphy of gene expression through melanin induction. *Radiology 204*, 425-429.
102. Moats, R.A., Fraser, S.E., Meade, T.J. (1997) A "smart" magnetic resonance imaging agent that reports on specific enzymatic activity. *Angew. Chem. Int. Ed. Engl. 36*, 726-728.
103. Louie, A.Y., Hüber, M.M., Ahrens, E.T., et al. (2000) In vivo visualisation of gene expression using magnetic resonance imaging. *Nature Biotechnology 18*, 321-325.
104. McMurry, T.J., Dunham, S.U., Dumas, S.A., et al. (1998) Bioactivated MRI contrast agents: preliminary results in sensing alkaline phosphatase via changes in protein binding. Proceedings of the International Society for Magnetic Resonance in Medicine, Sixth Annual Meeting, Sydney, Australia: 636.
105. Li, W.-H., Fraser, S.E., Meade, T.J. (1999) A calcium-sensitive magnetic resonance imaging contrast agent. *J. Am. Chem. Soc. 121*, 1413-1414.

RADIOACTIVE MICROSPHERES FOR MEDICAL APPLICATIONS

URS HÄFELI
Cleveland Clinic Foundation, Radiation Oncology Department T28
9500 Euclid Ave., Cleveland, OH 44195

Summary

This paper reviews the preparation and application of radioactive microspheres for medical purposes. It first discusses the properties of relevant radioisotopes and then explores the diagnostic uses of gamma-emitter labelled microspheres, such as blood flow measurement and imaging of the liver and other organs. The therapeutic uses of alpha- and beta-emitting microspheres, such as radioembolization, local tumour therapy and radiosynovectomy, are then described, and the recent developments in neutron capture therapy using gadolinium microspheres and boron liposomes discussed. The review concludes with some considerations in radiopharmaceutical kit preparations and radioisotope generator use, as well as with some radiobiological and dosimetric concerns.

1. Definition of microspheres

Many different kinds of microparticles are used for both diagnostic and therapeutic medical applications. In the broadest terms, as the name implies, microparticles or microspheres are defined as small spheres made of any material and sized from about 10 nm to about 2000 μm. The term nanospheres is often applied to the smaller spheres (sized 10 to 500 nm) to distinguish them from larger microspheres. Ideally, microspheres are completely spherical and homogeneous in size (Figure 1A), although particles less homogeneous in size and shape are generally termed microspheres as well (Figure 1B). Depending on the preparation method and material used, microspheres show a typical size distribution which often deviates from the mono-sized ideal (Figure 1C). The category of microparticles also includes colloids which are crystallised, insoluble conglomerates of defined chemical composition, liposomes which are phospholipid vesicles, and naturally occurring particles such as red blood cells or leukocytes. When discussing general points in this review, the entire group of microparticles will simply be referred to as "microspheres". Larger molecules such as antibodies or peptides are also occasionally included in this group but will not be

M. De Cuyper and J.W.M. Bulte (eds.), Physics and Chemistry Basis of Biotechnology, 213–248.

considered in this review. Wilder [1] and Papatheofanis [2] have nicely described the therapy of tumours using radiolabelled antibodies, i.e. radioimmunotherapy, or radiolabelled peptides.

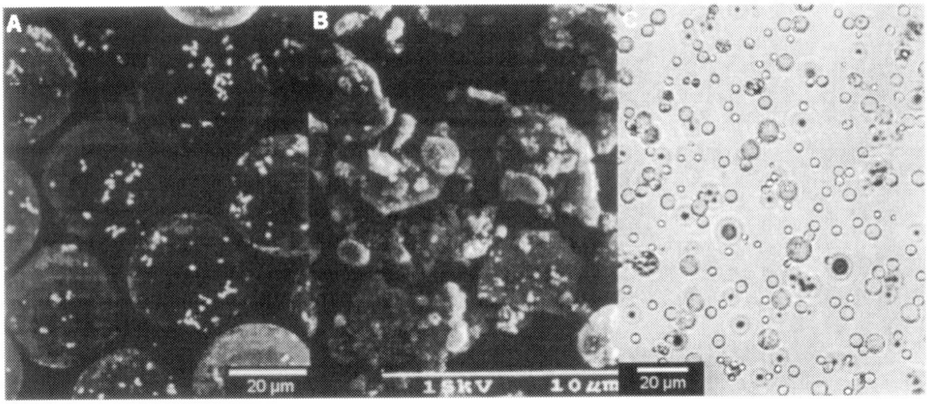

***Figure 1**. Microspheres for the delivery of radioactive isotopes. A) Spherical glass microspheres containing the two β-emitters ^{186}Re and ^{188}Re. B) Carbon-iron microspheres labelled with radioactive ^{99m}Tc. C) Poly(lactic acid) microspheres labelled with ^{188}Re.*

2. Applications and in vivo fate of microspheres

The largest application for microspheres in medicine is drug delivery. Sales of advanced drug delivery systems in the U.S. alone exceeded $13 billion in 1997, and are expected to increase. The medical uses of particulate drug delivery systems cover all areas of medicine such as cardiology, endocrinology, gynaecology, immunology, pain management and oncology. Most of the advanced drug delivery systems utilise microspheres or microcapsules for the encapsulation of drugs and proteins (see Table 1). The drug-loaded microspheres can be applied locally or delivered to the target area after intravenous injection by either passive means (e.g., trapping by size) or active means (e.g., magnetic targeting). From the target area, the encapsulated drug is slowly released over the desired time period, the length of which is determined mainly by the drug's biological half-life and its release kinetics from the microsphere matrix. This type of encapsulated drug delivery system has the advantage of protecting the encapsulated drug from the *in vivo* environment until time of release. Even very unstable substances such as growth hormones, interferon [3], or neuroactive peptides, [4] can be given in one daily dose instead of in several daily injections. Oral applications of very sensitive drugs such as insulin are also possible, as shown by Mathiowitz and others [5].

***Table 1**: Medical applications of microspheres*

Mechanism	Application (Examples)	Microsphere Matrix (Examples)	Ref.
Controlled drug delivery after local application	Release of proteins, hormones and peptides over extended times	PLA, PLGA, chitosan, polycyanoacrylate, polyanhydride	[6]
Oral drug delivery of easily degraded drugs	Gene therapy with DNA plasmids; delivery of insulin, LHRH	PLGA, styrene, polymethylmethacrylate	[7]
Vaccine delivery	Hepatitis, influenza, pertussis, ricin toxoid, diphteria toxoid; birth control	PLGA, chitosan	[8,9]
Drug targeting after intravenous / intra-arterial application	Passive targeting of leaky tumour vessels, active targeting of tumour cell antigens, magnetic targeting with microspheres	Any biocompatible material; liposomes or erythrocytes	
Drug delivery without toxic side effects	Tumour targeting with doxorubicin, treatment of Leishmaniasis	PLA, PLGA, starch cyanoacrylates, etc.; (PEG-) liposomes	[10,11]
Specific cell labelling	Stem cell extraction, bone marrow purging	Magnetic polystyrene microspheres	[12]
Affinity chromatography	Isolation of antibodies in immunology, cell separation, toxin extraction	Polymer resins such as Agarose-polyacrolein, Sephadex (polymer supports)	[13]
Adsorption of harmful substances from blood	Haemoperfusion	Agarose-PA, Sepharose, activated carbon, polyvinyl alcohol, polyacrylamide	[14]

Table 1: Medical applications of microspheres

Mechanism	Application (Examples)	Microsphere Matrix (Examples)	Ref.
Particle agglutination tests (qualitative and quantitative)	Diagnostic tests for infectious diseases (bacterial, viral, fungal...); other tests in human diagnostics (growth hormones, FDP,...)	Polystyrene (latex), silica, superparamagnetic particles	[15]
Endovascular embolization	Complex arteriovenous malformations in the brain; liver and other tumour treatment; management of life threatening haemoptysis and haematemesis	Poly(vinylalcohol), glass, polyurethane, poly(2-hydroxyethyl methacrylate)	[16]
Structure for cell growth	Cell culture of adherent cells in large amounts, 3D tissue structures possible	Gelatine, Sephadex, dextran, cellulose, collagen	[17]

Abbreviations: PLA = poly(lactic acid), PLGA = poly(lactide-co-glycolide), PEG = polyethylene glycol

The biodistribution and final fate of intravenously injected microspheres is highly dependent on their size and surface charge. Microspheres sized 10 to 30 μm are larger than capillaries and will be trapped in the first capillary bed that they encounter. This effect is used for radioembolization therapy in which microspheres are injected into the artery that leads to the tumour of interest. Positively charged microspheres sized in the micrometer range are quickly taken out of the blood pool by the reticuloendothelial cells of the liver and spleen [18]. Particles smaller than 0.1 μm are able to pass the fenestration in the liver and may be able to target the hepatocytes, although most are still taken up by the liver's Kupffer cells. Negatively charged or neutral nanospheres such as small PEG-coated nanospheres or liposomes can evade this fast uptake and circulate in the blood system for up to several days [19]. Over time, these long-circulating nanospheres will be concentrated in the tumour area because of the leaky capillary system of the newly growing tumour vasculature, which allows for extravasation of the nanospheres [20]. A more active way of increasing the concentration of nano- or microspheres in the target tissue is to bind antibodies against the target cells on the nanospheres' surface [21]. Alternatively, nanospheres, microspheres and colloids with a high affinity for white blood cells can be prepared. Such particles are rapidly taken up by the white blood cells and then concentrate in inflammatory regions because of chemotaxis and phagocytosis [22]. Made radioactive, such nanospheres are useful for diagnostic (imaging) purposes, as well as for therapy.

3. General properties of radioactive microspheres

The subgroup of microspheres that is radioactive behaves and is generally used in a similar fashion to non-radioactive microspheres. But in addition to the matrix substance, which defines the microsphere and gives it its targeting properties in a desired tissue or organ, the radioactive microsphere also contains one or more radionuclide(s) that are intimately bound to it.

Even in small concentrations, radioactive microspheres are able to deliver high radiation doses to a target area without damaging the normal surrounding tissue. The radioactivity, unlike drugs, is never released from the microspheres but acts from within over a radioisotope-typical distance. The effective treatment range in tissue is up to about 90 μm (10 cell layers) for α-emitters, never more than 12 mm for β-emitters and up to several centimetres for γ-emitters.

3.1. ALPHA-EMITTERS

Alpha particles are positively charged ions consisting of two protons and two neutrons, emitted during the radioactive decay of many nuclei with high atomic numbers. During decay, energy is released mainly as the kinetic energy of the α-particles. Since the path length of an α-particle with an energy of 5 to 8 MeV is on the order of 40 to 80 ☞m, the effective treatment radius is limited to several cell diameters from the atom that emits the particle, and nonspecific irradiation of distant tissues is eliminated. [23] The high linear energy transfer (LET) of such energetic particles (~100 keV/μm) and the limited ability of cells to repair the damage to DNA from α-particle irradiation contribute to their extraordinary cytotoxicity. At low doses in the range of 1 to 2 Gy, α-radiation is about 5 to 100 times more toxic than γ- or β-radiation. Furthermore, α-particle mediated cell killing is insensitive to conditions of hypoxia, which are often found in necrotic tumours and may compromise the clinical efficacy of β-, γ- or x-ray radiation.

The dosimetry of α-emitters is special since the dose deposition from the low-range α-particles must be considered on a cell by cell level. The normal approach of prescribing activity per gram of (tumour-)tissue will not lead to meaningful results, because it is very difficult to distribute radioactive microspheres absolutely homogeneously. Microdosimetry with α-emitters has been expertly described by Humm [24].

Most research with α-emitting radiopharmaceuticals and the first clinical trials in 1996 have involved antibodies labelled with ^{213}Bi, ^{211}At, ^{212}Bi, ^{225}Ac, ^{212}Pb, ^{255}Fm, ^{223}Ra, and ^{149}Tb (see Table 2). This work will not be covered here, but a comprehensive review of the so-called radioimmunotherapy with α-emitters is given by McDevitt [25].

***Table 2**. Alpha-emitters useful for delivery in particulate radiopharmaceuticals*

Radioisotope	Half-life	α-yield	Other radiation (keV)	Range in tissue in μm	Production
^{149}Tb	4.13 h	17%	β^- (400 max.), γ (165, 352, 511)	28	Accelerator
^{211}At	7.2 h	100%	γ (77-92, 500-900)	65	Accelerator
^{212}Bi	60 min	36%	β (2246 max, 64%) γ (727, 12%)	70, 42, 87	^{224}Ra-generator
^{213}Bi	46 min	100%	γ (440, 28%), β (1420 max, 98%)	43	^{225}Ra-generator
^{223}Ra	11.4 d	300%	γ (0.031-0.45)	43	^{227}Ac-generator
^{225}Ac	10.0 d	400%	γ (0.037-0.187)	48	^{225}Ra-generator
^{255}Fm	20.1 h	93%		63	^{255}Es-generator

3.2. BETA-EMITTERS

In 1896, Henri Becquerel discovered β-decay, which is the commonly used name for β^- - or negatron-decay. During β-decay, a neutron in the unstable nucleus is transformed into a proton, an electron and a neutrino, which is an uncharged particle with undetectable small mass. Additionally free energy is produced and released in the form of kinetic energy and given to the electron and the neutrino. Since the free energy is distributed in an isotope-characteristic but random fashion to the β-electron and the neutrino, we will always measure a spectrum of electrons with different energies. An electrons maximum energy E_{max} is measured when no energy transfer to the neutrino takes place (see Table 3). Each β-decay has its characteristic energy-spectrum, and the average energy is typically about a third of E_{max}. Passing through tissue, the ejected β-electrons interact with other (mainly water) atoms and lose energy, leading to excited and ionised atoms. These activated species (e.g., radicals) are responsible for therapeutic effects (e.g., DNA damage of cancer cells), but also for toxicity (damage to normal cells nearby).

One of the first β-emitters used in particulate form for the treatment of lung tumours was ^{198}Au-labeled microspheres (size 30-50 μm) [26]. Unfortunately, ^{198}Au also emits

***Table 3**. Beta-emitters useful in particulate radiopharmaceuticals*

Radio-isotope	Half-life	Average / max. beta-energy* in keV	Max. range in tissue in mm	X_{90}^ (mm)	Gamma-lines in keV (%)	Production
^{3}H	12.3 y	5.7 / 18.0			none	^{6}Li(n,α)-^{3}H
^{14}C	5730 y	49.5 / 156.0			none	^{14}N(n,p)-^{14}C
^{32}P	14.3 d	694.9 / 1710.2	8.7	2.2	none	^{32}S(n,p)-^{32}P or ^{31}P(n,γ)-^{32}P
^{90}Y	64.1 h	933.6 / 2280.0	12.0	2.8	none	^{90}Sr/^{90}Y generator
^{131}I	8.0 d	181.7 / 806.9	2.4		364.5 (81.2%)	^{131}Te (β^-)-^{131}I
^{153}Sm	46.5 h	224.2 / 808.2	3.0	0.7	103.2 (29.8%)	^{152}Sm(n,γ)-^{153}Sm
^{165}Dy	2.3 h	440.2 / 1286.7	6.4	1.3	94.7 (3.6%)	^{164}Dy(n,γ)-^{165}Dy
^{166}Ho	26.8 h	665.1 / 1853.9	10.2	2.1	80.6 (6.7%)	^{165}Ho(n,γ)-^{166}Ho
^{169}Er	9.4 d	99.6 / 350.9 keV	1.0		< 0.2%	^{168}Er(n,γ) ^{169}Er
^{177}Lu	6.7 d	133.3 / 497.8	1.7		113.0 (6.4%) 208.4 (11.0%)	^{176}Lu(n,γ)-^{177}Lu
^{186}Re	89.2 h	346.7 / 1069.5	5.0	1.0	137.2 (9.42%)	^{185}Re(n,γ)-^{186}Re
^{188}Re	17.0 h	764.3 / 2120.4	11.0	2.1	155.0 (15.1%)	^{188}W/^{188}Re generator
^{198}Au	2.7 d	311.5 / 960.7	4.4	0.9	411.8 (95.5%)	^{197}Au(n,γ)-^{198}Au

**NuDat database [29]; ^ Distance in tissue within which 90% of dose is deposited [30]*

high energy γ-rays. This led to higher than necessary radiation doses to the other organs as well as to hospital personnel. To avoid this exposure, the pure β-emitters ^{32}P and ^{90}Y have been favoured during the last decade and have become the predominant radioactive isotopes for many therapeutic applications. Recently, however, it has been shown that a certain amount of low-energy γ-radiation can actually be useful for imaging, either during or after the application of the radioactive microspheres [27]. During infusion, with the help of a γ-camera or γ-detector, the surgeons are able to a) direct the radioactive microspheres and b) adjust the necessary amounts of radioactivity.

Short-lived radioisotopes (Table 3) can be used to optimise radiobiological aspects of therapy. Specifically, it has been shown that not only the total dose, but the dose-rate is very important for the treatment outcome in radiotherapy [28]. Short-lived radioisotopes such as ^{165}Dy or ^{188}Re pack the "punch" into a much shorter time-period, allowing less time for tumours to recover and grow back. Although much more research is required in this area, many of the radioactive β-emitting lanthanides are seen as promising candidates for local or directed radiotherapy, with microspheres serving as the delivery system. The dosimetry of β-emitting radioactive microspheres depends on the application. In the simplest case, when the radiopharmaceutical is distributed homogeneously throughout the target (tumour) area, the MIRD scheme is used [31,32]. MIRD calculations can be done on a PC [using a program provided free to the interested user] [33]. Harbert's calculations are used when the radiopharmaceutical is in a plane from which it irradiates the tissue [34] (Appendix K). This approach is, for example, appropriate in the treatment of cystic brain tumours or in radiosynovectomy (see below). Dosimetric modelling using Monte Carlo simulations can be used for microspheres of different sizes and different β-emitters [35-37].

3.3. GAMMA-EMITTERS

A large group of radioisotopes emits γ-rays during decay. Gamma rays represent excess energy that is given off as the unstable nucleus breaks up and decays in its efforts to reach a stable form. The energy is emitted in the form of electromagnetic radiation (photons), with a radioisotope-characteristic photon energy typically expressed in kilo-electronvolt (keV). Photons are absorbed in biological material by both the photoelectric and Compton process, and then indirectly ionise the surrounding atoms, producing chemical and biological changes. Most γ-emitters are used primarily for diagnostic purposes and those used in nuclear medicine (Table 4) were chosen so that a) their γ-ray energy is not too high (radiation safety concerns) and matches the γ-camera, b) their half-life is practical and logistically feasible, c) they are easily available and inexpensive, and d) they can be bound to microspheres in an easy (kit) and stable fashion.

Table 4. Gamma-emitters used in particulate radiopharmaceuticals

Radioisotope	Half-life	Gamma lines (Efficiency)	Production
^{51}Cr	27.7 d	320 keV (10%)	^{50}Cr (n,γ) ^{51}Cr
^{67}Ga	78.2 h	93 keV (40%) 184 keV (20% 300 keV (17%) 393 keV (4%)	^{68}Zn (n,p) ^{67}Ga
^{99m}Tc	6.0 h	140 keV (89%)	$^{99}Mo/^{99m}Tc$-generator
^{111}In	2.8 d	171 keV (90%) 245 keV (94%)	^{111}Cd (p,n) ^{111}In
^{123}I	13.2 h	159 keV (83%) 528 keV (1%)	^{121}Sb (α,2n) ^{123}I
^{125}I	60 d	35 keV (7%) 27-32 keV x-ray (140%)	^{124}Xe (n,γ) ^{125}Xe ^{125}Xe (EC^*,β^+) ^{125}I

*EC = Electron capture

4. Preparation of radioactive microspheres

Microspheres can be made radioactive (= radiolabelled) either during or after their preparation. Although the former method is still more commonly used in medicine, the latter is preferred, especially for shorter-lived radioisotopes, because it is compatible with kit formulation. In this case, the microspheres can be stored for extended periods of time as part of a sterile nonradioactive kit and then be radiolabelled by the radiopharmacist in the nuclear medicine department shortly before use. Radiochemical stability problems are in this way minimised and logistical problems inherent to the use of radiopharmaceuticals avoided.

Depending on the particles, it is possible to enclose the activity, label throughout the entire volume, or label only certain structures, such as the surface, the outer or inner wall, the lipophilic or hydrophilic liposome compartment (Figure 2). The binding of radioactivity to particles can be done by covalent bonds, by chelation, by adsorption processes or by indirect means as, for example, avidin-biotin bonds which can bridge the microsphere and the radioisotope. In all these cases, *in vivo* biodegradation processes and reversible isotope exchange processes can lead to instability and release of the radioisotope into the immediate surrounding. Regarding biodegradation processes, the microspheres building material might undergo rapid ester-bond cleavage depending on the target organ's enzymatic activity or the locally produced radiation. The cleavage of covalent bonds can be beneficial in drug targeting if it takes place in the target organ. For example, the lysosomal *in vivo* activation of an inactive prodrug into the effective drug as shown with polymers binding daunomycin or puromycin [38], is a precondition for pharmacological action. In the case of radioactive microspheres, however, absolutely no degradation is wanted until complete decay of the radioisotope.

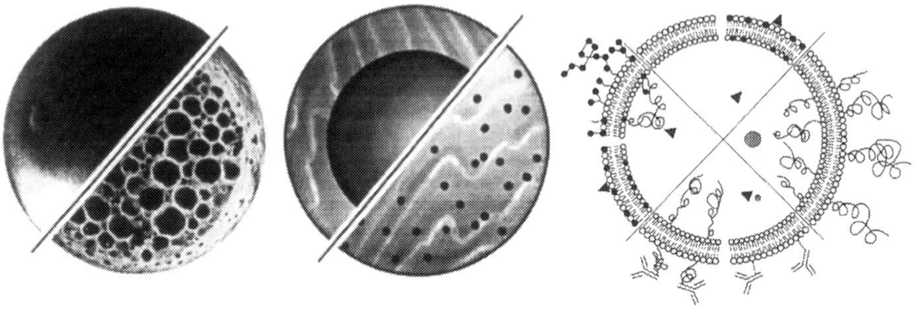

***Figure** 2. Depending on the particle type shown in this schematic drawing different compartments of the particles can be radiolabelled. A) Nonporous and porous microspheres or nanospheres. B) Reservoir and monolithic (matrix) microcapsules, C) Different types of liposomes.*

4.1. RADIOLABELING DURING THE MICROSPHERE PREPARATION

Colloids were the first microspheres to be radiolabelled during preparation. They are a unique class of radioactive microspheres (Table 5) that consist entirely of the defined inorganic compounds of a radioisotope which have precipitated into relatively homogeneous particles. The size range of colloids depends mostly on the preparation conditions such as temperature and pH and on the form of the precipitating agent. Sulphur colloids with ^{99m}Tc, for example, can be made in the size of 80 to 100 nm by treating a boiling acidic ^{99m}Tc pertechnetate solution with H_2S gas. Alternatively, sodium thiosulphate can be added to the radioactive solution, but the size distribution of the mixed colloid of Tc_2S_7 and sulphur is then much broader, from 80 to about 2000 nm [39]. Other useful colloids are the hydroxides and oxides of ^{99m}Tc and ^{113m}In prepared by coprecipitation with ferric hydroxide [40] and the oxides of ^{99m}Tc prepared by coprecipitation with Sn(II) [41].

One method of microsphere preparation which facilitates the production of homogeneously sized albumin microspheres that incorporate many different kinds of radioactive colloids was first described by Zolle et al [42]. The method consists of first transforming a radioactive substance into a precipitate, mixing it with an aqueous solution of albumin and then injecting it into a stirred solution of cottonseed oil. The fine dispersion of albumin droplets then forms spherical and stable albumin particles tightly enclosing the radioactive compounds after heating the mixture above 100 °C. Microspheres made from or with proteins, such as the above albumin microspheres, always contain tyrosine and histidine. Their phenol- and imidazole-rings can be easily iodinated using methods such as the chloramine T, the iodogen, the Bolton-Hunter or the iodo-bead method, to name just a few. A very good review that describes these techniques in detail is available from Amersham [43]. Another way of achieving iodinated microspheres is to radioiodinate the compound that will be incorporated into the microspheres during their formation. Yang et al. made radioactive PLA-microspheres by first labelling the contrast agents ethyliopanoate and ethyldiatrizoate, which were to be incorporated, with ^{131}I, dissolving them together with the polymer PLA in methylene chloride and then preparing the microspheres by a solvent evaporation method [44].

As with radioactive microspheres, radioactive liposomes can be made by adding radioactive compounds during their formation. Unilamellar liposomes of 70 nm diameter have been prepared by mixing the lipid-soluble radioactive complex oxodichloroethoxy-bis-(triphenylphosphine)^{186}rhenium(V) with phospholipids and the detergent sodium deoxycholate, followed by detergent removal on a small gel filtration column [45]. Such biocompatible ^{186}Re-liposomes can be used to deliver therapeutic radiation doses for radiosynovectomy (see below).

Table 5*. Methods of preparing radioactive microspheres in which radiolabelling is done during formation of the microspheres.*

Method of Labelling	Examples	Ref.
Colloid precipitation	^{99m}Tc sulphur colloid	[39]
	^{113m}In ferric hydroxide colloids	[40]
	^{165}Dy-FHMA (~5 μm)	[46]
	Chromic ^{32}Phosphate (1-2 μm)	[47]
Inclusion of radiolabelled compound	^{99m}Tc-HSA-gelatin microcapsules	[48]
	^{131}I-ethyldiatrizoate-PLA microspheres	[44]
	^{125}I-iododeoxyuridine-PLGA microspheres	[49]
	^{125}I-HSA magnetic albumin microspheres	[50]
Isotope exchange	^{211}At-microspheres	[34]
	^{14}C-, ^{35}S- and ^{3}H-labeling	[51]
Lipophilic inclusion	^{186}Re/^{188}Re-triphenylphosphine-liposomes	[45]
In situ production	^{99m}Tc-Buckminster fullerenes (C_{60} or C_{80}) or aggregates thereof (Technegas)	[52]

Abbreviations: FHMA = ferric hydroxide macroaggregates, HSA = human serum albumin

A relatively recent development in the preparation of radioactive particles is the *in situ* production of ^{99m}Tc-particles in a Technegas generator [52]. The ^{99m}Tc-pertechnetate is pyrolised together with carbon at 2500 °C and forms not only Buckminster fullerenes (C_{60} to C_{80}) each enclosing a technetium atom, but also agglomerates of graphite and technetium.

4.2 RADIOLABELING AFTER THE MICROSPHERE PREPARATION

Compared to radiolabelling during microsphere preparation, methods of radiolabelling already prepared microspheres are conceptually more straightforward. Spherical anion or cation exchange resins of different sizes are examples of microspheres which can be radiolabelled with ionic radionuclides (Table 6). The resins can be loaded with labelling efficiencies generally exceeding 95% by simple incubation in saline or aqueous buffer solutions containing the radioisotope. Their stability, however, has to be evaluated carefully, since not all resins have the capacity or the binding affinity necessary to bind radioisotopes such as $^{90}Y^{3+}$. Yttrium is a radioisotope that will, in its ionic form, be taken up easily by the bone marrow where it will remain until complete decay, leading to severe toxicity (myelosuppression). It is thus of utmost importance that bound ^{90}Y not be released *in vivo*. It has been found that of the cation-exchange resins Bio-Rex 70, Sephadex SP, Chelex 100, AG 50W-X8 or Cellex-P, only Bio-Rex 70 was able to provide the stability needed for ^{90}Y-radioembolization *in vivo* [53]. Other ion-exchange resins have been used for the adsorption of negatively charged radioisotopes. Pertechnetate, $^{99m}TcO_4^-$, for example, has been adsorbed to 300 μm -large Dowex 1-X4 beads [54]. Even larger 1 mm Amberlite 410 resin pellets were labelled with pertechnetate in the same way [55] and have been used for GI transit studies. Chromate, $^{51}CrO_4^{3-}$, has been adsorbed to Dowex 1-X8 sized 10 to 50 μm and used for the measurement of mucociliary functions [56].

Many different functional groups such as -OH, $-NH_2$, -SH and -COOH are used to bind specific drugs, radiolabelled chemicals, and chelators to microspheres, and to introduce other functional groups for further derivatisation. These chemical modifications are possible before microsphere preparation, but are more commonly performed afterwards. For example, the chelator DTPA (Figure 3) has been bound via an amide bond to albumin microspheres using one of the DTPA's carboxyl groups [58]. Such microspheres are quite versatile, since DTPA is able to chelate not only ^{111}In, but also ^{90}Y, ^{99m}Tc, ^{166}Ho and many other lanthanides. Currently, the two most stable and most often used chelators able to bind diagnostic and therapeutic radioisotopes are DOTA and MAG_3 (see Figure 3). DOTA (= 1,4,7,10-tetra- azacyclododecane N,N',N'',N'''-tetraacetic acid) is able to complex ^{212}Bi [72] and has also been shown to chelate ^{90}Y and ^{111}In with better than 99% stability over 2 weeks [73]. MAG_3 (= mercaptoacetylglycylglycylglycine) is able to complex the radioisotopes from group VIIB, ^{186}Re, ^{188}Re and ^{99m}Tc [74] at almost 100% stability in serum over 24 hours [75].

Table 6. *Methods of preparing radioactive microspheres from preformed, non-radioactive microspheres*

Method of Labelling	Examples	Ref.
Radiolabelling by ion exchange	Anion- and cation-exchange resins:	
	BioRex 70 loaded with ^{90}Y	[57]
	Dowex 1-X4 loaded with $^{99m}TcO_4^-$	[54,55]
	Dowex I-X8 loaded with $^{51}CrO_4^{3-}$	[56]
Chelation (complex formation) of the radioisotope	^{111}In-DTPA-albumin microspheres	[58]
	^{68}Ga-DTPA-albumin microspheres	[59]
	^{99m}Tc-polystyrene latex microspheres	[60]
	^{186}Re-polycysteine/polylysine microspheres	[61]
Isotope exchange with ^{131}I, ^{125}I and ^{211}At	^{131}I-Mitomycin C gelatine microspheres	[62]
	^{131}I-albumin microspheres	[58]
	^{211}At-methacrylate microspheres	[63]
Neutron activation (typically n,γ-reaction)	^{90}Y-glass and ^{32}P-glass microspheres	[64]
	^{186}Re/^{188}Re-glass microspheres	[65]
	^{166}Ho-glass microspheres	[66]
	^{166}Ho-PLA microspheres	[67,68]
	^{186}Re/^{188}Re-PLA microspheres	[69]
Reduction to insoluble, colloidal compounds	^{99m}Tc-Sn PLA microspheres	[70]
Affinity to microsphere material	^{186}Re-HEDP bound to hydroxyapatite microspheres	[71]
	^{153}Sm-citrate bound to hydroxyapatite microspheres	[71]

Abbreviations: DTPA = Diethylenetriamine pentaacetic acid

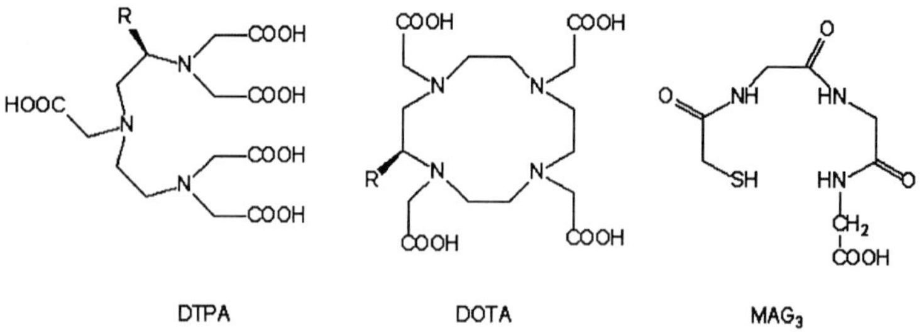

Figure 3. *Typical chelators used to complex diagnostic and therapeutic radioisotopes* ^{111}In, ^{90}Y, ^{212}Bi, ^{186}Re, ^{188}Re *and* ^{99m}Tc, *among many others.*

The radiolabelling of microspheres with chelator-groups on the surface typically involves an incubation with the radioisotope of between 5 and 60 minutes, at temperatures of 20 to 100 °C. The labelling of DOTA or DTPA with ^{90}Y, ^{111}In or many other $^{3+}$-charged ions occurs directly at the optimal pH. The complexation of the +VII pertechnetate or perrhenate, however, additionally involves the reduction of technetium and rhenium to the +V or +IV state. Many different reducing agents such as $NaBH_4$, $Na_2S_2O_4$, H_3PO_2, hydrazine, ascorbic acid or electric reduction have been used, but the most common method is the use of $SnCl_2$. The reduction and complexation of technetium, together with ways of developing it into kit form, has been well reviewed by Eckelman et al. and can be directly applied to many chelator-containing microspheres [76].

Microspheres made from appropriate materials can also be labelled using functional groups such as reduced sulfhydryl-groups, alone or in combination with nearby carboxyl- and amine-groups. This method has been termed the "direct method" by chemists using it for the radioactive labelling of antibodies [77] and works especially well for microspheres made from proteins, such as the human serum albumin microspheres labelled with ^{188}Re after reduction using Sn(II) [78]. Other microspheres that bind radioactivity with sufficient stability for therapy are $^{90}Y^{+3}$-labeled magnetic PLA microspheres with native carboxylic groups [79] and ^{99m}Tc-labeled polystyrene microspheres derivatised with poly(acryclic acid) in order to introduce carboxylic groups [60].

4.3. RADIOLABELING BY NEUTRON ACTIVATION OF PRE-MADE MICROSPHERES

A very effective way of preventing leakage of the radioactive isotope(s) from the microsphere is to seal the radioisotope into the microsphere matrix. The pre-made microspheres enclose the non-radioactive precursor of the radioisotope and are activated in a nuclear reactor by bombardment with thermal neutrons shortly before use (Table 6). The most stable matrix for this kind of microsphere activation is glass. Day and Ehrhardt pioneered such therapeutic radioactive microspheres (Figure 4) from aluminosilicate glass containing 17 mol% Y_2O_3 [80]. The glass mixture was melted in a platinum crucible at 1600 °C the annealed glass crushed and the splinters spheroidised by sprinkling them from above through an oxygen flame. During neutron-activation in the reactor, the non-radioactive ^{89}Y captured a neutron and became the radioactive β-emitter ^{90}Y. The leakage rate of the ^{90}Y enclosed in the glass matrix was extremely low. Not more than 92 Bq were released from 50 mg of microspheres when activated to therapeutic activities of 11.1 GBq. Very similar glass microspheres have also been prepared enclosing rhenium, resulting in ^{186}Re/^{188}Re microspheres after neutron activation [65]. Advantages of glass microspheres are their excellent stability, radiation resistance, insolubility and non-toxicity. Disadvantages include their high density (3.3 g/ml) which makes the complete injection through syringes and intravenous lines difficult, and their non-biodegradability which can lead to immunologic reactions. Research is ongoing, however, in the preparation of glass microspheres from biodegradable glass material such as lithium boride [81].

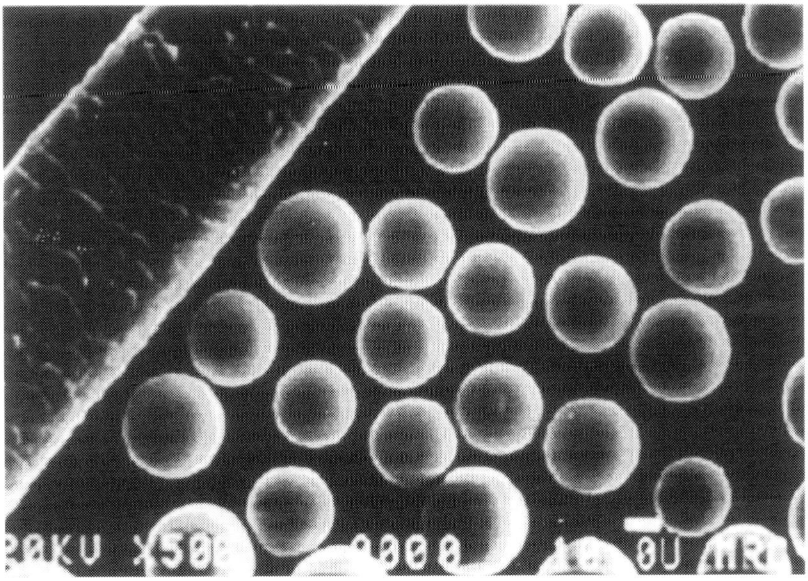

***Figure 4**. Yttrium glass microspheres for neutron activation in a nuclear reactor in comparison with the size of a hair.*

The disadvantages of glass were overcome by the preparation of PLA-microspheres containing either an acetylacetone-complex of ^{165}Ho [67,68] or small particles of metallic rhenium in its native form, ^{185}Re and ^{187}Re [69] (Table 6). The stability of the activated ^{166}Ho and ^{186}Re/^{188}Re-microspheres was sufficient for therapy (less than 1% of activity released within a week). The activation time of these poly(lactic acid) microspheres, however, is limited due to the radiolytic breakdown of ester bonds and must be characterised for each polymer-microsphere composition. Specifically, it has been found that activation of rhenium microspheres made from PLA with a molecular weight of 2000 for 1 hour at a neutron-flux of $5 \cdot 10^{12}$ n/cm^2/sec produced 450 MBq ^{188}Re and 78 MBq of ^{186}Re. Longer activation times led to melting and polymer breakdown [69]. The therapy of liver tumours which requires high specific activities is thus not possible with these $^{188}Re/^{186}Re$-PLA-microspheres, but they could be used in local treatment of brain metastases or applied to incompletely resected tumour tissue after surgery. Recently, ^{166}Ho-acetylacetonate-microspheres made from PLA with a molecular weight of 20,000 have been described [68]. The authors reported that up to 1 hour of neutron-activation at a flux of $5 \cdot 10^{13}$ n/cm^2/sec was possible, yielding an activity of 20 GBq in 400 mg of microspheres. This activity would be sufficient to allow for their transport to the hospital and use in liver tumour patients on the following day.

4.4. IN SITU NEUTRON CAPTURE THERAPY USING NON-RADIOACTIVE MICROSPHERES

Neutron capture therapy is an exciting bimodal tumour treatment concept originally proposed in 1936 by Locher [82]. The first component of this therapy is the delivery to tumour cells of non-radioactive atoms or molecules either alone or packed into carriers such as liposomes or microspheres. The target nuclei have large thermal and/or epithermal neutron capture cross-sections and a resulting reaction having a large positive Q value. The aim is to attain a higher concentration of these nuclei in the tumour than in the surrounding normal tissue cells. The second component is the exposure of a selected patient volume to a neutron beam. During neutron capture *in situ*, excessive energy between the initial and final state of the reactive nuclei (the positive Q-value) is released either as the recoil energy of heavy particles (^{6}Li, ^{10}B) and α-particles, or as γ-rays (^{155}Gd, ^{157}Gd) (see Table 7). In the case of boron or lithium neutron capture, most of this energy is deposited in the tumour cell, since the range of the produced particles is less than 10 μm. In the case of gadolinium neutron capture, energy is spread out more because of photonic interactions. Although the first clinical trials with neutron capture therapy were completed in the 1950's, it took developments of the next 40 years to make this therapeutic approach very promising for cancer therapy [83]. Developments included the synthesis of superior targeting compounds such as the sulphur containing boron compounds mercaptoundecahydrododecaborate (= BSH) or boronated porphyrins (= BOPP).

In neutron capture therapy, boron and gadolinium are generally delivered intravenously, although their delivery is also possible via microparticles. Tokumitsu et al developed Gd-DTPA loaded chitosan microspheres of 4.1 ± 1.3 μm for intratumoural injection [84,85]. Boron can be delivered in a similar way by packaging BSH into liposomes with anti-carcinoembryonic antigen antibodies on their surface [86]. Different boron compounds have been encapsulated not only in the aqueous compartment of liposomes, but also in their bilayer and attached to their surface and work in this area is ongoing [87].

Table 7. *Isotopes useful for neutron capture therapy. Their half-life is zero.*

Isotope	Reaction	Q [MeV]	Cross section σ_{th} [barn]	Mode of energy deposition	Range in tissue in μm	Delivery vehicles
^{6}Li	$^{6}Li(n,\alpha)^{3}H$	4.784	940	α - 2.105 MeV ^{3}H - 2.734 MeV	< 10	
^{10}B	$^{10}B(n,\alpha)^{7}Li$	2.790	3837	α_0 - 1.775 MeV α_1 - 1.47 MeV $^{7}Li_0$ - 1.015 MeV $^{7}Li_1$ - 0.84 MeV	7.2, 8.9	Liposomes containing BSH [86,88]
^{155}Gd	$^{155}Gd(n,\gamma)^{156}Gd$	11.452	61000			
^{157}Gd	$^{157}Gd(n,\gamma)^{158}Gd$	7.937	254000			Chitosan microspheres; microcapsules containing Gd-DTPA [84,85]

5. Diagnostic uses of radioactive microspheres

Diagnostic studies with radiopharmaceuticals include dynamic and static imaging and *in vivo* function tests. Dynamic imaging provides information about the biodistribution and pharmacokinetics of drugs in organs. Performed with a γ-camera, dynamic studies are generally carried out over a pre-set length of time and provide clues to the functioning of the organ being examined. Static imaging, on the other hand, provides morphological information about an organ such as its shape, location and size. Furthermore it allows the exact location of tumours to be determined. Static imaging, unlike dynamic imaging, is normally done at a single point in time, with the imaging time being dependent on the organ activity. In contrast to dynamic and static imaging, *in vivo* function tests do not require imaging. Instead they are evaluated by comparing an injected or swallowed amount of radioactivity to the measured radioactivity in urine or blood.

All three types of diagnostic studies can be performed with radioactive microspheres which contain one or several -emitters that can be detected by a γ-camera. The first such "microspheres" in clinical use were red and white blood cells, which were taken from a patient, labelled with ^{111}In or ^{51}Cr, and then re-injected. Red blood cells labelled with ^{51}Cr are commonly used for the measurement of red blood cell mass and for imaging of the spleen. For the latter purpose, the red blood cells are denatured by heating, which renders them spheroidal and nondeformable, and makes them easy to take up by the spleen. Another common application of radiolabelled red blood cells is the accurate determination of total systemic arterial blood flow or venous return, as well as, for blood flow determination within specific organs [89]. These blood flow parameters are important when drugs for the treatment of cardiovascular diseases are evaluated. White blood cells labelled with ^{111}In-oxine are used for the detection of inflammatory diseases, abscesses or other infections. A less expensive method has been developed in which the neutral and lipophilic ^{99m}Tc-HMPAO complex is prepared from a kit and then incubated with the leukocytes [90]. Platelets labelled with ^{111}In are also used to detect actively forming deep vein thrombi, to measure blood flow, and to detect regions of infection [91]. Radiolabelled blood cells are still used today, although pre-made radioactive microspheres containing several different γ-emitters (see Table 7) are easier to use and do not require time-consuming labelling procedures [92]. Unfortunately, the radioactive microspheres of homogenous size are made from polystyrene and thus are not biodegradable, making them inappropriate for clinical use.

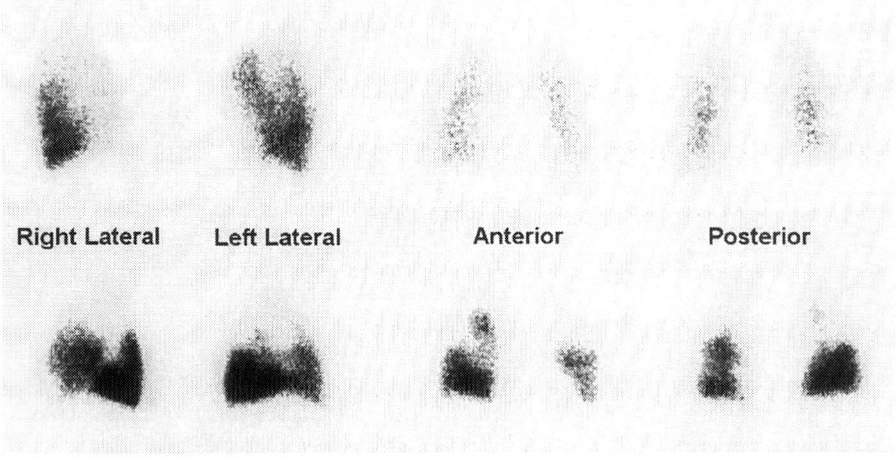

***Figure** 5. Diagnostic lung imaging obtained after the injection of ^{99m}Tc-labeled macroaggregated albumin in different projections. The top row shows a normal lung and the bottom row the lung of a patient with multiple pulmonary emboli in both lobes of the lungs.*

For the diagnosis of pulmonary embolism, both the inhalation of small, radioactive ^{99m}Tc-carbon particles (Technegas) and the perfusion of the lung with ^{99m}Tc-labeled albumin particles are used. In the first case, Technegas behaves, due to the small particle size of less than 100 nm, much more like a gas than a radioaerosol and diffuses into the entire accessible lung volume. In the second case, macroaggregated albumin is mainly used for the quantification of shunts associated with intrapulmonary arteriovenous malformations and the diagnosis of pulmonary diseases such as cancer and hypertension (Figure 5). The diagnostic determination of shunts within an organ is generally done prior to using radioactive microspheres in radioembolization therapy (see below) [93] in order to prevent radiotoxicity to the lungs [94]. The biological half-life of the albumin particles is only 1 to 3 hours, so any therapeutic interventions can easily be performed afterwards.

***Table** 7: Radioactive microspheres for diagnostic applications*

Application	Type of radioactive microspheres used	Particle size
Gated blood pool study	^{111}In- or ^{51}Cr-labeled red blood cells	6-8 μm
Thrombus imaging in deep vein thrombosis	^{111}In-labeled platelets ^{99m}Tc-macro-aggregated human serum albumin (MAA) ^{99m}Tc-sulfur colloid	0.5-1 μm 10-90 μm 0.05-0.6μm
Blood flow measurements	Polystyrene-microspheres labelled with theγγ-emitters ^{141}Ce, ^{57}Co, ^{114m}In, ^{85}Sr, ^{51}Cr, and others (animal experiments)	10, 15 μm (other sizes)
Investigation of biodistribution and fate of (drug-loaded) microspheres	^{3}H, ^{14}C-labeled microspheres (animal experiments) ^{141}Ce-polystyrene microspheres	all sizes 11.4 μm
Lung scintigraphy	^{99m}Tc-impregnated carbon particles (= Technegas) ^{99m}Tc-macro-aggregated human serum albumin (MAA)	50 nm 10-90 μm
Diagnostic radioembolization	^{99m}Tc-macro-aggregated human serum albumin (MAA)	10-90 μm
Liver and spleen imaging	^{99m}Tc-macro-aggregated human serum albumin (MAA) ^{99m}Tc-sulfur colloid ^{99m}Tc-tin colloid	10-90 μm 0.05-0.6μm 0.05-0.6μm
Bone marrow imaging	^{99m}Tc-sulfur colloid ^{99m}Tc-antimony sulphide colloid	0.05-0.6μm 0.05-0.6μm

Table 7: Radioactive microspheres for diagnostic applications

Application	Type of radioactive microspheres used	Particle size
Infection localisation	^{111}In-labeled leukocytes	12-20 μm
	^{111}In-labeled liposomes	20 nm-1μm
	^{99m}Tc-labeled liposomes	20 nm-1μm
	^{99m}Tc-albumin nanocolloid	<80 nm
Tumour imaging	^{99m}Tc-labeled liposomes	20 nm-1μm
	^{67}Ga-NTA- or ^{111}In-NTA-labeled liposomes	65 nm
Gastrointestinal transit studies	^{99m}Tc-sulfur colloid	0.05-0.6 μm
	^{111}In-labeled ion exchange resins	
Local restenosis prevention in coronary arteries	^{141}Ce microspheres (preliminary imaging tests)	11.4 μm

For liver, spleen, bone marrow and lymphatic system imaging, colloidal microparticles, such as ^{99m}Tc-sulfur colloids, are most useful (Table 7). To illustrate, Figure 6 shows the changes in a cirrhotic patient made visible by ^{99m}Tc-sulfur colloid. The lymphatic system can also be imaged or targeted with drugs through the use of the poly(lysine) nanospheres [95]. The ideal nanospheres for this purpose are 10-30 nm, contain carbohydrate groups on the surface, and are able to bind the γ-emitter ^{111}In via the covalently bound chelator DTPA.

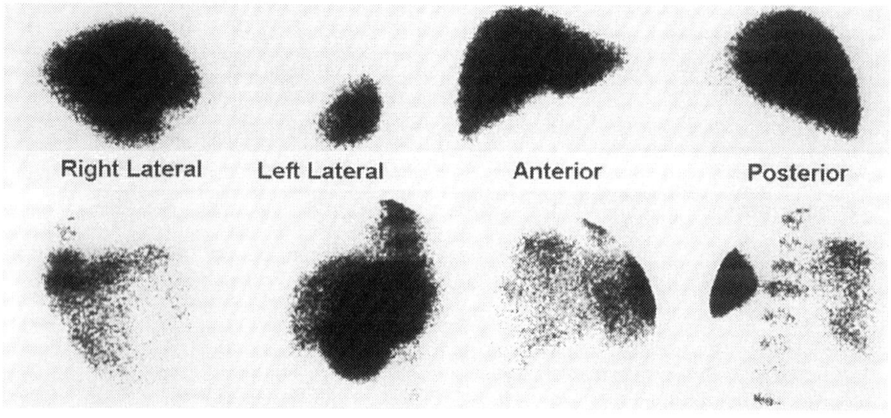

***Figure 6**. Liver scintigraphy performed with ^{99m}Tc-sulfur colloid in different projections. The top row shows a normal liver, the bottom row the corresponding views of a liver from a patient with cirrhosis.*

The radiopharmaceutical ^{99m}Tc-sulfur colloid is also used for gastrointestinal blood loss studies, for the preparation of a ^{99m}Tc-labeled egg sandwich for gastric emptying studies [39], and for the determination of oesophageal transit and gastro-oesophageal reflux. For colonic transit studies, radioisotopes with a half-life longer than ^{99m}Tc are more appropriate, and ^{111}In-labeled ion exchange resins, but also ^{131}I-cellulose are utilised [96]. Latex-particles of 2.5 μm size and labelled with ^{99m}Tc have also been shown to give excellent abdominal images [97]. In all these gastrointestinal transit time studies, the size of the radiolabelled microspheres does not influence the measured times.

Radiolabelled liposomes, another diagnostic class of radioactive particles, has been used for tumour imaging since 1977. In order to prolong the blood residence time and maximise tumour uptake, neutral, positively and negatively charged small unilamellar vesicles (= SUV's) of 65 nm encapsulating ^{111}In were made and their biodistribution measured in mice [98]. The highest uptake of 18.5% of injected dose per gram of tumour was measured with the neutral liposomes. A further attempt to minimise the high blood-background radioactivity levels was to inject ^{67}Ga- or ^{111}In-labeled liposomes containing biotin groups on their surface and then chasing the non-tumour bound liposomes 2 hours later with avidin [99]. This chase removed the unbound liposomes effectively from the circulation and the blood concentration of the radioactive liposomes dropped to a tumour-to-blood ratio of about 15 to 1 shortly after the avidin chase. The avidin-biotin-liposome conglomerates accumulated in the liver and increased the liver activity about 2.5 fold.

Radiolabelled microspheres can also be used to image cancer lesions. An interesting application is the use of PLA-microspheres labelled with ^{131}I-iopanoic acid derivatives for the imaging of liver tumours [44]. The normal liver parenchyma lined with Kupffer cells takes up the microspheres, but the cancer lesions do not possess fixed macrophages and therefore exclude the radioactive microspheres, showing the focal lesions as defects. The recently introduced ^{99m}Tc-PLA microspheres which were radiolabelled in a $SnCl_2$-containing kit could be used for the same application [100], although the stability described as "more than 80% bound after 6 hours" is not optimal yet.

The *in vivo* faith of microspheres after intra-arterial catheter-mediated delivery through a porous balloon to a rabbit's femoral artery has been investigated with radioactive ^{141}Ce-microspheres [101]. Although only 0.14% to 0.16% of the microspheres were delivered to the vessel wall, an average of 92% of these microspheres was still present 7 days later. In addition, much higher amounts of the microspheres were found in the periadventitia (the vessel's "outside") and the overlying musculature and are believed to be caused by the increase of vaso vasora present in atherosclerotic patients. Although the targeted amounts of microspheres are small, they can lead to drug concentrations a few hundred times higher than the serum concentrations, allowing for effective restenosis therapy with microspheres containing cytostatic or antiproliferative agents, especially from the periadventitia side.

6. Therapeutic uses of radioactive microspheres

Many radiolabelled particles, microspheres and liposomes are appropriate for therapy once the encapsulated diagnostic radioisotope has been exchanged for a therapeutic one from the α- or β-emitter group. Typical uses in the last 20 to 40 years include local applications for the treatment of rheumatoid arthritis, liver tumours and cystic brain tumours. However, their use remains experimental because of smaller than expected target uptake, unwanted toxicity and insufficient treatment effects that have resulted from radiochemical instability and suboptimal biodistribution of the radiopharmaceutical. In addition, there exists a general negative attitude towards the use of radioactive substances in spite of proven superior results of many radiation therapies [102-104]. What follows is a review of a few α-emitter applications as well as the more established β−emitter therapies.

6.1. THERAPY WITH ALPHA-EMITTING MICROSPHERES

Different α-emitters have been tested in ovarian cancer mouse models. Microsphere-bound ^{211}At, for example, was applied in mice with ovarian cancer metastases and was found to be more effective than the β-emitting ^{32}P- and ^{90}Y-microspheres [63]. It was, however, also shown that the amount of radioactivity had to be tailored carefully. More than 1 MBq of ^{211}At per animal led to shorter survival times of the treated mice. This effect is very likely due to the instability of ^{211}At which is highly toxic to the lymphatic tissue and thyroid gland when leakage occurs. First clinical trials with the same α-emitter bound to albumin microspheres have been reported by Wunderlich et al. [105]. The authors injected the microspheres into the arteries leading to tongue and larynx tumours. After 4 hours, 80% of the radioactivity was bound to the tongue and 12% to the lungs. The rest was found in the abdomen. The tongue tumour was completely ablated, and no side effects or recurrences were observed at 2 year follow-up. Another α-emitter, ^{212}Pb, in the form of radioactive colloids [106] was also investigated in an ovarian cancer mouse model. Tumour necrosis and decrease in ascites was observed in a dose-related manner, with acute gastro-intestinal toxicity developing at the highest doses. The therapeutically effective radioisotope in these experiments was ^{212}Pb's daughter nuclide ^{212}Bi (Table 2).

To increase the limited range of α-emitters (see general properties of α-emitters), the radiopharmaceutical should be delivered close to the tumour from where it releases the radioisotope, allowing it to diffuse into the surrounding area. Ideally, the released radioactivity binds to the tumour's cell surface and not to the surrounding normal tissue, something that could be accomplished, for example, by pre-targeting the cancer cells with an antibody metallothionein bioconjugate. This approach has been tested *in vitro* with the biodegradable polymer mixture of PHEA (= α,β-poly(hydroxyethyl)-D,L-aspartamide) and Pluronic enclosing ^{212}Bi. Within 1 hour, the polymer began to resemble Swiss cheese, with its many small holes of about 1 μm in diameter (Figure 7). The size of the holes further increased, and after 2 days, more than 75% of the radioactivity had been set free [107]. The polymer tested was in the form of a paste, but microspheric radiopharmaceutical delivery forms are also possible. This approach is limited by the diffusion distances of the radioisotope [108]. In a chopped meat model, ^{212}Bi has been shown to diffuse a maximum distance of 10 mm. It thus may only be useful for the treatment of very small tumours, metastases or leftover tumour cells from incompletely resected tumours.

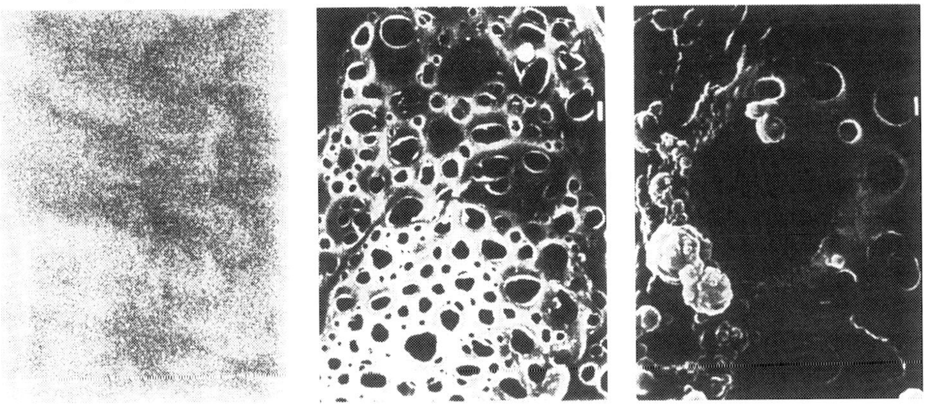

Figure 7. Scanning electron microscopy picture of the biodegradable polymer PHEA/P immediately after preparation (left), after 1 hour (middle) and after 24 hours (right) at 37 °C in PBS pH 7.4. The bar to the right represents 1 μm.

6.2. THERAPY WITH BETA-EMITTING MICROSPHERES

One of the first applications of α-emitting microspheres was the treatment of inaccessible tumours [109]. In this approach termed radioembolization therapy, 20 to 50 μm microspheres are injected into the artery leading to the tumour. Since the microspheres are larger than the newly formed capillaries in the tumour, they are trapped and become lodged in the tumour from where they irradiate the surrounding cancerous tissue with radiation doses 20 to 30 times higher than what is achievable with external radiation therapy. This approach, pioneered with ^{65}Zn- and ^{198}Au-microspheres by Müller and Rossier in Switzerland [26], was further investigated with ^{198}Au- and ^{90}Y-microspheres in many types of tumours by Ariel [110-112] and with ^{32}P-resin microspheres by Caldarola and Dogliotti [113]. Turner et al. investigated ^{166}Ho-labeled cation exchange resins [114] and Häfeli et al. $^{186}Re/^{188}Re$-labeled glass microspheres for the same application [65,115]. Currently, radioembolization therapy is primarily used for the treatment of liver tumours, both hepatomas and liver metastases [116]. Since liver tumours get most of their blood supply from their hepatic artery [117], the radioactive microspheres injected into this artery are preferentially flushed into the tumour. Radiochemically highly stable glass ^{90}Y microspheres sized 25 to 35 μm (Figure 4) are commercially available for the treatment of liver tumours in Canada and since June 1999 also in the United States (Theraspheres™; Nordion, Kanata, Ontario, Canada).

Prior to radioembolization, a diagnostic step is generally performed in order to prevent arterial shunting in the liver. Arterial shunts can divert large amounts of the highly cytotoxic microspheres to the lungs and thus lead to pneumonitis [118]. The diagnostic step consists of determining the "shunt index" by injecting ^{99m}Tc-labeled macro-aggregated albumin microspheres sized between 1 to 10 μm and imaging their biodistribution. If less than 5% of the radioactivity shunts to the lungs, then β-emitting microspheres such as ^{90}Y-glass microspheres [64] or ^{90}Y-resin microspheres [27] are injected into the hepatic artery of the patient. The first results in a disease which carries a grave prognosis with a survival rate of less than 50% after 1 year are very encouraging. After intra-hepatic injection, the microspheres are preferentially taken up by the tumour at an average ratio of about 4 to 1 (tumour to normal liver ratio) [119,120]. Very high radiation doses without side effects can thus be given. Treating 7 patients with doses of 50 to 100 Gy, Houle showed that the larger doses are necessary for successful treatment results [121], and doses between 80 and 150 Gy are now recommended. Further improvements in treatment outcome are possible by injecting the vasoconstricting agent Angiotensin II immediately after the microsphere injection. Normal hepatic vessels are able to react by constriction, but the developing tumour capillaries are not. As a result, larger amounts of the microspheres are diverted to the tumour bed [113]. The clinical results regarding radioembolization therapy have been described in detail by Harbert [109].

Another therapeutic application of β-emitting colloids and microspheres is the radioactive ablation of inflamed synovia in arthritic joints, which has been termed radiosynovectomy or sometimes radiosynoviorthesis. Fellinger and Schmid [122] reported in 1952 the first use of ^{198} Au gold colloid for the treatment of rheumatoid arthritis in knees. Their results were not very encouraging probably due to underdosing, but they did not give up and later confirmed the value of this treatment [123]. Therapy with ^{198}Au has the drawback of a 411 keV γ-emission. To overcome this drawback other β-emitters such as ^{186}Re [124], ^{90}Y [125], ^{165}Dy [46] and ^{32}P-colloids [126] have been investigated. Today, the choice of the radioisotope is entirely based on the size of the joint and the radioisotopes' treatment range (for example, ^{90}Y and ^{188}Re for knee and shoulder, and ^{186}Re and ^{169}Er for finger or elbow) (Table 8).

***Table 8**: Radioactive microspheres for therapeutic applications*

Application	Type of radioactive microspheres used	Particle size
Radioembolization of liver and spleen tumours	^{90}Y-glass microspheres (Theraspheres™)	25-35 μm
	^{186}Re/^{188}Re-glass microspheres	25-35 μm
	^{188}Re-Aminex A27 microspheres	20-50 μm
	^{166}Ho-Aminex A-5 microspheres	13 μm

***Table 8**: Radioactive microspheres for therapeutic applications*

Application	Type of radioactive microspheres used	Particle size
Radiosynovectomy of arthritic joints	^{35}S-colloid	0.05-0.6 μm
	^{90}Y-resin microspheres	20-50 μm
	^{90}Y-silicate, ^{90}Y-citrate	0.01-1 μm
	^{165}Dy-ferric hydroxide macroaggregates	2-5 μm
	^{169}Er-citrate	0.1-1 μm
	^{186}Re-sulfur-colloid	30-50 nm
	^{188}Re-macro-aggregated albumin	10-20 μm
Local radiotherapy	^{90}Y-labeled poly(lactic acid) microspheres	1-5 or 10-50 μm
	^{165}Dy-acetylacetone poly(lactic acid) microspheres	1-5 or 10-50 μm
	^{166}Ho-acetylacetone poly(lactic acid) microspheres	1-5 or 10-50 μm
	$^{186}Re/^{188}Re$-labeled poly(lactic acid) microspheres	1-5 or 10-50 μm
	^{211}At-microspheres	1.8 μm, 3-10 μm
	^{212}Pb-sulfur colloid	<1 μm
	^{212}Pb-ferrous(ferric) hydroxide	<1 μm
Intracavitary treatment (peritoneal ovarian tumour metastases, cystic brain tumours)	chromic ^{32}P-phosphate	1-2 μm
	^{90}Y-silicate, ^{90}Y-citrate	0.01-1 μm
	^{198}Au suspensions	5-25 nm

Traditionally used radioactive colloids are not ideal because their small particle size and large size distribution lead to radiation leakage from the joint [126,127]. Higher than desired leakage has also been measured in liposomes filled with ^{99m}Tc [128] and in liposomes that contain the chelating DTTA-group covalently bound to cholesterol [129]. In the second case, the chelator was incorporated into the liposomes' phospholipid-wall during preparation and was then able to bind different radioisotopes such as the β-emitter ^{177}Lu (Table 3) and the γ-emitter ^{67}Ga (Table 4).
More radiochemically stable and better-defined microspheres of about 5 μm seem to be optimal for retention in joints. Many of the recently developed microspheres such as biodegradable glass microspheres containing ^{153}Sm, ^{166}Ho, ^{90}Y, ^{165}Dy, ^{186}Re or ^{188}Re

[130], ^{188}Re-labeled albumin microspheres [78], ^{166}Ho- or ^{165}Dy-enclosing biodegradable poly(lactic acid) microspheres [67,131] and ^{90}Y- or $^{186}Re/^{188}Re$-enclosing biodegradable poly(lactic acid) microspheres [79,132] can be produced in the appropriate size, will biodegrade after complete decay and can easily be made radioactive. More information about radiosynovectomy is available in an extensive review written by Harbert [133]. It covers the medical applications and procedures in detail. The radiation dosimetry of radiosynovectomy is covered by Johnson et al. [30].

Another important area for β-emitting microspheres is their use in the local treatment of tumours. The delivery of these radioactive microspheres has been attempted in several ways. In one of them, radioactive microspheres are directly injected into the tumour. Wang et al., for example, radiolabelled ion exchange resin microspheres with ^{188}Re and injected them directly into rat hepatomas [134]. Twelve out of 15 rats survived longer than 60 days in the treatment group, as compared to 5 out of 15 rats in the control group. In another novel method for the treatment of solid tumours, Order et al. combined embolization therapy and local radiotherapy, injecting first non-radioactive macro-aggregated albumin microspheres followed by colloidal ^{32}P-chromic phosphate [135]. The blockage of the capillaries induced before the ^{32}P-injection resulted in a 3-fold increase of colloid uptake, an effect that lasted for at least 48 hours. This technique has been tested in a first clinical phase I trial for the treatment of non-resectable pancreatic cancer [136]. Four patients had a complete response with a duration ranging from 2-57 weeks and 5 patients had a partial response with a duration ranging from 4-21 weeks, corresponding to an objective response of 53% (9 of 17 patients). Six of these patients were alive 33-57 weeks after treatment.

At the current time, there is only one approved application for radioactive microspheres in the United States [137]. It is the use of ^{32}P-chromic phosphate colloid for the treatment of cystic brain tumours such as craniopharyngiomas and astrocytomas. The radiocolloid is typically instilled using stereotaxic equipment, either with or without surgical resection or drainage of the cyst. There exists persuasive evidence that this therapeutic approach is as or more efficacious than conventional methods not only for patients with recurrent malignancies, but also for patients receiving primary radiocolloid therapy [138]. Radioactive glass- and poly(lactic acid)microspheres containing a mixture of ^{186}Re and ^{188}Re have recently been incorporated into a bioadhesive gel of either carboxymethylcellulose or fibrin glue and applied to the surface of growing rat 9L-glioblastomas [139]. The control group's survival was 18 days, whereas 4 out 6 of the treated animals were still alive on day 35, which represented the end of the experiment (Figure 8). The amount of radioactive ^{186}Re and ^{188}Re injected was less than 50 μCi combined. The surviving animals showed no signs of toxicity and had not lost any weight. Such microspheres are now planned for a clinical phase I trial of the treatment of recurrent brain tumour metastases intraoperatively after debulking. This therapeutic approach looks especially promising because the likelihood of local recurrence in these patients is very high [140], and the local radiation with γ-emitters could be done in addition to chemotherapy or whole brain irradiation without risking undue toxicity.

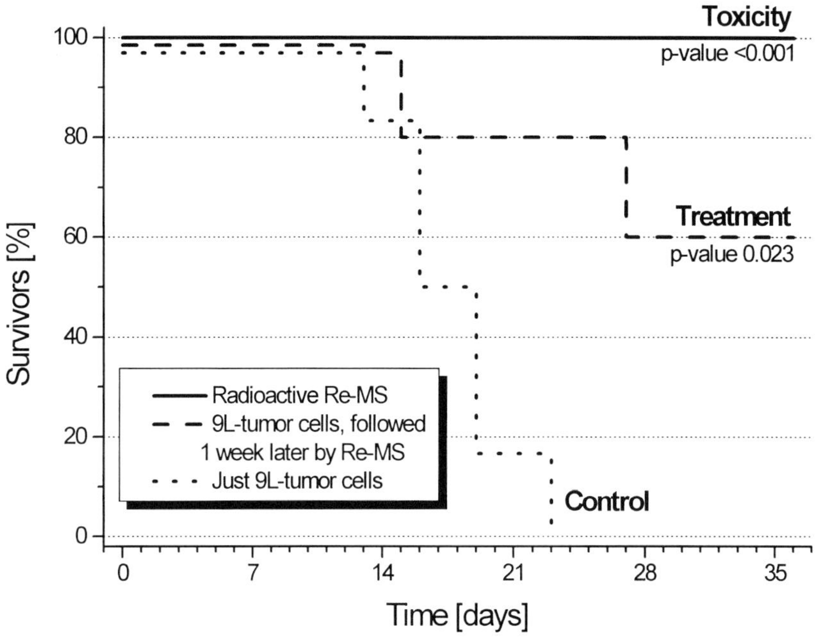

Figure 8. *Treatment of 9L-glioblastoma brain tumours in Sprague Dawley rats. The treatment and toxicity group received 50 μCi* ^{186}*Re and* ^{188}*Re in 0.5 mg glass microspheres contained in 30 μl of fibrin glue.*

Radioactive microspheres filled with magnetite and radiolabelled with the β-emitter ^{90}Y can also be used for targeted cancer therapy. This has been shown with 30% magnetite-containing poly(lactic acid) microspheres sized 20 to 30 μm that were injected intraperitoneally into C57BL6/N mice and targeted to a subcutaneously growing EL-4 murine lymphoma of about 0.5 g [141]. The injection of microspheres took place inside the peritoneal cavity as far from the tumour as possible. After injection, a round, 2 mm thick rare earth magnet with a diameter of 10 mm was taped directly above the tumour. The magnetic field on top of the magnet was 0.12-0.16 Tesla. A dose dependent decrease in tumour size was observed after the 7 day treatment period (Figure 9). Close examination revealed that 3 out of 4 tumours in the 80 Gy group and 2 out of 4 tumours in the 120 Gy group were completely eradicated, but that the remaining 1 or 2 tumours, respectively, had grown. It was precisely these tumours that had initially been found to be oblong or flattened out, thus causing the magnetic microspheres to be concentrated farther than 5 mm away from the edges of the tumour. Considering that 90% of the dose of ^{90}Y is deposited within 2.8 mm [30], it follows that the tumour cells farther away were undertreated with the applied amount of radioactivity. The tumours which were not eradicated were therefore local treatment failures.

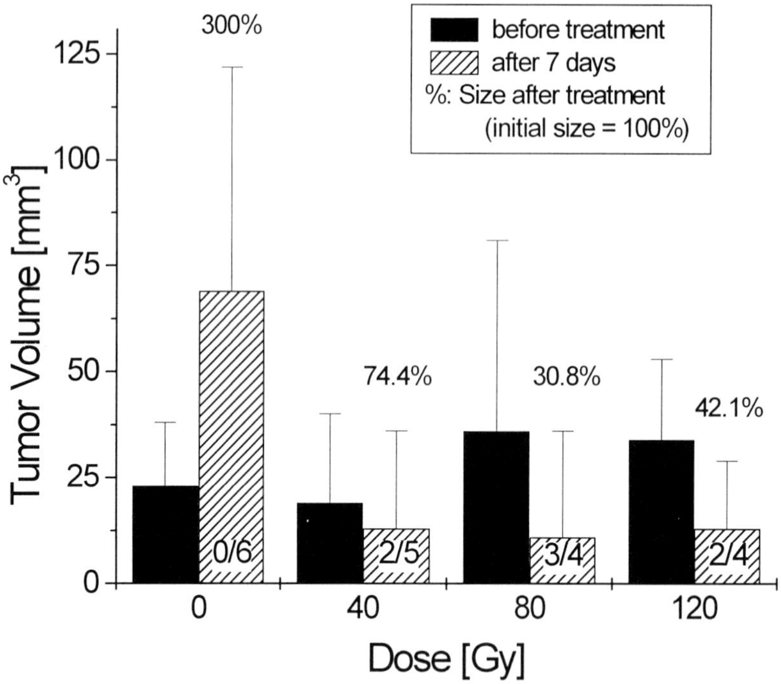

Figure 9. Treatment results of subcutaneous EL-4 lymphomas in mice after magnetic targeting of ^{90}Y-PLA microspheres (n = 6). The numbers inside the bars represent the ratio of completely eradicated tumours to the total number of tumours.

7. Considerations for the use of radioactive microspheres

The recent surge in the evaluation and clinical testing of radiopharmaceuticals is closely related to the recent development of user-friendly kits which allow the user to prepare radioactive microspheres or other radiolabelled agents in a hospital's radiopharmacy. These kits have served to reduce concerns about the safety, cost and handling of radioactive pharmaceuticals. Current manipulations needed in most kit preparations typically include the addition of a radioisotope, incubation for a predetermined length of time between 5 and 60 minutes, verification of the activity of the radiopharmaceutical by a simple measurement in a dose calibrator and, sometimes, a thin layer chromatogram for quality control. Ideally, the kit preparation leads to highly stable radioactive microspheres with no purification needed. Saha gives additional information in an excellent up-to-date introduction into the currently used radionuclides, radionuclide generators and radiopharmaceuticals in a nuclear pharmacy [142]. Saha covers not only all technical aspects of a nuclear pharmacy, but also the radiation regulations and radiation protection aspects.

Increased interest in radiopharmaceuticals is also explained by easier access to generator-produced α- and β-emitting radioisotopes. The β-emitter ^{188}Re, for example (Table 3), can now be inexpensively obtained by any hospital radiopharmacy, in the form of a $^{188}W/^{188}Re$ generator from Oak Ridge National Laboratories. This generator contains the parent nuclide ^{188}W with a half-life of 69.4 days permanently bound to an alumina column. Because of ongoing decay into the daughter-nuclide ^{188}Re, hundreds of mCi of a sterile ^{188}Re-solution can be eluted from the column every day over the course of about 3 months [143]. Rhenium-188 is currently being tested in clinical trials for the radioactive treatment of restenosis [144], for local cancer therapy [134,145,146], for radioembolization therapy [147] and for radioimmunotherapy [148].

Two of the most important parameters for the in vivo use of radiopharmaceuticals are their stability and target specificity. The stability can be improved by using more specific chelators and by attaching the chelators to the microspheres through a linker that does not interfere with their metal-binding properties. Many of these stability issues have already been optimised for radiolabelled antibodies [149] and can thus be directly applied to radioactive microspheres and their kit preparation. The second parameter, target specificity, can be addressed by using additional surface chemistry to modify and functionalise the microspheres' surface. This allows for circulation time optimisation and foreign body response minimisation. In addition, more specific targeting of microspheres to areas other than the reticuloendothelial system (mainly liver and spleen) is possible, as well as the modification of the microspheres' adsorption behaviour to blood proteins [150].

Also important for the application of therapeutic radioactive microspheres is that the radioisotope be chosen for radiobiological and dosimetric reasons. The target size should, for example, be matched with the radiation range of the radioisotope, thus maximising the therapeutic effect and minimising the toxicity [151]. Also, dose rates should be taken into account and radioisotopes chosen so that dose rate and total dose deposited are optimal for the target lesion [152]. These parameters are not yet well established, but are nevertheless important and should be investigated further.

Another area for the optimisation of radioactive microspheres is research into the most appropriate size of microspheres for *in vivo* application. It has long been known that differently sized microspheres of identical composition show different biodistribution profiles [153,154]. In addition, for both diffusion and erosion release mechanisms, the release rate of the encapsulated drugs is theoretically dependent on the available surface area. Smaller microspheres with a much larger surface area should thus release their contents at a much faster rate with all the other parameters being equal. Unfortunately, no homogenous, mono-sized microspheres made from biodegradable materials are currently available.

The use of radioactive microspheres is the basis of a large variety of well-established and original concepts for future biomedical, diagnostic and therapeutic applications. Optimally, radioactive microspheres should be combined with biologically active molecules such as proteins, peptides, hormones, lectins, and antibodies. This will allow for the diagnosis and treatment of many different diseases with microsurgical precision and will lead to better treatment concepts with fewer side effects.

References

1. Wilder RB, DeNardo GL, and DeNardo SJ. Radioimmunotherapy: Recent results and future directions. *J. Clin. Oncol.* **14:** 1383-1400 (1996).
2. Papatheofanis FJ and Munson L. Peptide radiopharmaceutical imaging. *Appl. Radiol.* **June:** 11-17 (1994).
3. Cleland JL and Jones AJS. Stable formulations of recombinant human growth hormone and interferon-ô for microencapsulation in biodegradable microspheres. *Pharmaceut. Res.* **13:** 1464-1475 (1996).
4. Mehta RC, Jeyanthi R, Calis S, Thanoo BC, Burton KW, and DeLuca PP. Biodegradable microspheres as depot system for parenteral delivery of peptide drugs. *J. Contr. Rel.* **29:** 375-384 (1994).
5. Mathiowitz E, Jacob JS, Jong YS, Carino GP, Chickering DE, Chaturvedi P, Santos CA, Vijayaraghavan K, Montgomery S, Bassett M, and Morrell C. Biologically erodable microspheres as potential oral drug delivery systems. *Nature* **386:** 410-414 (1997).
6. Langer R. Drug delivery and targeting. *Nature* **392:** 5-10 (1998).
7. Chen H and Langer R. Oral particulate delivery: status and future trends. *Adv. Drug Del. Rev.* **34:** 339-350 (1998).
8. Muir W, Husband AJ, Gipps EM, and Bradley MP. Induction of specific IgA responses in rats after oral vaccination with biodegradable microspheres containing a recombinant protein. *Immunol. Lett.* **42:** 203-207 (1994).
9. Hanes J, Chiba M, and Langer R. Polymer microspheres for vaccine delivery. *Pharm. Biotech.* **6:** 389-412 (1995).
10. Smith OP, Hann IM, Cox H, and Novelli V. Visceral leishmaniasis: rapid response to AmBisome treatment. *Arch. Dis. Childhood* **73:** 157-159 (1995).
11. Codde JP, Lumsden AJ, Napoli S, Burton MA, and Gray BN. A comparative study of the anticancer efficacy of doxorubicin carrying microspheres and liposomes using a rat liver tumour model. *Anticancer Research* **13:** 539-544 (1993).
12. Treleaven JG. Bone marrow purging: An appraisal of immunological and non-immunological methods. *Adv. Drug Del. Rev.* **2/3:** 253-269 (1988).
13. Arshady R. Polymer supports, reagents and catalysts. *In* Arshady R (Ed.). Microspheres, microcapsules and liposomes. Citus Books, London, 1999, pp. 197-235.
14. Mikhalovsky SV. Microparticles for haemoperfusion and extracorporeal therapy. *In* Arshady R (Ed.). Microspheres, microcapsules and liposomes. Citus Books, London, 1999, pp. 133-169.
15. Bangs LB. Microspheres for medical diagnostics: Specific tests and assays. *In* Arshady R (Ed.). Microspheres, microcapsules and liposomes. Citus Books, London, 1999, pp. 71-96.
16. Flandroy PMJ, Grandfils C, and Jerome RJ. Clinical applications of microspheres in embolization and chemoembolisation: A comprehensive review and perspectives. *In* Rolland A (Ed.). Pharmaceutical particulate carriers: Therapeutic applications. Marcel Dekker Inc., New York, 1993, pp. 321-366.
17. Boschetti E and Schwarz A. Polymer microbeads: Biological applications. *In* Arshady R (Ed.). Microspheres, microcapsules and liposomes. Citus Books, London, 1999, pp. 191-224.
18. Papisov MI. Modelling in vivo transfer of long-circulating polymers (two classes of long circulating polymers and factors affecting their transfer in vivo). *Adv. Drug Del. Rev.* **16:** 127-139 (1995).
19. Torchilin VP and Trubetskoy VS. Which polymers can make nanoparticulate drug carriers long-circulating? *Adv. Drug Del. Rev.* **16:** 141-155 (1995).
20. Ackerman NB. The blood supply of experimental liver metastases. IV. Changes in vascularity with increasing tumour growth. *Surgery* **75:** 589-596 (1974).
21. Gupta PK. Review article: Drug targeting in cancer chemotherapy: A clinical perspective. *J. Pharm. Sci.* **79:** 949-962 (1990).
22. Roser M, Fischer D, and Kissel T. Surface-modified biodegradable albumin nano- and microspheres. Part II: Effect of surface charges on in vitro phagocytosis and biodistribution in rats. *Eur. J. Pharm. Biopharm.* **46:** 255-263 (1998).
23. Macklis RM, Kinsey BM, Kassis AI, Ferrara JLM, Atcher RW, Hines JJ, Coleman CN, Adelstein SJ, and Burakoff SJ. Radioimmunotherapy with alpha-particle-emitting immunoconjugates. *Science* **240:** 1024-1026 (1988).
24. Humm JL, Macklis RM, Bump K, Cobb LM, and Chin LM. Internal dosimetry using data derived from autoradiographs. *J. Nucl. Med.* **34:** 1811-1817 (1993).

25. McDevitt MR, Sgouros G, Finn RD, Humm JL, Jurcic JG, Larson SM, and Scheinberg DA. Radioimmunotherapy with alpha-emitting nuclides. *Eur. J. Nucl. Med.* **25:** 1341-1351 (1998).
26. Muller JH and Rossier PH. A new method for the treatment of cancer of the lungs by means of artificial radioactivity. *Acta Radiologica* **35:** 449-468 (1951).
27. Rösler H, Triller J, Baer HU, Geiger L, Beer HF, Becker C, and Blumgart LH. Superselective radioembolization of hepatocellular carcinoma: 5-year results of a prospective study. *Nucl. Med.* **33:** 206-214 (1994).
28. Hall EJ and Brenner DJ. The dose-rate effect in interstitial brachytherapy: a controversy resolved. *Brit. J. Radiology.* **65:** 242-247 (1992).
29. Kinsey RR. National Nuclear Data Center: Nuclear Data from NuDat at Brookhaven National Laboratory [http://www.nndc.bnl.gov/nndc/nudat/]. : (1998).
30. Johnson LS, Yanch JC, Shortkroff S, Barnes CL, Spitzer AI, and Sledge CB. Beta-particle dosimetry in radiation synovectomy. *Eur. J. Nucl. Med.* **22:** 977-988 (1995).
31. Loevinger R, Budinger TF, and Watson EE. MIRD primer for absorbed dose calculations. Society of Nuclear Medicine, New York, 1991.
32. Russell JL, Carden JL, and Herron L. Dosimetry calculations for Yttrium-90 used in the treatment of liver cancer. *Endocurietherapy/Hyperthermia Oncology* **4:** 171-186 (1988).
33. Stabin MG. MIRDOSE: Personal computer software for internal dose assessment in nuclear medicine. *J. Nucl. Med.* **37:** 538-546 (1996).
34. Harbert JC, Eckelman WC, and Neumann RD. Nuclear medicine: Diagnosis and therapy. Thieme Medical Publishers, New York, 1996.
35. Bardies M, Lame J, Myers MJ, and Simoen JP. A simplified approach to beta dosimetry for small spheres labelled on the surface. *Phys. Med. Biol.* **35:** 1039-1050 (1990).
36. Akabani G, Poston JW, and Bolch WE. Estimates of beta absorbed fractions in small tissue volumes for selected radionuclides. *J. Nucl. Med.* **32:** 835-839 (1991).
37. Siegel JA and Stabin MG. Absorbed fractions for electrons and beta particles in spheres of various sizes. *J. Nucl. Med.* **35:** 152-156 (1994).
38. Duncan R, Kopeckova-Rejmanova P, Strohalm J, Hume I, Cable HC, Pohl J, Lloyd JB, and Kopecek J. Anticancer agents coupled to N-(2-hydroxypropyl)methacrylamide copolymers. I. Evaluation of daunomycin and puromycin conjugates in vitro. *Br. J. Cancer* **55:** 165-174 (1987).
39. Nelp WB. Evaluation of colloids for RES function studies. *In* Subramanian G, Rhodes B, Cooper JF, and Sodd VJ (Eds.). . Society of Nuclear Medicine, New York, NY, 1975, pp. 349-356.
40. Goodwin DA, Stern HS, and Wagner HN. Ferric hydroxide particles labelled with indium In-113m for lung scanning. *JAMA* **206:** 339-343 (1968).
41. Lin MS and Winchell HS. A "kit" method for the preparation of technetium-tin(II) colloid and a study of its properties. *J. Nucl. Med.* **13:** 58-65 (1972).
42. Zolle I. Method for incorporating substances into protein microspheres. US Patent No. 3937668, 1976.
43. Amersham. Guide to radioiodination techniques: Iodine-125. Amersham International, Little Chalfont, England, 1993.
44. Yang DJ, Kuang LR, Li C, Kan Z, and Wallace S. Computed tomographic liver enhancement with poly(d,l-lactide)-microencapsulated contrast media. *Invest. Radiol.* **29 Suppl. 2:** S267-S270 (1994).
45. Häfeli U, Tiefenauer LX, Schubiger PA, and Weder HG. A lipophilic complex with ^{186}Re/^{188}Re incorporated in liposomes suitable for radiotherapy. *Nucl. Med. Biol. Int. J. Rad. Appl. Instr. Part B* **18:** 449-454 (1991).
46. Sledge CB, Noble J, Hnatowich DJ, Kramer RT, and Shortkroff S. Experimental radiation synovectomy by ^{165}Dy ferric hydroxide macroaggregate. *Arthritis Rheum.* **20:** 1334-1342 (1977).
47. Howson MP, Shepard NL, and Mitchell NS. Colloidal chromic phosphate P-32 synovectomy in antigen-induced arthritis in the rabbit. *Clin. Orthopaed. Rel. Res.* **229:** 283-293 (1988).
48. Gürkan H, Yalabik-Kas HS, Hincal AA, and Ercan MT. Streptomycin sulphate microspheres. Formulation and in vivo distribution. *J. Microencapsulation* **3:** 101-108 (1986).
49. Reza MS and Whateley TL. Iodo-2'-deoxyuridine (IUdR) and I-125-IUdR loaded biodegradable microspheres for controlled delivery to the brain. *J. Microencapsulation* **15:** 789-801 (1998).
50. Senyei AE and Widder KJ. Drug Targeting: Magnetically responsive albumin microspheres - a review of the system to date. *Gynecol. Oncol.* **12:** 1-13 (1981).

51. Teder H, Johansson CJ, d'Argy R, Lundin N, and Gunnarsson PO. The effect of different dose levels of degradable starch microspheres (Spherex) on the distribution of a cytotoxic drug after regional administration to tumour-bearing rats. *Europ. J. Cancer* **31A:** 1701-1705 (1995).
52. Burch WM, Sullivan PJ, and McLaren CJ. Technegas - a new ventilation agent for lung scanning. *Nuclear Medicine Communications* **7:** 865-871 (1986).
53. Schubiger PA, Beer HF, Geiger L, Rösler H, Zimmermann A, Triller J, Mettler D, and Schilt W. ^{90}Y-resin particles - Animal experiments on pigs with regard to the introduction of superselective embolization therapy. *Nucl. Med. Biol. Int. J. Rad. Appl. Instr. Part B* **18:** 305-311 (1991).
54. Quinlan MF, Salman SD, Swift DL, Wagner HN, and Proctor DF. Measurement of mucociliary function in man. *Am. Rev. Respir. Dis.* **99:** 13-23 (1969).
55. Stivland T, Camilleri M, Vassallo M, Proano M, Rath D, Brown M, Thomforde G, Pemberton J, and Phillips S. Scintigraphic measurement of regional gut transit in idiopathic constipation. *Gastroenterology* **101:** 107-115 (1991).
56. Simon H, Drettner B, and Jung B. Messung des Schleimhauttransportes in der menschlichen Nase mit Cr-51 markierten Harzkügelchen. *Acta Otolaryngol.* **83:** 378-390 (1977).
57. Zimmermann A, Schubiger PA, Mettler D, Geiger L, Triller J, and Rösler H. Renal pathology after arterial Y-90 microsphere administration in pigs: A model for superselective radioembolization therapy. *Invest. Radiol.* **30:** 716-723 (1995).
58. Willmott N, Murray T, Carlton R, Chen Y, Logan H, McCurrach G, Bessent RG, Goldberg JA, Anderson J, McKillop JH, and McArdle CS. Development of radiolabelled albumin microspheres: A comparison of gamma-emitting radioisotopes of Iodine (^{131}I) and Indium (^{111}In/^{113m}In). *Nucl. Med. Biol. Int. J. Rad. Appl. Instr. Part B* **18:** 687-694 (1991).
59. Wagner SJ and Welch MJ. Gallium-68 labelling of albumin and albumin microspheres. *J. Nucl. Med.* **20:** 428-433 (1979).
60. Ercan MT, Tuncel SA, Caner BE, and Piskin E. Tc-99m-labeled monodisperse latex particles with amine or carboxylic functional groups for colon transit studies. *J. Microencapsulation* **10:** 67-76 (1993).
61. Day DE, Ehrhardt GJ, and Zinn KR. radiolabelled protein composition and method for radiation synovectomy. U.S.A. Patent No. 5403573, 1995.
62. Yan C, Li X, Chen X, Wang D, Zhong D, Tan T, and Kitano H. Anticancer gelatine microspheres with multiple functions. *Biomaterials* **12:** 640-644 (1991).
63. Vergote I, Larsen RH, de Vos L, Nesland JM, Bruland O, Bjorgum J, Alstad J, Trope C, and Nustad K. Therapeutic efficacy of the á-emitter ^{211}At bound on microspheres compared with ^{90}Y and ^{32}P colloids in a murine intraperitoneal tumour model. *Gynecol. Oncol.* **47:** 366-372 (1992).
64. Ehrhardt GJ and Day DE. Therapeutic use of 90Y microspheres. *Nucl. Med. Biol. Int. J. Rad. Appl. Instr. Part B* **14:** 233-242 (1987).
65. Conzone SD, Häfeli UO, Day DE, and Ehrhardt GJ. Preparation and properties of radioactive rhenium glass microspheres intended for in-vivo radioembolization therapy. *J. Biomed. Mat. Res.* **42:** 617-625 (1998).
66. Brown RF, Lindesmith LC, and Day DE. ^{166}Holmium-containing glass for internal radiotherapy of tumours. *Nucl. Med. Biol. Int. J. Rad. Appl. Instr. Part B* **18:** 783-790 (1991).
67. Mumper RJ and Jay M. Poly(L-lactic acid) microspheres containing neutron-activatable Holmium-165: A study of the physical characteristics of microspheres before and after irradiation in a nuclear reactor. *Pharmaceut. Res.* **9:** 149-154 (1992).
68. Nijsen JFW, Zonnenberg BA, Woittiez JRW, Rook DW, Swildens-van Woudenberg IA, van Rijk PP, and van het Schip AD. Holmium-166 poly lactic acid microspheres applicable for intra-arterial radionuclide therapy of hepatic malignancies: effects of preparation and neutron activation techniques. *Eur. J. Nucl. Med.* **26:** 699-704 (1999).
69. Häfeli UO, Roberts WK, Pauer GJ, Kraeft SK, and Macklis RM. Preparation and stability of biodegradable radioactive rhenium microspheres (Re-186 and Re-188) for use in radiotherapy. *J. Pharm. Sci.* **submitted:** (1999).
70. Ercan MT. Rapid determination of hydrolysed-reduced Technetium-99m in particulate radiopharmaceuticals. *Appl. Radiat. Isot. - Int. J. Radiat. Appl. Instrum. Part A* **43:** 1175-1177 (1992).

71. Chinol M, Vallabhajosula S, Goldsmith SJ, Klein MJ, Deutsch KF, Chinen LK, Brodack JW, Deutsch EA, Watson BA, and Tofe AJ. Chemistry and biological behaviour of Samarium-153 and Rhenium-186-labeled hydroxyapatite particles: Potential radiopharmaceuticals for radiation synovectomy. *J. Nucl. Med.* **34:** 1536-1542 (1993).
72. Junghans RP, Dobbs D, Brechbiel MW, Mirzadeh S, Raubitschek AA, Gansow OA, and Waldmann TA. Pharmacokinetics and bioactivity of 1,4,7,10-tetra-azacyclododecane N,N',N'',N'''-tetraacetic acid (DOTA)-bismuth-conjugated anti-Tac antibody for á-emitter 212Bi therapy. *Cancer Res.* **53:** 5683-5689 (1993).
73. Camera L, Kinuya S, Garmestani K, Wu C, Brechbiel MW, Pai LH, McMurry TJ, Gansow OA, Pastan I, Paik CH, and Carrasquillo JA. Evaluation of the serum stability and in vivo biodistribution of CHX-DTPA and other ligands for Yttrium labelling of monoclonal antibodies. *J. Nucl. Med.* **35:** 882-889 (1994).
74. Fritzberg AR. Radioimmunotherapy with Rhenium-186 and Rhenium-188. *In* Bryskin BD (Ed.). Rhenium and Rhenium Alloys. TMS (Minerals, Metals and Materials Society), Warrendale, PA, 1997, pp. 479-487.
75. Fritzberg AR, Abrams PG, Beaumier PL, Kasina S, Morgan AC, Rao TN, Reno JM, Sanderson JA, Srinivasan A, Wilbur DS, and Vanderheyden JL. Specific and stable labelling of antibodies with Tc-99m with a dialled dithiolate chelating agent. *Proc. Natl. Acad. Sci. USA* **85:** 4025-4029 (1988).
76. Eckelman WC, Steigman J, and Paik CH. Radiopharmaceutical chemistry. *In* Harbert JC, Eckelman WC, and Neumann RD (Eds.). Nuclear medicine: Diagnosis and therapy. Thieme Medical Publishers, New York, 1996, pp. 213-266.
77. Griffiths GL, Goldenberg DM, Diril H, and Hansen HJ. Technetium-99m, Rhenium-186, and Rhenium-188 direct labelled antibodies. *Cancer* **73:** 761-768 (1994).
78. Wunderlich G, Pinkert J, and Franke WG. Studies on the processing and in vivo stability of Re-188 labelled microspheres. *In* Nicolini M and Mazzi U (Eds.). Technetium, rhenium and other metals in chemistry and nuclear medicine. SGE Ditoriali, Padova, Italy, 1999, pp. 709-712.
79. Häfeli UO, Sweeney SM, Beresford BS, Sim EH, and Macklis RM. Biodegradable magnetically directed ^{90}Y-microspheres: Novel agents for targeted intracavitary radiotherapy. *J. Biomed. Mat. Res.* **28:** 901-908 (1994).
80. Day DE and Day TE. Radiotherapy Glasses. *In* Hench LL and Wilson J (Eds.). An Introduction to Bioceramics. World Scientific, New Jersey, 1993, pp. 305-317.
81. Conzone SD. Glass microspheres for medical applications. *Ph.D. thesis*, University of Missouri, Rolla, 1999.
82. Locher GL. Biological effects and therapeutic possibilities of neutrons. *AJR* **36:** 1-13 (1936).
83. Mehta SC and Lu DR. Targeted drug delivery for boron neutron capture therapy. *Pharmaceut. Res.* **13:** 344-351 (1996).
84. Akine Y, Tokita N, Tokuuye K, Satoh M, Fukumori Y, Tokumitsu H, Kanamori R, Kobayashi T, and Kanda K. Neutron capture therapy of murine ascites tumour with gadolinium-containing microcapsules. *J. Cancer Res. Clin. Oncol.* **119:** 71-73 (1992).
85. Tokumitsu H, Ichikawa H, Fukumori Y, and Block LH. Preparation of gadoptentetic acid-loaded chitosan microparticles for gadolinium neutron capture therapy of cancer by a novel emulsion-droplet coalescence technique. *Chem. Pharm. Bull.* **47:** 838-842 (1999).
86. Yanagie H, Tomita T, Kobayashi H, Fujii Y, Nonaka Y, Saegusa Y, Hasumi K, Eriguchi M, Kobayashi T, and Ono K. Inhibition of human pancreatic cancer growth in nude mice by boron neutron capture therapy. *Br. J. Cancer* **75:** 660-665 (1997).
87. Rawls RL. Bringing boron to bear on cancer. *C&EN* **March 22:** 26-29 (1999).
88. Hawthorne MF and Shelly K. Liposomes as drug delivery vehicles for boron agents. *J. Neuro-Oncol.* **33:** 53-58 (1997).
89. Heymann MA, Payne BD, Hoffman JI, and Rudolph AM. Blood flow measurements with radionuclide-labelled particles. *Progress in Cardiovascular Diseases* **20:** 55-79 (1977).
90. Peters AM, Danpure HJ, Osman S, Hawker RJ, Henderson BL, Hodgson HJ, Kelly JD, Neirinckx RD, and Lavender JP. Clinical experience with Tc-99m-hexamethyl propylene amine oxime for labelling leukocytes and imaging inflammation. *The Lancet* **2:** 946-949 (1986).
91. Knight L. Thrombus-localising radiopharmaceuticals. *In* Fritzberg AR (Ed.). Radiopharmaceuticals: Progress and clinical perspectives. CRC Press, Boca Raton, Florida, 1986, pp. 23-40.

92. Marcus ML, Heistad DD, Ehrhardt JC, and Abboud FM. Total and regional cerebral blood flow measurement with 7- 10-, 15-, 25-, and 50-ìm microspheres. *J. Appl. Physiol.* **40(4):** 501-507 (1976).
93. Triller J, Rösler H, Geiger L, and Baer HU. Methodik der superselektiven Radioembolisation von Lebertumoren mit Yttrium-90-Resin-Partikeln. *Fortschr. Röntgenstr.* **160:** 425-431 (1994).
94. Lin M. Radiation pneumonitis caused by Yttrium-90 microspheres: Radiologic findings. *AJR* **162:** 1300-1302 (1994).
95. Papisov MI and Brady TJ. System of drug delivery to the lymphatic tissues. U.S.A. Patent No. 5582172, 1996.
96. Frier M and Perkins AC. Radiopharmaceuticals and the gastrointestinal tract. *Eur. J. Nucl. Med.* **21:** 1234-1242 (1994).
97. Caner BE, Ercan MT, Kapucu LO, Tuncel SA, Bekdik CF, Erbengi G, and Piskin E. Functional assessment of human gastrointestinal tract using Tc-99m-latex particles. *Nuclear Medicine Communications* **12:** 539-544 (1991).
98. Proffitt RT, Williams LE, Presant CA, Tin GW, Uliana JA, Gamble RC, and Baldeschwieler JD. Tumour-imaging potential of liposomes loaded with 111In-NTA: Biodistribution in mice. *J. Nucl. Med.* **24:** 45-51 (1983).
99. Ogihara-Umeda I, Sasaki T, and Nishigori H. Development of a liposome-encapsulated radionuclide with preferential tumour accumulation - the choice of radionuclide and chelating ligand. *Nucl. Med. Biol. Int. J. Rad. Appl. Instr. Part B* **19:** 753-757 (1992).
100. Diaz RV, Mallol J, Delgado A, Soriano I, and Evora C. Tc-99m microspheres based on biodegradable synthetic polymers (MSP). A new patented radiopharmaceutical. *Poster* **:** (1997).
101. Wilensky RL, March KL, Gradus-Pizlo I, Schauwecker D, Michaels MB, Robinson J, Carlson K, and Hathaway DR. Regional and arterial localisation of radioactive microparticles after local delivery by unsupported or supported porous balloon catheters. *Am. Heart J.* **129:** 852-859 (1995).
102. Molho P, Verrier P, Stieltjes N, Schacher JM, Ounnoughene N, Vassilieff D, Menkes CJ, and Sultan Y. A retrospective study on chemical and radioactive synovectomy in severe haemophilia patients with recurrent haemarthrosis. *Haemophilia* **5:** 115-123 (1999).
103. Keys HM, Bundy BN, Stehman FB, Muderspach LI, Chafe WE, Suggs CL, Walker JL, and Gersell D. Cisplatin, radiation, and adjuvant hysterectomy compared with radiation and adjuvant hysterectomy for bulky stage IB cervical carcinoma. *The New England Journal of Medicine* **340:** 1154-1161 (1999).
104. Rose PG, Bundy BN, Watkins EB, Thigpen JT, Deppe G, Maiman MA, Clarke-Pearson DL, and Insalaco S. Concurrent cisplatin-based radiotherapy and chemotherapy for locally advanced cervical cancer. *The New England Journal of Medicine* **340:** 1144-1153 (1999).
105. Wunderlich G, Franke WG, Doberenz I, and Fischer S. Two ways to establish potential At-211 radiopharmaceuticals. *Anticancer Research* **17:** 1809-1814 (1997).
106. Rotmensch J, Atcher RW, Schlenker R, Hines J, Grdina D, Block BS, Press MF, Herbst AL, and Weichselbaum RR. The effect of the á-emitting radionuclide Lead-212 on human ovarian carcinoma: a potential new form of therapy. *Gynecol. Oncol.* **32:** 236-239 (1989).
107. Häfeli U, Atcher RW, Morris CE, Beresford B, Humm JL, and Macklis RM. Polymeric radiopharmaceutical delivery systems. *Radioactivity & Radiochemistry* **3:** 11-14 (1992).
108. Macklis RM, Atcher R, Morris C, Beresford B, Häfeli U, and Humm J. Controlled release biodegradable radiopolymers for intracavitary radiotherapy using a Pb-212 alpha emitting generator system. **:** (1992).
109. Harbert JC. Therapy with intra-arterial radioactive particles. *In* Harbert JC, Eckelman WC, and Neumann RD (Eds.). Nuclear medicine: Diagnosis and therapy. Thieme Medical Publishers, New York, 1996, pp. 1141-1155.
110. Ariel IM. The treatment of metastases to the liver with interstitial radioactive isotopes. *Surgery, Gynecology & Obstetrics* **110:** 739-745 (1960).
111. Ariel IM. Treatment of inoperable primary pancreatic and liver cancer by the intra-arterial administration of radioactive isotopes (Y-90 radiating microspheres). *Ann. Surg.* **162:** 267-278 (1965).
112. Ariel IM. Radioactive isotopes for adjuvant cancer therapy. *Arch. Surg.* **89:** 244-249 (1964).
113. Burton MA, Gray BN, Kelleher DK, Klemp P, and Hardy N. Selective internal radiation therapy: Validation of intra-operative dosimetry. *Radiology* **175:** 253-255 (1990).
114. Turner JH, Claringbold PG, Klemp PFB, Cameron PJ, Martindale AA, Glancy RJ, Norman PE, Hetherington EL, Najdovski L, and Lambrecht RM. Ho-166-microsphere liver radiotherapy: a preclinical SPECT dosimetry study in the pig. *Nuclear Medicine Communications* **15:** 545-553 (1994).

115. Häfeli UO, Casillas S, Dietz DW, Pauer GJ, Rybicki LA, Conzone SD, and Day DE. Radioembolization of Novikoff hepatomas using radioactive rhenium (Re-186/Re-188) glass microspheres. *Int. J. Radiat. Oncol. Biol. Phys.* **44:** 189-199 (1999).
116. Andrews JC, Walker SC, Ackermann RJ, Cotton LA, Ensminger WD, and Shapiro B. Hepatic radioembolization with Yttrium-90 containing glass microspheres: Preliminary results and clinical follow up. *J. Nucl. Med.* **35:** 1637-1644 (1994).
117. Ackerman NB, Lien WM, Kondi ES, and Silverman NA. The blood supply of experimental liver metastases. I. The distribution of hepatic artery and portal vein blood to "small" and "large" tumours. *Surgery* **66:** 1067-1072 (1969).
118. Leung TWT, Lau WY, Ho SKW, Ward SC, Chow JHS, Chan MSY, Metreweli C, Johnson PJ, and Li AKC. Radiation pneumonitis after selective internal radiation treatment with intraarterial Y-90-microspheres for inoperable hepatic tumours. *Int. J. Radiat. Oncol. Biol. Phys.* **33:** 919-924 (1995).
119. Ho S, Lau WY, Leung TWT, Chan M, Chan KW, Lee WY, Johnson PJ, and Li AKC. Tumour-to-normal uptake ratio of Y-90 microspheres in hepatic cancer assessed with Tc-99m macroaggregated albumin. *Brit. J. Radiology.* **70:** 823-828 (1997).
120. Blanchard RJ, Grotenhuis I, LaFave JW, Frye CW, and Perry JN. Treatment of experimental tumours. *Arch. Surg.* **89:** 406-410 (1964).
121. Shepherd FA, Rotstein LE, Houle S, Yip TCK, Paul K, and Sniderman KW. A phase I dose escalation trial of Yttrium-90 microspheres in the treatment of primary hepatocellular carcinoma. *Cancer* **70:** 2250-2254 (1992).
122. Fellinger K and Schmid J. Die lokale Behandlung der rheumatischen Erkrankungen. *Wien Z. Inn. Med.* **33:** 351 (1952).
123. Delbarre F, Cayla J, Roucayrol JC, et al. Synoviortheses (synoviotherapie par les radioisotopes). Etude de plus de 400 traitements et perspectives d'avenir. *Ann. Med. Interne* **121:** 441 (1970).
124. Delbarre F, Roucayrol JC, Ingrand J, Sanchez A, Menkes CJ, and Aignan M. Une nouvelle preparation radioactive pour la synoviorthese: le rhenium 186 colloidal: Avantages par rapport au colloide d'or 198. *Nouvelle presse medicale* **2:** 1372 (1973).
125. Gumpel JM, Beer TC, Crawley JCW, and Farran HEA. Yttrium 90 in persistent synovitis of the knee - a single centre comparison. The retention and extra-articular spread of four 90Y radiocolloids. *Brit. J. Radiology.* **48:** 377-381 (1975).
126. Winston MA, Bluestone R, Cracchiolo A, and Blahd WH. Radioisotope synovectomy with P-32-chromic phosphate - kinetic studies. *J. Nucl. Med.* **14:** 886-889 (1973).
127. Noble J, Jones AG, Davies MA, Sledge CB, Kramer RI, and Livni E. Leakage of radioactive particle systems from a synovial joint studied with a gamma camera. *J. Bone Joint Surg.* **65A:** 381-389 (1983).
128. Zalutsky MR, Noska MA, Gallagher PW, Shortkroff S, and Sledge CB. Use of liposomes as carriers for radiation synovectomy. *Nucl. Med. Biol. Int. J. Rad. Appl. Instr. Part B* **15:** 151-156 (1988).
129. Knight CG, Bard DR, and Page Thomas DP. Liposomes as carriers of antiarthritic agents. *Ann. N. Y. Acad. Sci.* **:** 415-428 (1988).
130. Day DE and Ehrhardt GJ. Composition and method for radiation synovectomy of arthritic joints. U.S.A. Patent No. 4889707, 1989.
131. Mumper RJ, Mills BJ, Yun Ryo U, and Jay M. Polymeric microspheres for radionuclide synovectomy containing neutron-activated Holmium-166. *J. Nucl. Med.* **33:** 398-402 (1992).
132. Häfeli U, German R, Pauer G, Casillas S, and Dietz D. Production of Rhenium-Powder with a jet mill and its incorporation in radioactive microspheres for the treatment of liver tumours. *In* Bryskin BD (Ed.). Rhenium and Rhenium Alloys. TMS (Minerals, Metals and Materials Society), Warrendale, PA, 1997, pp. 469-477.
133. Harbert JC. Radionuclide therapy in joint diseases. *In* Harbert JC, Eckelman WC, and Neumann RD (Eds.). Nuclear medicine: Diagnosis and therapy. Thieme Medical Publishers, New York, 1996, pp. 1093-1109.
134. Wang SJ, Lin WY, Chen MN, Chi CS, Chen JT, Bo WL, Hsieh BT, Shen LH, Tsai ZT, Ting G, Mirzadeh S, and Knapp FF. Intratumoural injection of rhenium-188 microspheres into an animal model of hepatoma. *J. Nucl. Med.* **39:** 1752-1757 (1998).
135. Order SE, Siegel JA, Lustig RA, Principato R, Zeiger LS, Johnson E, Zhang H, Lang P, Pilchik NB, Metz J, DeNittis A, Boerner P, Beuerlein G, and Wallner PE. A new method for delivering radioactive cytotoxic agents in solid cancers. *Int. J. Radiat. Oncol. Biol. Phys.* **30:** 715-720 (1994).

136. Westlin JE, Andersson-Forsman C, Garske U, Linne T, Aas M, Glimelius B, Lindgren PG, Order SE, and Nilsson S. Objective responses after fractionated infusional brachytherapy of unresectable pancreatic adenocarcinomas. *Cancer* **80:** 2743-2748 (1997).
137. Harbert JC. Radiocolloid therapy of cystic brain tumours. *In* Harbert JC, Eckelman WC, and Neumann RD (Eds.). Nuclear medicine: Diagnosis and therapy. Thieme Medical Publishers, New York, 1996, pp. 1083-1091.
138. Backlund EO. Colloidal radioisotopes as part of a multi-modality treatment of craniopharyngiomas. *J. Neurosurg. Sci.* **33:** 95-97 (1989).
139. Häfeli UO and Pauer GJ. Brachytherapy of brain tumours using rhenium microspheres in fibrin glue. *J. Natl. Cancer Inst.* in preparation (1999).
140. Wallner KE, Galicich JH, Krol G, Arbit E, and Malkin MG. Patterns of failure following treatment for glioblastoma multiforme and anaplastic astrocytoma. *Int. J. Radiat. Oncol. Biol. Phys.* **16:** 1405-1409 (1989).
141. Häfeli UO, Pauer GJ, Roberts WK, Humm JL, and Macklis RM. Magnetically targeted microspheres for intracavitary and intraspinal Y-90 radiotherapy. *In* Häfeli U, Schütt W, Teller J, and Zborowski M (Eds.). Scientific and clinical applications of magnetic carriers. Plenum, New York, 1997, pp. 501-516.
142. Saha GB. Fundamentals of nuclear pharmacy. Springer, New York, 1998.
143. Knapp FF, Beets AL, Guhlke S, Zamora PO, Bender H, Palmedo H, and Biersack HJ. Availability of Re-188 from the alumina-based W-188/Re-188 generator for preparation of Re-188-labeled radiopharmaceuticals for cancer treatment. *Anticancer Research* **17:** 1783-1795 (1997).
144. Waksman R. Clinical trials in radiation therapy for restenosis: Past, present and future. *Vascular radiotherapy monitor* **1:** 10-18 (1998).
145. Bender H, Zamora PO, Rhodes BA, Guhlke S, and Biersack HJ. Clinical aspects of local and regional tumour therapy with Re-188-RC-160. *Anticancer Research* **17:** 1705-1712 (1997).
146. Blower PJ, Lam ASK, O'Doherty MJ, Kettle AG, Coakley AJ, and Knapp FF. Pentavalent rhenium-188 dimercaptosuccinic acid for targeted radiotherapy: synthesis and preliminary animal and human studies. *Eur. J. Nucl. Med.* **25:** 613-621 (1998).
147. Wang SJ, Lin WY, Chen MN, Hsieh BT, Shen LH, Tsai ZT, Ting G, and Knapp FF. radiolabelling of lipiodol with generator-produced Re-188 for hepatic tumour therapy. *Appl. Radiat. Isot.* **47:** 267-271 (1996).
148. Chen JQ, Strand SE, Tennvall J, Lindgren L, Hindorf C, and Sjögren HO. Extracorporeal immunoadsorption compared to avidin chase: Enhancement of tumour-to-normal tissue ratio for biotinylated rhenium-188 chimeric BR96. *J. Nucl. Med.* **38:** 1934-1939 (1997).
149. Schubiger PA, Alberto R, and Smith A. Vehicles, chelators, and radionuclides: Choosing the "building blocks" of an effective therapeutic radioimmunoconjugate. *Bioconj. Chem.* **7:** 165-179 (1996).
150. Lück M, Pistel KF, Li YX, Blunk T, Müller RH, and Kissel T. Plasma protein adsorption on biodegradable microspheres consisting of PLGA, PLA or ABA triblock copolymers containing poly(oxyethylene). Influence of production method and polymer composition. *J. Contr. Rel.* **55:** 107-120 (1998).
151. O'Donoghue JA, Bardies M, and Wheldon TE. Relationships between tumour size and curability for uniformly targeted therapy with beta-emitting radionuclides. *J. Nucl. Med.* **36:** 1902-1909 (1995).
152. Hall EJ and Brenner DJ. The dose-rate effect revisited: Radiobiological considerations of importance in radiotherapy. *Int. J. Radiat. Oncol. Biol. Phys.* **21:** 1403-1414 (1991).
153. Ogawara KI, Yoshida M, Higaki K, Kimura T, Shiraishi K, Nishikawa M, Takakura Y, and Hashida M. Hepatic uptake of polystyrene microspheres in rats: Effect of particle size on intrahepatic distribution. *J. Contr. Rel.* **59:** 15-22 (1999).
154. Papisov MI, Savelyev VY, Sergienko VB, and Torchilin VP. Magnetic drug targeting. In vivo kinetics of radiolabelled magnetic drug carriers. *Int. J. Pharm.* **40:** 201-206 (1987).

RADIATION-INDUCED BIORADICALS:
Physical, chemical and biological aspects

WIM MONDELAERS[*] AND PHILIPPE LAHORTE[*†]
[*] *Laboratory of Subatomic and Radiation Physics, Radiation Physics group, Ghent University, Proeftuinstraat 86, B-9000 Ghent, BELGIUM. Member of IBITECH.*[†] *Division of Nuclear Medicine, Ghent University Hospital, De Pintelaan 185, B-9000 Ghent, BELGIUM.*

Abstract

This chapter is part one of a review in which the production and application of radiation-induced bioradicals is discussed. Bioradicals play a pivotal role in the complex chain of processes starting with the absorption of radiation in biological materials and ending with the radiation-induced biological after-effects. The general aspects of the four consecutive stages (physical, physicochemical, chemical and biological) are discussed from an interdisciplinary point of view. The close relationship between radiation dose and track structure, induced DNA damage and cell survival or killing is treated in detail. The repair mechanisms that cells employ, to insure DNA stability following irradiation, are described. Because of their great biomedical importance tumour suppressor genes involved in radiation-induced DNA repair and in checkpoint activation will be treated briefly, together with the molecular genetics of radiosensitivity. Part two of this review will deal with modern theoretical methods and experimental instrumentation for quantitative studies in this research field. Also an extensive overview of the applications of radiation-induced bioradicals will be given. A comprehensive list of references allows further exploration of this research field, characterised in the last decade by a substantial advance, both in fundamental knowledge and in range of applications.

1. Introduction

The discovery of X-rays in 1895 by Roentgen set in motion a long era of intense research on the fundamental and applied aspects of ionising radiation. Soon after the discovery of X-rays it was widely recognised that ionising radiation could modify the properties of matter and that mankind could take advantage of this. In the early days it

M. De Cuyper and J.W.M. Bulte (eds.), Physics and Chemistry Basis of Biotechnology, 249–276.

was believed that irradiation would be the ideal tool for successful medical therapy and diagnostics, and for the synthesis of new and exotic materials and products. However, these ideas, which excited the imagination of many researchers and industrialists, have not developed to the extent anticipated in these pioneering days. The major reasons were twofold: an insufficient fundamental insight in the basic processes governing the complex reaction chains initiated by radiation and the unavailability of powerful instruments and techniques for the production and study of the radiation-induced species. During the last decade however there has been a virtual explosion of advances in the field of radiation research, mainly driven by a fruitful cross-fertilisation of the multidisciplinary research community involved. Scientists from disciplines as radiation, nuclear and atomic physics, nuclear medicine and radiotherapy, quantum chemistry, radiobiology, biological modelling, organic chemistry, genetics, accelerator technology,... contribute to this exciting evolution.

Free radicals are situated at the crossroads of these research disciplines. The authors were asked to present an up-to-date review in which various aspects of radiation-induced radicals in biological matter are discussed. The author's goal in this respect has been to span the gap between the underlying physical sciences and the world of biological research and applications. Obviously, any attempt to cross the boundaries of traditional disciplines in a restricted amount of space is done at risk of losing scientific rigor and depth. We have tried to counterbalance this by incorporating numerous review articles and reference works. The comprehensive literature citations will help the interested reader to pursue their search to broader scopes and deeper levels.

The scientific interest in the physical and chemical aspects of the radiation-induced formation of bioradicals is important because it allows to an understanding of the primary processes from which originate many complex reaction chains with significant biological consequences. The macroscopic biological effects that can be induced by minimal radiation energy transfers are applied for the benefit of man in several domains of human activity such as radiotherapy, medical imaging, sterilisation, polymer chemistry, food processing, waste management. Advanced knowledge of radiation-induced bioradicals will lead to a better understanding of radiation damage, its after-effects and potential safeguards.

This contribution is divided over two chapters of this volume.

In this first chapter, Radiation-induced bioradicals: physical, chemical and biological aspects, we will concentrate on the basic concepts of the creation of radiation-induced bioradicals. We will give a short description of the interaction mechanisms of ionising radiation with biological systems and of the immediate and long-term consequences of this interaction. Knowledge of the basic physics of radiation interaction and energy transfer is fundamental to understand the consecutive chemical and ultimate biological effects in living matter. The subsequent stages of the radiation energy deposition processes and their effects will be elaborated in detail.

The second chapter, Radiation-induced bioradicals: technologies and research, will deal with modern theoretical methods and experimental instrumentation for quantitative studies in this research field. Also an extensive overview of the applications of radiation-induced bioradicals will then be given.

2. The interaction of ionising radiation with matter

In this chapter we will concentrate on bioradicals induced by ionising radiation. Radiation in general can excite and ionise molecules while getting absorbed by matter. Ionising radiation includes those types of radiation that are capable of producing ions by ejecting electrons from their atomic or molecular structure. In contrast, visible or near ultraviolet photons interact with matter by predominantly producing excited states. They are called non-ionising radiations. The chemical reactions induced by them are in the domain of photochemistry. Besides the undeniable similarities between the chemical and biological effects of both types of radiation, there are profound differences. The energy absorption process of ionising radiation is almost entirely dependent on the atomic composition while non-ionising radiation absorption is mainly influenced by molecular binding properties of the irradiated medium. While the energy of the quanta of non-ionising radiation is about the same as that of the chemical bond (2-5 eV) [1], the energy of ionising radiation is several orders of magnitude higher, enabling ionisation of many different molecules of the substance during a chain of interactions. This will generate a wide variety of reactive species, especially radicals, which lead to chemical and biological products in the irradiated specimen. The interaction of non-ionising radiation has a more selective nature, because low-energy photons with energies corresponding to the absorption spectrum of atoms and molecules are selectively absorbed. The result is a well-defined photochemical reaction. Such selectivity will not be observed with ionising radiations, subject of this chapter.

Ionising radiation includes high-energy atomic particles (electrons, protons, neutrons, α-particles...) as well as high-energy electromagnetic radiation (X-rays, γ-radiation). High energy in this context refers to energies greater than the ionisation energies of atoms and molecules, but, in practice, energies in the range of kilo-electronvolt (keV) or Mega-electronvolt (MeV) are used for radiobiological research and applications.

Although the major part of the following discussion is applicable to matter in general, we will focus mainly on biological systems. A complete and detailed description of the interaction of ionising radiation with matter is very complex. The radiation transport in matter is governed by a combination of many possible interaction processes, each having energy-dependent interaction probabilities. To deal with the statistical nature of these events and the many parameters involved (radiation type, medium, geometric and energetic characteristics, reaction mechanisms, enzymatic repair), radiation physics, chemistry and biology have to rely on Monte Carlo techniques (Andreo 1991; Ballarini 1999; Begusova 1999; Briesmeister 1993; De Marco 1998; Halbeib 1984; Hill 1999; Ma 1999; Michalik 1995a,b; Nikjoo 1998; Rogers 1991; Tomita 1997). In these 'roulette-type' theoretical techniques primary and secondary radiation species are followed in space and time, the type and probability of

---

[1] One electronvolt (1 eV) is the kinetic energy an electron, or another singly charged particle, gains on being accelerated by a potential difference of one volt (1 eV = 1.602 10^{-19} J). Most covalent chemical bonds in organic materials have bond dissociation energies between 3 and 4.5 eV.

interaction being sampled by random numbers. Within the scope of this chapter, however, we have to limit ourselves to some general fundamental aspects (Turner 1995, Attix 1998). The understanding of these basic processes will provide a very good insight in the parameters controlling the immediate chemical and the long-term biological consequences of the interaction of radiation with biological systems.

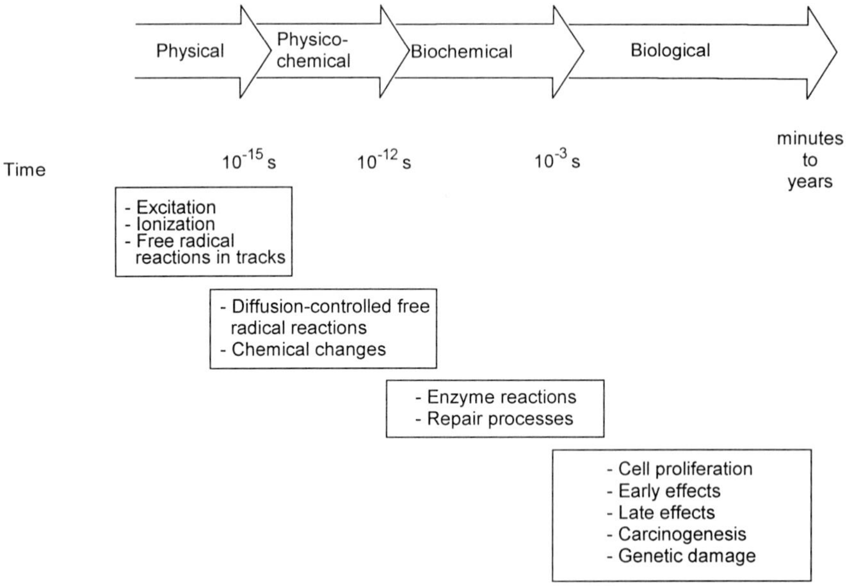

Figure 1. *Approximate time scale of events in the interaction of ionising radiation with matter.*

When radiation penetrates (biological) matter a succession of processes is generated. Figure 1 gives an approximate time-scale of events following the interaction of ionising radiation with matter. There is a possible overlap between the individual stages, both in time and with respect to the character of the elementary processes. As an introduction we will give a short overview of the chain of consecutive processes set in motion by radiation impinging on a biological system. A detailed description of the different stages will be given in paragraphs 2 to 5 of this chapter.

Radiation, be it a photon or a particle, traverses a molecular dimension of a few angstroms in 10^{-18} to 10^{-17} s. The energy of the radiation is initially distributed among a large number of atoms and molecules through the interaction of the radiation with their orbital electrons, giving rise to ionised and electronically excited atoms and molecules. These species are concentrated in tracks along the path of the ionising species. The spatial distribution of the electron-excited or ionised molecules depends on the properties of the irradiated medium and of the radiation. In covalently bonded systems, such as organic and biological systems, a large proportion of the ionised and excited molecules react or dissociate with the formation of free radicals, concentrated in tracks distributed in a manner similar to their parent molecules. The electrons produced by the absorption of radiation are rapidly thermalised and consequently solvated in most liquid media. The duration of these processes is very short (10^{-15} s).

By the end of the energy deposition an irradiated molecular system contains ions, electrons, excited molecules and free radicals. As we will see further, these species will in general be the same in a particular material, regardless of the type or energy of the radiation. All ionising radiations will therefore give rise to qualitatively similar chemical effects. The quantitative differences of chemical effects of distinct radiations stem from the different spatial distributions of the reactive species.

The chemical reactivity of positive and negative ions is not high, but if they recombine with other ions they can form free radicals. Because of their unpaired electron free radicals are very reactive. Their reactions are usually very fast, in particular the radical-radical reactions, so radicals play a predominant role during this stage. Radicals have been shown to be transient intermediates also in many non-radiation-induced chemical reactions. However, the initial high concentration of radiation-induced radicals along the radiation tracks can lead to a completely different radical behaviour as compared to systems where the radicals are more uniformly distributed. During the initial stage there exists an enormous variety of intermediates leading to a complicated set of reactions (Klassen 1987). It is the production of this large number of free radicals that accounts for the fact that high-energy radiation is much more effective in inducing chemical changes than, for example, an equivalent amount of thermal energy. It is also the basis of the great variety of applications of radiation-induced processes (Woods 1994).

About 10^{-12} s after the initial events, any radicals that have not reacted within the tracks, have diffused from these and become essentially homogeneously distributed in the medium. Chemical changes in the material being irradiated are generally the result of further free radical reactions, that are completed within approximately 1 ms in gaseous and liquid systems. In solids the reactions proceed much slower, due to the reduced mobility of the free radicals. Trapped radicals may be detected even weeks or months after irradiation.

When ionising radiation is absorbed in living material, there is a possibility that it will act directly on critical targets in the cell. The molecules may be ionised or excited, thereby initiating a chain of events that leads to biological change and cell death if the change is critical. In contrast to this direct effect, radiation may also interact with other atoms or molecules in the cell, particularly water, to produce free radicals which can diffuse far enough to reach and damage the DNA.

Radiation effects to living species (e.g. loss of viability, sterility, cancer, and genetic damage) can occur over longer time scales, from a few hours to many years, depending on the irradiation conditions. In each case, however, the changes to the living system are the result of the chemical changes brought about in the first fractions of a second after irradiation.

Based on this description, the complex chain of processes starting with the absorption of radiation in biological materials and ending with the radiation-induced biological after-effects can be subdivided in four stages: 1° the absorption of radiation energy in the substance (physical stage); 2° energy transfer among the intermediates (physicochemical stage); 3° restoration of the chemical equilibrium (biochemical stage) and 4° long-term biological processes taking place due to the applied radiation (biological stage). We will treat these four stages now more in detail.

3. The physical stage

High-energy radiations used to initiate radiochemical reactions include charged particles (α and β-particles or electrons), photons (γ-radiation and X-rays) and neutrons. At some point in the radiation absorption process almost all the energy of the ionising radiations is transferred to fast-moving charged particles, ionising or exciting nearby molecules of the material, as they are slowed down. These charged particles may represent the primary radiation itself, as in the case of charged particle irradiation, secondary electrons in the case of photon irradiation, or protons and other ionising particles produced by neutron interactions. Therefore, high-energy charged particles are called direct ionising particles, while photons and neutrons are indirect ionising radiations.

3.1. DIRECT IONIZING RADIATIONS

A charged particle traversing matter exerts electromagnetic forces on atomic electrons and imparts so-called collision energy to them (ICRU 46 1992; Knoll 1989). The energy transferred may be sufficient to knock an electron out of an atom and thus ionise it. Alternatively, it may leave the atom in an excited non-ionised state. When an atom is ionised, the secondary electron produced may have enough energy to cause several more ionisations or excitations along a branched track before being thermalised. The energy releases by particles are discrete events, the spacing of which will depend on the energy and type of the particles.

The average rate of collision energy loss of particles in a medium –dE/dx or collision stopping power can be derived, using relativistic quantum mechanics, from the Bethe formula (1):

$$-\frac{dE}{dx}=\frac{4\pi N_A z^2 e^4}{m_e v^2}\rho\frac{Z}{A}B \tag{1}$$

In this relation:

N_A = Avogadro's number
z = charge of the particle
e = magnitude of the electron charge
m_e = electron rest mass
v = speed of the particle

ρ = density of the irradiated medium
Z = atomic number of the absorbing atom
A = atomic mass of the absorbing atom
B = correction factor, dependent on the medium, particle type and energy
As can be seen from this formula, the stopping power increases with the density of the irradiated medium.

If they have the same energy, heavy particles such as α-particles and protons, are much slower than electrons. Therefore the average rate of energy loss of heavy particles is much greater. Because they are heavier than the atomic electrons with which they collide their energy loss per collision is small and their deflection is almost negligible. Therefore they are gradually slowed down as a result of a large number of small energy losses. They travel along almost straight paths through matter leaving a dense track of ionised and excited atoms in their wake (ICRU 49, 1993).

In contrast, electrons and positrons can lose a large fraction of their energy in a single collision with an atomic electron, thereby suffering relatively large deflections. The energy releases are widely spaced along the particle tracks. Because of their small mass, electrons are frequently scattered through large angles by nuclei. Electrons and positrons generally do not travel through matter in straight lines (ICRU 39, 1984).

The v^{-2}-dependence of the stopping power, (see the Bethe formula above), indicates that for low velocities at the end of the particle trajectories, the deposited energy increases sharply (the v^{-2}-dependence is slightly compensated by a smaller decrease of the factor B with energy). For electrons this absorbed dose increase is substantial only over the last nm of the trajectory, where clusters of ionisations occur. For high-energy protons this region extends over a few mm (the so-called Bragg peak). The Bragg peak of protons forms the basis of the application of these particles in proton radiotherapy, allowing more precise dose distributions in tumours (Courdi 1993; Rosenwald J. 1993).

A high-speed particle can also be sharply decelerated and deflected by an atomic nucleus, causing it to emit the energy lost as electromagnetic radiation in a process called "bremsstrahlung" (braking radiation). The rate of energy loss by this process is proportional with z^2Z^2/m^2, where z and Z are the charge of the particle and the atomic number of the nucleus, and m the mass of the particle. For particles with a heavier mass than electrons bremsstrahlung production is negligible at the energies used for the irradiation of biological samples. For electrons bremsstrahlung production is a second-order effect in the radiolysis of biological materials, because they are low-Z materials. The bremsstrahlung process is in this context only relevant for the production of X-rays with electron accelerators (see next chapter).

The range of a charged particle, i.e. the distance that a particle can penetrate into matter, depends on the initial energy of the particle and the density of the absorber. The reciprocal of the stopping power (including collision and radiative processes) gives the distance travelled per unit energy loss. Therefore, the range R(T) of a particle of kinetic energy T is the integral of this quantity down to zero energy (2):

$$R(T) = \int_0^T \left(\frac{dE}{dx} \right)^{-1} dE \qquad (2)$$

Because the common unit for stopping powers is MeV cm² g^{-1}, the range is expressed in g cm^{-2}. To obtain the range in cm R has to be divided by the density ρ of the absorbing material.

The range of 1 MeV α-particles, protons and electrons in water are respectively 4.6 μm, 39 μm and 4.37 mm. For 10 MeV these values are 0.1 mm, 1.2 mm and 49.8 mm. For other materials the range is roughly inversely proportional with the density.

3.2. INDIRECT IONIZING RADIATIONS

Unlike charged particles, which generally lose energy through a large number of small energy transfers, photons (X-rays and γ-rays) tend to lose a large amount of energy when they interact with matter. However, the interaction probability of photons is rather low, so that many photons will pass through a finite thickness of material without change in energy or direction. A simple exponential law (3) can describe the reduction of the number of photons transmitted through a sample with thickness d:

$$I = I_0 e^{-\mu d} \qquad (3)$$

I_0 and I are the radiation intensities before and after the sample and μ is the absorption coefficient taking into account the several processes that contribute to the attenuation and scattering of a photon beam (Gerward 1993; Henke 1993; Hubbell 1999). Their relative importance depends on the photon energy and on the nature of the irradiated material. There are three important photon interactions, all three producing fast electrons causing subsequently many excitations and ionisations: photoelectric effect, Compton scattering and pair-production.

When the photon energy is below 0,5 MeV, the photoelectric effect is predominant. The total energy, i.e. the entire photon, is used up in the ejection of an electron from an atom shell. Subsequently this fast electron causes many excitations and ionisations.

Compton scattering arises predominantly when photons in the energy range 0.5 – 5 MeV collide with free or loosely bound electrons in the absorber. Part of the photon energy is transferred to the electron as kinetic energy, and the photon is deflected from its initial direction (Cooper 1997; Harding 1997).

When a photon has an energy of 1.02 MeV or higher it may extinct in the proximity of an atomic nucleus of the absorber, giving rise to an electron-positron pair. This process is called "pair-production".

Neutrons, being particles without charge, gradually lose energy by direct collisions with nuclei of matter. Ion pairs are produced by these collisions, the hit nucleus losing one or more of its orbital electrons (ICRU 46, 1992). The energy transfer in these elastic (billiard-ball like) collisions are most effective when the mass of the nucleus is comparable with the mass of the neutrons, as is the case for light elements, especially in

H-rich media, such as biological materials. Other, less important, neutron interactions also generate predominantly secondary ionising particles and photons as reaction products.

3.3. LINEAR ENERGY TRANSFER (LET)

As described, radiation of different types and energy will lose energy in discrete events at different rates. Excitation and ionisation occurs along the track of the primary radiation in single events, the spacing along the track depending on the type and the velocity of the primary radiation and the density of the medium. The spatial distribution differences are related to the linear rate at which radiation loses energy in a medium, referred to as linear energy transfer (LET), usually expressed in unit keV μm^{-1}. High LET or rapid energy loss will lead to particle tracks densely populated with ions and excited molecules, while low LET radiation gives widely separated spurs (Goodhead 1988, 1989). Values of LET range from 0.2 keV μm^{-1} to 40-50 keV μm^{-1}. To have a general picture, the different radiation types can be arranged in order of increasing LET: high-energy electrons and X-rays, γ-radiation; low-energy X-rays and β-particles; protons; deuterons; α-particles; heavy ions, and fission fragments from nuclear reactions.

High LET values affect radiolysis yields by increasing the probability of reactions between the reactive species that are formed in the radiation tracks. One reaction product may predominate with low LET-radiation and another with high LET-radiation. As we will see further, a wide variety of biological effects are induced by ionising radiation. The biological effectiveness may be strongly dependent on LET, i.e. of the nature of the radiation tracks. For example for fast electrons the discrete energy releases are widely spaced and even though a track passes through a DNA molecule, there is a chance that no energy releases will occur in it. The track left by an α-particle on the other hand is so dense that if the α-particle passes through the DNA, there will enough energy releases and clustering of damage to destroy it (Hill 1999; Nikjoo 1998).

3.4. DOSE AND DOSE EQUIVALENT

The primary physical quantity to characterise irradiation processes is the amount of energy transferred from the radiation to the absorbing medium. The energy absorbed per unit mass from any kind of ionising radiation in any target is defined as the absorbed dose D. The unit of the absorbed dose is called Gray (symbol Gy). One Gray is equal to an energy absorption of one Joule per kilogram of the irradiated material. The older unit, frequently encountered in publications, is the rad. 1 Gy is equal to 100 rad.

The penetration of radiation energy into a sample can be represented by depth-dose curves in which the relative absorbed dose at a particular point is plotted against the distance (i.e. the depth) of that point from the irradiated face of the sample. The shape of these depth-dose curves depends on the nature of the radiation, the energy, the density of the irradiated material and the irradiation geometry. Depth-dose distributions have to be taken in account when the distribution of physical, chemical or biological effects are important in an irradiated medium (e.g. radiotherapy, sterilisation of pharmaceuticals...).

As already mentioned, the range of heavy particles is very small compared to the electron range. For heavy particles there is only substantial absorbed dose along the first micrometers of their trajectory. Therefore, electron beams are preferred for the production of radicals in biological systems. Because photons have even a larger penetration depth, they are chosen for the irradiation of bulky samples. However, photon beams generally produce lower dose rates.

LET reflects the deposited energy distribution in a material on a microscopic scale (nm-range). This localised spatial distribution is rather inhomogeneous, dependent on the radiation type. The absorbed dose is a macroscopic quantity that 'smears out' the energy deposition over volume elements in the mm-range. Absorbed dose is radiation-independent, 1 Gy always corresponding to 1 J/kg.

It has long been recognised that the absorbed dose needed to achieve a given level of biological change (e.g. cell killing) is often different for different kinds of radiation. This is closely related to the initial track structure or the microscopic spatial distributions of the reactive species during the physicochemical stage. To account for different biological effectiveness of different kinds of radiation, the concept of dose equivalent is introduced. The dose equivalent H is given by $H = DQ$, where D is the absorbed dose and Q is the quality factor which is related to the LET of the radiation. Values of Q range from unity for high-energy photon and electron radiation to 25 for neutrons, protons and heavier particles. When the dose is expressed in Gy, the unit of dose equivalent is the Sievert (Sv). With the dose in rad, the equivalent dose unit is rem. 1 Sv is equal to 100 rem.

3.5. INDUCED RADIOACTIVITY

When ionising radiation impinges on matter, energy may be imparted to some nuclei of the atoms. Under certain conditions, this may be sufficient to induce an atomic nucleus to become so unstable that its emits one or more nuclear particles together with γ-radiation. This reaction changes the nucleus into that of a different element or an isotope of the original one. By these nuclear transformations ionising radiation may induce radioactivity in matter which previously showed none. The risk of radioactivity production depends on the properties of the matter irradiated, and on the energy and type of the ionising radiation employed. It is essential that treatment with ionising radiation of living matter or in a manufacturing process leading to human consumption does not produce radioactivity. Therefore a maximum energy limit to the beams is prerequisite. Following more than 30 years experience, especially in food processing, the energy limits are fixed at 10 MeV for electron irradiation and 5 MeV for photon beams. As described in the next chapter, appropriate radiation sources, recommended by the WHO, IAEA and FAO, can be chosen, that do not induce any noticeable radioactivity in biological specimens. It has been amply demonstrated that there is no danger of inducing radioactivity with such selected sources (Diehl 1995; FAO/IAEA 1996; WHO 1984).

4. The physicochemical stage

Regardless the type of ionising radiation, finally the absorbed energy is transferred to atomic electrons. When inner-shell electrons are ejected they interact with other less-firmly-bound atomic electrons, so that the absorbed energy is rapidly distributed over the least-strongly-bound electrons; for example the nonbonding electrons on oxygen or nitrogen and the π electrons of unsaturated compounds. During the physicochemical stage very fast reactions occur in the direct neighbourhood of the radiation tracks (Glass 1991). The tracks may be densely or sparsely populated with the active species, dependent on the radiation LET. Therefore, depending on the LET, the relative proportion of the chemical products formed and their distribution may differ appreciably. The resulting biological effects will change accordingly.

To understand radiation-induced effects in biological materials one has to recall that the critical molecules such as DNA, RNA, or protein in the living cell are irradiated in an aqueous environment. Damage to these molecules can be imparted either by a direct hit of the molecule or by means of an indirect mechanism by the free radicals induced in water. Therefore the study of the chemical changes induced by ionising radiation in liquid water is very important (Ferradini 1999; von Sonntag 1991). The different types of radiation interact with water by distinctly different processes but the overall result will be the formation of the ionised and excited water molecules, H_2O^+ and H_2O^*, and subexcitation electrons. These species, so-called primary products, are produced in local regions of the tracks by the above-mentioned physical processes in less than 10^{15} s.

At room temperature, a water molecule can move an average distance comparable to its diameter (2.9 Å) in about 10^{-12} s. Thus, 10^{-12} s after passage of an ionising particle or photon marks the beginning of the ordinary, diffusion-controlled chemical reactions that take place around the radiation track. During the physicochemical stage, from 10^{-15} s to 10^{-12} s, the three primary products induce changes in their direct environment. First an ionised water molecule reacts with a neighbouring molecule (4), forming a hydronium ion and a hydroxyl radical:

$$H_2O^+ + H_2O \rightarrow H_3O^+ + OH^\circ \qquad (4)$$

Second, an excited water molecule dissipates energy by ejecting an electron (5) and proceeding according to the previous reaction or by molecular dissociation (6):

$$H_2O^* \rightarrow H_2O^+ + e^- \qquad (5)$$

$$H_2O^* \rightarrow H^\circ + OH^\circ \qquad (6)$$

Third, the subexcitation electrons migrate, losing energy by rotational and vibrational excitation of water molecules, and become thermalised. The thermalised electrons orient the permanent dipole moments of neighbouring water molecules, forming a cluster, called a hydrated electron e^-_{aq}, which is a radical that has a significant lifetime in water. The hydrated electron has a lifetime under physiological conditions of a few microseconds. The net result of these fast reactions is to produce three important highly reactive species: the hydrated electron e^-_{aq}, OH° and H° (Buxton 1988; Fulford 1999; Laverne 1993).

For high LET radiation the concentration of the radicals in the tracks is very high and molecular products are formed through the recombination of radicals (7):

$$H° + H° \rightarrow H_2 \qquad (7.1)$$

$$OH° + OH° \rightarrow H_2O_2 \qquad (7.2)$$

The last reaction produces hydrogen peroxide H_2O_2, a very active oxidising agent.

Reactions occurring within a track in pure materials are often similar to what is observed in aqueous solutions. If a simplified representation of an organic compound is given by RH_2, the following fast reactions with ions RH_2^+ and excited molecules RH_2^* can take place: dissociation of ions or excited molecules giving radical or molecular products (reactions 8.1 to 8.4); dissipation of excitation energy without chemical reaction (reaction 8.5); recombination of radicals (reaction 8.6) and ion-molecule reactions (reaction 8.7).

$$RH_2^+ \rightarrow RH^+ + H° \qquad (8.1)$$

$$RH_2^+ \rightarrow R^+ + H_2 \qquad (8.2)$$

$$RH_2^* \rightarrow LH° + MH° \qquad (8.3)$$

$$RH_2^* \rightarrow R + H_2 \qquad (8.4)$$

$$RH_2^* \rightarrow RH_2 \qquad (8.5)$$

$$RH° + H° \rightarrow RH_2 \qquad (8.6)$$

$$RH_2^+ + RH_2 \rightarrow RH_3^+ + RH° \qquad (8.7)$$

Free radicals can be formed directly when a molecule dissociates at a covalent bond, so that one bonding electron remains with each fragment. In organic compounds bond scission tends to be almost random in straight-chain hydrocarbons such as for example hexane, giving a variety of different radicals and a relatively large number of radiolysis products (17 different reaction products for hexane). However scission tends to be more specific if the molecule has weaker covalent bonds such as those at branches in the carbon skeleton (Woods 1994). The effects on non-hydrocarbon organic materials are determined largely by the presence of functional groups.

The process of molecule dissociation, producing radicals, is reversible. During the physicochemical stage, there is a substantial chance that the radicals, not separated within the track, will recombine. Recombination will often give back the original molecule, although alternative reactions may deliver different products. Radicals that do not react with other radicals in this region of high radical concentration diffuse into the bulk of the medium and generally react with products in the medium during the chemical stage, where the usual chemical kinetics apply. However the initial high concentration of radicals close to the radiation tracks can lead to a completely different radical behaviour for radiation-induced reactions than usually encountered in systems where the radicals are more uniformly distributed.

All radiation types produce qualitatively the same reactive species in the local track regions. The chemical and biological differences that result at later time are due entirely to the different spatial distribution of the patterns of initial energy deposition.

5. The chemical stage

At about 10^{-12} s after passage of the ionising radiation the reactants begin to migrate randomly in thermal motion around their tracks. The chemical kinetics are now diffusion-controlled (von Sonntag 1987).

The possible reactions of radicals are manifold. They can be uni-molecular (rearrangement and dissociation) and bimolecular. The latter can be subdivided into reactions that include a radical among the products (addition, abstraction) and those that terminate the radical reaction, leading to molecular or ionic species (combination, disproportionation, electron transfer). The most typical examples of reactions in systems of biological interest are: hydrogen abstraction by a radical (reactions 9.1 and 9.2); ion neutralisation (reaction 9.3); radical combination (reaction 9.4) and radical disproportionation (reaction 9.5).

$$RH_2 + OH^\circ \rightarrow RH^\circ + H_2O \qquad (9.1)$$

$$RH_2 + H^\circ \rightarrow RH^\circ + H_2 \qquad (9.2)$$

$$RH_3 + e^-_{aq} \rightarrow RH^- + nH_2O \qquad (9.3)$$

$$2RH^\circ \rightarrow RH—RH \qquad (9.4)$$

$$2RH^{\circ} \rightarrow RH_2 + R \quad (9.5)$$

In an aqueous biological environment the water radicals and hydrogen peroxide attack organic molecules. The reactions are of a statistical nature so that randomly distributed large biomolecules in the medium undergo rather indiscriminate bond scissions, with some preference concentrated on those parts of the molecule having the greatest variation in electron density or where the weaker covalent bonds are present. Radicals tend to react with the functional groups of these molecules. Complicated biological macromolecules containing a great number of reactive functional groups are very sensitive to irradiation. Even breakage of single hydrogen bridges may cause lasting effects in these materials. Certain functional groups are particularly susceptible to one or more of the radicals, but the situation is very diversified and complex. For example, H° and OH° will react rapidly with alkenes, while the reaction with e^{-}_{aq} will proceed slowly. Halogen compounds are particularly susceptible to attack by e^{-}_{aq} but not by OH°. Hydroxyl radicals react with virtually all organic compounds, either by adding to a multiple bond or by hydrogen abstraction. Halogen radicals, classed as electron acceptors, tend to attack preferentially points of high electron density in the surrounding medium. In contrast, methyl radicals have a tendency to lose electrons and seek electron-deficient centres. During the first millisecond after radiation exposure there is a competition between radical scavenging and damage-fixing reactions in biologically important molecules. The radicals react with other radicals by combination or disproportionation, or react with other molecules, to give a wide range of products.

5.1. RADICAL REACTIONS WITH BIOMOLECULES

The study of the chemistry of bioradicals is important for the understanding of later biological effects. As we will discuss further damage to the genetic material DNA is the most critical event in radiation exposure of biological systems. Radiation exposure of cells in living organisms may result in cell replication failure or in chromosome aberrations, leading to mutagenesis and carcinogenesis. Radiation damage to proteins, lipids and carbohydrates is relevant for effects such as enzyme inactivation and also has applications in the radiation treatment of foods and drugs where toxicity of radiation products is a point of major concern. Amino acids and sugars are important for dosimetric applications. The amino acid L-α-alanine is currently used as a reference dosimeter suitable over a wide dose range (Callens 1996; McLaughlin 1993; Van Laere 1993). The study of radiation-induced effects on various sugars (e.g. glucose, fructose, and sucrose) is relevant for the detection of some irradiated foodstuff.

The reactivity with free radicals in general depends to a great extent on the structure of the reactants. The selectivity is determined largely by the energetics of the processes taking place. As we will see in the next chapter, ab-initio quantum chemical calculations in combination with EPR measurements are very important to interpret these energetics and to elucidate the possible structure of bioradicals (Lahorte 1999a,b). Also primary radiation-damage in DNA is studied through ab-initio molecular-orbital calculations (Colson 1995; Wetmore 1998a,b).

5.1.1. Radiation damage to DNA

The primary target for radiation-induced cell damage is the DNA molecule (Kiefer 1990; von Sonntag 1994). This is supported by many sources of evidence, including the following (McMillan 1993):

- microirradiation studies show that to kill cells by irradiation of cytoplasm requires far higher doses than irradiation of the nucleus
- isotopes with a short range emission, incorporated into cellular DNA, produce very effectively radiation cell killing and DNA damage
- the incidence of chromosomal aberrations following irradiation is closely linked to cell killing
- thymidine analogues, as IUdR and BrUdR (iodo- and bromo-deoxyuridine) incorporated in chromatin modify radiosensitivity.

DNA is a complex molecule, a long chain polymer composed of nucleotides. Each nucleotide contains a nitrogenous base (adenine, guanine, thymine and cytosine), linked through a sugar (deoxyribose) to a phosphoryl group. The backbone of the molecule consists of alternating sugar-phosphate groups.

LET and track structure play an important role in the production of DNA damage (Frankenburgschwager 1994; Hill 1999; McMillan 1993). The exposure of mammalian cells to 1 Gy of low-LET ionising radiation leads to the production of around 1000 tracks with 2×10^5 ion pairs per cell nucleus, roughly 2000 of which may be produced directly in the DNA itself. The same dose of high-LET radiation produces only about 4 tracks per cell nucleus, but the intense ionisation within each track leads to more severe damage where the track intersects the DNA. In addition to this direct effect, damage may result indirectly from free radicals produced in water close to DNA. Free radicals produced in a radius of 2 nm around a DNA molecule are believed to contribute to the radiation damage.

DNA may be affected by one or more of the following types of damage (Alberts 1994; Box 1995; Hildenbrand 1990):

- strand breaks, single or multiple
- modification of bases and sugars
- cross-linking and dimerisation

The amount of DNA damage that can be detected immediately after irradiation is substantial. Often quoted are the estimates for a clinical dose of 1 Gy of $>$ 1000 base damages, ~ 1000 single-strand and ~ 40 double-strand breaks, together with cross-links between DNA strands and with nuclear proteins (Ward 1990). During the first millisecond after radiation exposure, free radicals take part in a variety of competitive reactions, some of which lead to the fixation of damage, others to the scavenging and inactivation of radicals. Besides these chemical repair processes, enzymatic repair and rejoining of DNA breaks, further reduce the damage during the subsequent few hours. Irradiation at clinically used doses induces a vast amount of DNA damage, most of which is successfully repaired by the cell. The frequency of lethal lesions is typically 1 per Gy (Steel 1996). It is clear that cells generally have a remarkably high ability to repair radiation-induced DNA damage.

On the basis of a variety of experimental investigations it is generally believed that multiple-strand breaks are the critical lesions for radiation cell killing (McMillan 1990; Obe 1992; Ward 1990). There are two possible mechanisms for the formation of double-strand breaks. After the formation of a single-strand break, a deoxyribose radical is still present. Eventually it may be transferred to the complimentary DNA strand, generating another strand-breakage, close to the first. Secondly, as we have already mentioned, the ionising events are not homogeneously distributed. Clusters of ionisation and free radicals occur within a diameter of a few nanometer along the tracks of high-LET radiation and at the very end of the range of low-LET particles.

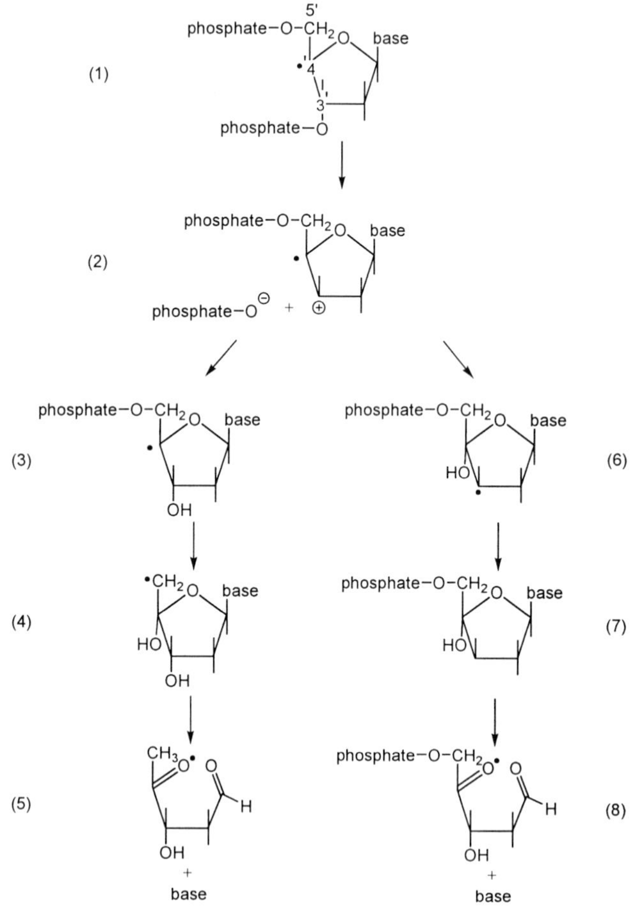

Figure 2. Reaction pathways for the formation of altered sugars in DNA leading to strand breakage.

Such an event may produce a particularly severe lesion, which might consist of one or more double-strand breaks, together with a number of single-strand breaks, base damage,... Such a Local multiply damaged site (Steel 1996) may be too complex to allow enzymatic repair. DNA damage is often investigated in theoretical (Michalik 1995) and experimental (Spotheim-Maurizot 1995) model systems, e.g. dried or frozen DNA to study the direct effect or in dilute aqueous systems to study the indirect effect. Under aerobic conditions, the main source of DNA damage are OH° radicals from water radiolysis (indirect effect) (Mee 1987; von Sonntag 1994) and one-electron oxidation (direct effect) (von Sonntag 1987). In Figure 2, reaction pathways for the formation of altered sugars in DNA leading to chain breakage are given as an example (Mee 1987). Radical 1 is formed due to hydrogen abstraction with a OH° radical. The C4' position of the sugar is considered as the main site of sugar attack by OH° radicals (Spotheim-Maurizot 1995). Abstraction at the C4' position is preferred over other positions due to sterical hindrance (Obe 1992). The radical 1 removes first the phosphate ester anion thus breaking the chain. Further interaction with OH° radicals produces either radicals 3 or 6. For radical 3 the second phosphate group can be eliminated. Radical 4 is terminated in disproportionation reactions with other radicals, followed by ring opening and the elimination of the unaltered base-forming sugar. For radical 6 the second phosphate group cannot be removed, but in an analogous series of reactions, the ring opens and an unaltered base is released. The base radicals may also lead to strand breaks by radical transfer to the sugar moieties (Becker 1993).

Fast neutron and photon-irradiation of double- and single-stranded synthetic oligonucleotides have shown that radiation-induced damages occur at all nucleotide sites (Isabelle 1995). In all structures, the probabilities of inducing a modification are almost identical for adenine, thymine and cytosine, but this probability is higher for guanine, in accordance with the known higher reactivity of guanine base towards the OH° radicals. An almost complete description of the radical reactions mediated by OH° radicals and also of direct one-electron oxidation has been developed recently for the guanine compounds, guanine having the lowest ionisation potential (Cadet 1999). The radiosensitivity of each nucleotide is the same in the single- and double-stranded DNA. The similarity of radiosensitivity of the single and of the double stranded DNA strongly suggests that the stacking and the helical conformation, governing the accessibility of the sites to radical attack, are the determinant factors and not the interstrand H bonding (Spotheim-Morizot 1995). The radiosensitivity of DNA is also modulated by its interactions with proteins, the presence of bound proteins protecting the DNA in regions of contact (Isabelle 1993).

5.1.2. Radiation damage to proteins

Proteins generally have very high rate constants for reaction with the reactive species of water. The possible sites of interaction with proteins can be predicted from pulse radiolysis experiments (see next chapter) on amino acids and peptides, but the intrinsic reactivity of these sites are influenced by the structure of the proteins, governing the accessibility of the sites to radical attack (Houée-Levin 1994; Sharpatyi 1995).

Based on known protein structural factors and kinetic constants of radical attachments it can be estimated that most hydrated electrons should react with the surface peptide carbonyl groups, and to a lesser extent with disulfide bridges and imidazole rings (Mee 1987). The reactions of OH° radicals with proteins are preferentially addition to aromatic rings, abstraction of hydrogen from the peptide backbone and abstraction of hydrogen from the side chains. In most cases the reactions of the H° radical are similar with reactions of either e^-_{aq} or the OH° radical. Pulse radiolysis experiments have established that the primary addition of reactive intermediates to sites are followed by an intramolecular chain of events, such as internal electron migration into a potential sink, either histidine or a disulfide bridge. Depending on the irradiation conditions also cross-linking and scission of the peptide backbone occur. The effects of the highly reactive intermediates result in irreversible chemical changes.

It is well known that the radiation resistance of enzymes is relatively high (Diehl 1981). Compared to DNA and RNA, their radiation resistance is at least an order of magnitude higher. The radiation sensitivity of different enzymes irradiated in dilute aqueous solutions varies by a factor of up to 200. If an active centre of an enzyme contains amino acids which are particularly sensitive to radiation, for example sulphur-containing amino acids, this enzyme is effectively inactivated. As an example the radiation sensitivity of papain, containing one SH-group which is essential for enzyme activity, is about 20 times higher than of glyceraldehyde-3-phosphate dehydrogenase, where 3 to 4 SH-groups have to be destroyed before activity is lost.

5.1.3. Radiation damage to lipids and polysaccharides

Interest in the irradiation of fats is related mainly to food irradiation. Up to even rather high doses the changes in common fat quality indices are slight (Mee 1989). Of primary importance is the fatty-acid composition of the irradiated fat, especially the degree of unsaturation (i.e. the presence of one or more double bonds). Polyunsaturated fatty acids, containing three or more double bonds are very sensitive and are readily destroyed by radiation. The initial radiation event is the formation of fatty-acid free radicals, either directly or, if water is present, by abstraction of a hydrogen atom from the fatty-acid chain by OH° radicals. Peroxy radicals are formed by reaction with oxygen. By abstraction of a hydrogen atom from another fatty-acid molecule, peroxy radicals are converted to hydroperoxides, thus promoting a chain reaction (10):

$$R^\circ + O_2 \rightarrow RO_2^\circ \qquad (10.1)$$

$$RO_2^\circ + RH \rightarrow ROOH + R^\circ \qquad (10.2)$$

Lipid hydroperoxides are very unstable and spontaneously decompose to form a mixture of mono- and dialdehydes and ketones of various chain lengths. Consequently irradiated foods containing unsaturated fats will have an increased content of these radiation products and a decreased concentration of (poly)unsaturated fatty acids.

In polysaccharides, glycosidic linkages join the basic monomer units. The polysaccharide cellulose is the most abundant organic compound in vegetation. There is a considerable interest in radiolysis of cellulose from the viewpoints of food sterilisation, textile improvement and viscose industry. The predominant effect of irradiation of cellulose is degradation, due to the splitting of the glycosidic bond. The reactions are similar to the reactions on the deoxysugars of DNA as described above.

5.2 RADIATION SENSITISERS AND PROTECTORS

The effects of radiation may be modified appreciably by the presence of oxygen or other compounds with sensitising or protecting properties (Schulte-Frohlinde 1995; Tomita 1997). Oxygen is in biological systems usually present in high concentrations. The presence of oxygen has an important influence on the radical induced reactions because molecular oxygen has a high affinity for free radicals. The H° radical is easily transformed (11) into long-living reactive oxygen containing radicals HO_2° and $O_2^{-\circ}$ and molecules H_2O_2.

$$H^\circ + O_2 \rightarrow HO_2^\circ \qquad (11.1)$$

$$H^\circ + HO_2^\circ \rightarrow H_2O_2 \qquad (11.2)$$

$$e^-_{aq} + O_2 \rightarrow O_2^{-\circ} + nH_2O \qquad (11.3)$$

With organic compounds, organic peroxide will be formed, leading to a chain reaction (12):

$$R^\circ + O_2 \rightarrow RO_2^{-\circ} \qquad (12.1)$$

$$RO_2^{-\circ} + RH \rightarrow RO_2H + R^\circ \qquad (12.2)$$

Oxygen gives rise to further reactive products and fixes the radiation damage. It plays a central sensitising role in radiation effects. These reactions explain for example the reduced radiosensitivity of hypoxic cells. In this way the biological effects of radiation therapy are enhanced appreciable in the presence of oxygen (Oxygen Enhancement Factors 2 to 3).

Many chemical agents have been found to alter radiation sensitivity. In general, oxidising agents are radiation sensitisers, whereas reducing agents are usually radiation protectors. The use of antioxidants, like vitamin C, vitamin E and beta-carotene, as radiation protecting agents is actually under investigation (Platzer 1998).

Compounds containing sulphydryl (-SH groups) have a particular affinity for free radicals. Their presence within the cell may decrease radiation effects. Thiols like cysteine or gluthatione, which are readily attacked by radicals, can often protect other solutes against radiation damage by scavenging radicals that would otherwise react with the solute. Cysteine and other molecules that contain weakly bound hydrogen atoms can also act as repair agents by replacing a hydrogen atom lost as the result of radical attack (13). Two cysteine radicals dimerise to cystine (14).

$$R^{\circ} + CySH \rightarrow RH + CyS^{\circ} \quad (13)$$

$$2CyS^{\circ} \rightarrow CyS{-}Scy \quad (14)$$

These two sulfhydryl compounds are notable radioprotectors. There is a considerable interest in manipulating the levels of cellular thiols either to protect normal tissue or to increase radiosensitivity of tumour cells (McMillan, 1993). The most successful research into radiation sensitizers is on nitroimidazoles and other sensitizers of hypoxic cells (Adams 1991). This class of sensitisers mimic oxygen and their consumption rate in tissue is low, so that they are able to diffuse into hypoxic tumour cells (Overgaard 1993). A clinical meta-analysis demonstrates clearly that an overall clinical gain can be obtained (Overgaard 1994).

6. The biological stage

The chemical changes brought about immediately after the radiation impact can initiate modifications of the biochemical functions in living species that emerge only after longer periods of time (Steel 1993). Since irradiation at normal levels of exposures changes only a very small portion of the molecules, biological effects have to be attributed to damage of critical molecules, such as enzymes and genetic material. Radical reactions with biological molecules are completed within roughly 1 ms. The biological phase includes all subsequent processes (Hall E. 1994, Oleinick 1990). Biochemical changes occur after seconds, while some cells functions are affected within minutes. After chemical repair reactions during the chemical stage, enzymatic reactions act on the residual chemical damage during subsequent hours (Biaglow 1987). The vast majority of the DNA lesions are successfully repaired. Some lesions fail to repair leading to cell death after a number of mitotic divisions. Current evidence suggests that double-strand breaks are the most closely linked to cell death (Cassoni 1992; Dealmodovar 1994; McMillan 1990; Nunez 1996; Obe 1992). Some damage induced by radiation may be insufficient to stop cell division but enough to lead to mutations with a loss of ability to provide normal progeny. Mutations arise by non-lethal alteration of the base sequence in DNA. A secondary effect of cell killing is compensatory cell proliferation. Depending on the delivered radiation dose clinical damage will become apparent in days (skin, gastrointestinal and nervous system), weeks (lung fibrosis, spinal cord and blood vessel damage) or years (cataracts and cancer). Genetic effects are seen in the next or subsequent generations. Radiation-induced genetic changes can result from gene mutations and from chromosome alterations. Some mutations involve a deletion of a portion of a chromosome. Broken chromosomes can rejoin in various ways, introducing errors into the normal arrangement. The death of most cells is associated with certain types of chromosomal aberrations, in particular those aberrations that lead to the creation of chromosome fragments without a centromere and to the loss of a substantial piece of the genome at a subsequent mitosis (Bedford 1991). The higher organisms, with more complex biochemical relationships, are more at risk from radiation damage than simples ones (Britt 1996).

6.1. DOSE-SURVIVAL CURVES

The effect of radiation on a biological function of a living species (cell, microorganism, plant, animal, human...) is usually studied by determining the survival of the function with increase in radiation dose. Already since many years, there is a vast amount of literature on dose-survival curves for numerous cell types and microorganisms (Fertil 1981, Pollard 1983, Deacon 1984). Survival curves of single cells allow to quantify the biological effects of radiation on a cellular basis. They also provide a better understanding of the influence of various physical, chemical and biological modifications on cell survival. The radiation response curves are conveniently represented by plotting the natural logarithm of the surviving fraction (e.g. enzyme molecules, viruses, bacterial colonies, tumour cells, plants...) as a function of the dose they receive (Niemann 1982). Radiation response curves of cells can have different shapes, as shown in Figure 3. A number of different mathematical models adequately simulate the shape of survival curves for biological systems (Joiner 1993).
A linear semilog survival (curve A) implies exponential survival. Exponential behaviour can be accounted for by a single-target single-hit model (if a species is hit, it will not survive). These simple straight survival curves, usually found for the inactivation of viruses and bacteria, are also useful in describing the radiation response of very sensitive human cells, the response to very low dose rates and to high LET radiations (Joiner 1993).

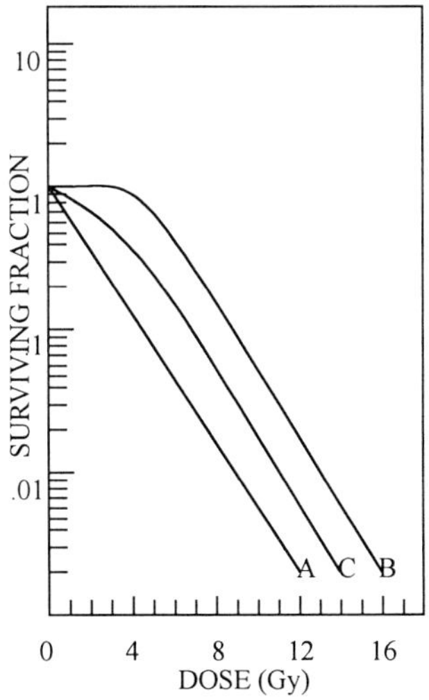

Figure 3. *Dose-survival curves: (A) single-target single-hit model, (B) multi-target or multi-hit models and (C) lesion-interaction or repair-saturation models.*

The radiation dose response curves for many biological systems are not straight lines, but instead have a 'shoulder' (curve B). This shoulder implies that for low doses the radiation effects are reduced. This can be explained by multi-target or multi-hit models (Katz 1994), describing different situations: (1) the systems may require hits in more than one sensitive volume before lethality (for example mammalian cells at high doses); (2) the systems may require more than one hit in the same area (e.g. bacterial colonies).

The main shortcoming of these models is the flat response for very low doses of low LET radiations. Much experimental cell survival curves show evidence for a finite initial slope (curve C) (Biaglow 1987). Two different types of model describe accurately the linear quadratic curves: lesion-interaction models (Curtis 1986) and repair-saturation models (Alper 1980). Lesion-interaction models postulate two classes of lesions. One class is directly lethal, while the other is only potentially lethal and may be repaired enzymatically or may interact with other potentially lethal lesions to form lethal lesions. Repair-saturation models propose a dose-dependent repair rate. Potentially lethal lesions are either repaired or fixed. Higher doses produce too much damage for the repair system to handle and it saturates. Therefore, proportionally less damage can be repaired before fixation. So far it has not been possible to determine which model has to be preferred (Steel 1999).

6.2. REPAIR MECHANISMS

As already mentioned, the majority of the damage induced in cells by radiation is satisfactorily repaired (Lehmann 1992). Evidence for this comes from studies of strand breaks in DNA, most of which disappear in a few hours (Olive 1999). Because all organisms are exposed since the beginning of life on earth to natural radioactivity and to cosmic radiation, enzymatic mechanisms for repairing damage to the genetic material have been developed during evolution. These processes may lead either to successful repair or to fixation of damage, so-called "misrepair" (Dahm-Daphi 1998; Sachs 1997). The nature of the repair processes depends to a large degree on the type of lesion. A simple excision-repair process can remove a change in one of the DNA bases. This involves nicking of the DNA on either side of the lesion, removal of a few bases around the point of damage, synthesis of new DNA and ligation to the original strand. Where both strands of the DNA are involved, there may be loss of information. A strand of DNA with a homologous sequence can be used as template in what is known a recombination repair (McMillan 1993). When recombinational processes have taken place some cells will have lost very large sections of the genome (Steel 1999) and if these encompass essential genes the cell is doomed.

6.3. RADIOSENSITIVITY AND THE CELL CYCLE

During the cell cycle radiation sensitivity changes. Mitosis and the G_2 phase were confirmed as the most sensitive, the S phase as most resistant. Because cells exhibit different degrees of radiosensitivity in different phases of the cell cycle, an asynchronous cell population will become partially synchronised by irradiation. The surviving cells will be those in the more resistant phases. As the cell population continues to grow the surviving cells become redistributed over the complete cycle, including the more sensitive phases. If now a given radiation dose is administered in many fractions instead of a single irradiation, this redistribution process will lead to an increased cell killing, because cells not killed by the first dose move from a less sensitive state to a more sensitive state in later cell cycles. Therefore dose fractionation is extremely important in radiotherapy (Thames 1987). It also allows cell recovery by repair of sub-lethal damage. It has the additional advantage that it allows sensitisation of radiation-resistant hypoxic tumour cell by reoxygenation of these cells if the tumour is shrinking between different fractions. The response of the normal oxygenated cells is unchanged by this procedure.

During the last decade there has been extensive research on the regulation of the cell cycle in relation to cancer therapy (Murray 1992; Weichselbaum 1991). A number of discrete checkpoints in the cell cycle have been detected and the biochemical processes that control them are now well understood (Dasicka 1999; Meek 1999; Rowley 1999). In mammalian cells the *p53* stress response gene plays a crucial role in regulating cell growth following exposure to various stress stimuli, in particular radiation, thereby maintaining genetic stability in the organism. The *p53* protein either inhibits the progress of cells through the checkpoints, inducing growth arrest which prevents the replication of damaged DNA (Gallagher 1999; Shackelford 1999) or it leads to programmed cell death (apoptosis, distinct from mitotic death), which is important for eliminating defective cells (Brown 1999; Lee 1995; Sionov 1999). *p53* inactivation allows damaged cells to cycle through a growth arrest. As *p53* is the most commonly mutated gene in human cancer there are some speculations that *p53*, when mutated, may constitute one step to malignancy (Hall P. 1994; Lane 1992; Lee 1994). It has attracted a great deal of interest as prognostic factor, diagnostic tool and therapeutic target (Steele 1998).

6.4. MOLECULAR GENETICS OF RADIOSENSITIVITY

Genetics of radiosensitivity is currently developing rapidly following two important developments. First, it was realised that a number of inherited human syndromes, especially Ataxia Telangiectasia (AT) have an associated increased radiosensitivity. Cells from these patients were intensively investigated and the critical AT gene has been identified and sequenced (Savitsky 1995). Secondly, there was the production of radiosensitive mutants of *in vitro* rodent cell lines, especially *xrs* mutants from Chinese hamster ovary cells. One of the genes that have been mutated was identified, *XRCC5* (Rathmell 1994). The human DNA sequence that complements this defect has been found and was the first human ionising-radiation-sensitivity gene to be identified, *XRCC1* (Thompson 1990). This search has gone in the context of the rapid identification of genes that control DNA repair. The unravelling of its molecular genetics is going on in bacterial, yeast and mammalian cell systems, facilitated by the remarkable homology among genes in these widely different species (Steel 1999, Cartwright 1993). Other studies have linked oncogenes with the development of radioresistance. Transformation of primary rat embryo cells with a combination of *c-*

ras and *v-myc* has been found to lead to radioresistance (McKenna 1992). Techniques of molecular genetics are beginning to elucidate the processes of cell killing and mutation.

7. Conclusion

This review has sought to give a general survey of the current knowledge of radiation-induced bioradicals and to indicate a number of research lines which have been developed during the last decade. The study of bioradicals is situated at the crossroads of different research disciplines. Extensive research in the fields of radiation physics chemistry and in radiobiology has delivered a good, albeit still incomplete description of the underlying phenomena during the various physical, chemical and biological stages of the reaction chain. These studies have revealed many effects on the radical yields of different intervening factors including irradiation (type and energy) and environmental conditions (presence of oxygen, temperature, phase, pressure, pH and solute concentration). The close relation between the initial energy deposition, radical yields and the ultimate biological outcome has been elucidated substantially. Of particular interest is the relationship between radiation dose and track structure, induced DNA damage and cell killing. In the last years considerable progress has been made in identifying genes that control the repair of radiation damage. Recent papers describe the mechanisms cells employ to insure DNA stability following irradiation. Several newly defined tumour suppressor genes are involved in radiation-induced DNA repair and in checkpoint activation. There is increasing evidence that the mutational status of the p53 controlling gene is an important determinant of the clinical outcome of cancer.
The powerful theoretical and experimental techniques, developed or perfected during the 90's and described in the next chapter, offer the best means to carry out studies on bioradicals and their influence on radiation-induced damage. It is not surprising that the majority of the information available in this area has been obtained with the help of these novel research methods. This leads to various attractive perspectives concerning the future development of this research area and its potential applications in many domains of human activity.

References

Adams G., Bremmer J., Stratford I., et al. 1991. Nitroheterocyclic compounds a radiosensitisers and bioreductive drugs. Radiother. Oncol. **20**: 85-91.
Alberts B., Bray D., Lewis J., Raff M., Roberts K., Watson J.D. 1994. *Molecular biology of the cell*, 3rd ed. Garland Publishing, New York.
Alper T. 1980. Keynote address: survival curve models. In Radiation biology in cancer research. (R. Meyn, H. Whiters, Eds.), pp. 3-18. Raven, New York.
Andreo P. 1991. Monte Carlo techniques in medical radiation physics. Phys. Med. Biol. **36**: 861-920.
Attix F.H. 1986. *Introduction to Radiological Physics and Radiation Dosimetry*. Wiley-Interscience, New York.
Ballarini F., Merzagora M., Monforti F., et al. 1999. Chromosome aberrations induced by light ions: Monte Carlo simulations based on a mechanistic model. Int. J. Radiat. Biol. **75**: 35-46.

Becker D., Sevilla M. 1993. The chemical consequences of radiation damage to DNA. Adv. Radiat. Biol. **17**: 121-180.

Bedford J. 1991. Sublethal damage, potentially lethal damage, and chromosomal aberrations in mammalian cells exposed to ionising radiations. Radiat. Oncol. Biol. Phys. **21**: 1457–1469.

Begusova M., Tartier L., Sy D., et al. 1999. Monte Carlo simulation of radiolytic attack to 5'-d[T(4)G(4)](4) sequence in a unimolecular quadruplex. Int. J. Radiat. Biol. **75**: 913-917.

Biaglow J. 1987. Ionising radiation in mammalian cells. In *Radiation Chemistry, Principles and Applications* (Farathaziz, Rodgers M., eds.). pp. 527-563.VCH Publishers, New York, 1987.

Box H., Freund H., Budzinski E., Wallace J., and Maccubbin A. 1995. Free radical –induced double-base lesions. Rad. Res. **141**: 91-94.

Brenner D. 1990. Track structure, lesion development, and cell survival. Radiat. Res. **124**: 29–37.

Briesmeister J. 1993. MCNP- a general Monte Carlo N-particle code, version 4A. Rep. LA-12625. Los Alamos National Laboratory, Los Alamos, New Mexico.

Britt A. 1996. DNA damage and repair in plants. Ann. Rev. Plant Phys. Plant Molec. Biol. **47**: 75-100.

Brown J., Wouters B. 1999. Apoptosis, p53, and tumour cell sensitivity to anticancer agents. Canc. Res. **59**: 1391-1399.

Buxton G., Greenstock C., Helman W., and Ross A. 1988. Critical-review of rate constants for reactions of hydrated electrons, hydrogen-atoms and hydroxyl radicals ($OH^{\circ}/0^{\circ-}$) in aqueous solution. J. Phys. Chem. Ref. Data **17**: 513-886.

Cadet J., Douki T., Gasparutto D., Gromova M., Pouget J.-P., Ravanat J.-L., Romieu A., and Sauvaigo S. 1999. Radiation-induced damage to DNA: mechanistic aspects and measurement of base lesions. Nucl. Inst. Meth. **B151**: 1-7.

Callens F., Van Laere K., Mondelaers W., Matthys P., and Boesman E. 1996. The study of the composite character of the ESR spectrum of alanine. Appl. Radiat. Isot. **47**: 1241-1250.

Cartwright R., and McMillan T. 1993. The isolation of repair genes. *In Molecular Biology for Oncologists* (Yarnold J., Stratton M., and McMillan T., Eds.). Elsevier Science Publishers, Amsterdam.

Cassoni A., McMillan T., Peackock J., and Steel C. 1992. Differences in the level of DNA-double-strand breaks in human tumour-cell lines following low dose-rate irradiation. Eur. J. Canc. **28A**: 1610-1614.

Colson A., Sevilla M. 1995. Elucidation of primary radiation-damage in DNA through application of ab-initio molecular-orbital theory. Int. J. Radiat. Biol. **67**: 627-645.

Cooper M. 1997. Compton scattering and the study of electron momentum density distributions. Radiat. Phys. Chem. **50**: 102–112.

Courdi A. 1993. Radiobiological rationale for protontherapy. Pathol. Biol. **41**: 117

Curtis S. 1986. Lethal and potentially lethal lesions induced by radiation - a unified repair model. Radiat. Res. **106**: 252-270.

Dahm-Daphi J., Dikomey E., Brammer I. 1998. DNA-repair, cell killing and normal tissue damage. Strahlentherapie und Onkologie **174**: 8-11.

Dasika G., Lin S., Zhao S., et al. 1999. DNA damage-induced cell cycle checkpoints and DNA strand break repair in development and tumourigenesis. Oncogene **18**: 7883-7899.

Deacon J., Peckham M., and Steel G. 1984. The radioresponsiveness of human tumours and the initial slope of the cell survival curve. Radiother. Oncol. **2**: 317-323.

Dealmodovar J., Nunez M., McMillan T., Olea N., Mort C., Villalobos M., Pedraza V., and Steel G. 1994. Initial radiation-induced DNA-damage in human tumour-cell lines - a correlation with intrinsic cellular radiosensitivity. Brit. J. Cancer **69**: 457-462.

De Marco J., Solberg T., and Smathers J. 1998. A CT-based Monte Carlo simulation tool for dosimetry planning and analysis. Med. Phys. **21**: 1943-1952.

Diehl J. 1981. *Radiolytic effects in foods, in Preservation of food by ionising radiation.* CRC Press Inc., Boca Raton, Florida.

Diehl J. 1995. *Safety of irradiated foods.* Marcel Dekker Inc., New York.

Evans H. 1994. The prevalence of multilocus lesions in radiation-induced mutants. Radiat. Res. **137**: 131–144.

FAO/IAEA. 1996. *Food irradiation Newsletter.* International Atomic Energy Agency, Vienna.

Ferradini C., Jay-Gerin J. 1999. Radiolysis of water and aqueous solutions – History and present state of the science. Can. J. Chem. – Rev. Can. Chim. **77**: 1542-1575.

Fertil B., and Malaise E.-P. 1981. Inherent cellular radiosensitivity as a basic concept for human tumour radiotherapy. Int. J. Radiat. Oncol. Biol. Phys. 7: 621-629.

Frankenburgschwager M., Harbich R., Beckonert S., and Frankenberg D. 1994. Half-live values for DNA double-strand break rejoining in yeast can vary by more than an order of magnitude depending on the irradiation conditions. Int. J. Rad. Biol. **66**: 543-547.

Fulford J., Bonner P., Goodhead D., Hill M., and O'Neill P. 1999. Experimental determination of the dependence of OH radical yield on photon energy: A comparison with theoretical simulations. J. Phys. Chem. A **103**: 11345-11349.

Gallagher W., Brown R. 1999. p53-oriented cancer therapies: Current progress. Ann. Oncol. **10**: 139-150.

Gerward L. 1993. X-ray attenuation coefficients: current state of knowledge availability. Radiat. Phys. Chem. **50**: 783–789.

Glass W., Varma M. 1991. *Physical and Chemical Mechanisms in molecular Radiation Biology*. Plenum Press, New York.

Goodhead D. 1988. Spatial and temporal distribution of energy. Health Phys. **55**: 231-240.

Goodhead D., Nikjoo H. 1989. Track structure analysis of ultrasoft X-rays compared to high- and low-LET radiations. Int. J. Radiat. Biol **55**: 513–529.

Halbeib J., and Mehlhorn T. 1984. ITS: the Integrated Tiger Series of coupled electron photon Monte Carlo transport codes. Rep. SAND84-0573. Sandia National Laboratories. Albuquerque, New Mexico.

Hall E. 1994. *Radiobiology for the Radiobiologist*. Lippincott, Philadelphia.

Hall P., and Lane D. 1994. Genetics of growth arrest and cell death: key determinants of tissue homeostasis. Eur. J. Cancer **30**: 2001-2012.

Harding G. 1997. Inelastic scattering effects and applications in biomedical science and industry. Radiat. Phys. Chem. **50**: 91–111.

Henke B., Gullikson E., Davis J. 1993. X-ray interactions: photoabsorption, scattering, transmission and reflection at E = 50 – 30000 eV, Z = 1 – 92. At. Data Nucl. Data Tables **54**: 181–342.

Hildenbrand K., Schultefrohlinde D. 1990. ESR-spectra of radicals of single-stranded and double-stranded DNA in aqueous-solution – implications of OH-induced strand breakage. Free Rad. Res. Comm. **11**: 195-206.

Hill M. 1999. Radiation damage to DNA: The importance of track structure. Radiat. Meas. **31**: 15-23.

Houée-Levin C. 1994. Radiolysis of proteins. J. Chim. Phys. Physico-Chem. Biol. **91**: 1107-1120.

Hubbell J. 1999. Review of photon interaction cross section data in the medical and biological context. Phys. Med. Biol. **44**: 1–22.

ICRU Report 37. 1984. *Stopping Powers for Electrons and Positrons*. International Commission on Radiation Units and Measurements, Bethesda, MD.

ICRU Report 46. 1992. *Photon, Electron, Proton and Neutron Interaction Data for Body Tissues*. International Commission on Radiation Units and Measurements, Bethesda, MD.

ICRU Report 49. 1993. *Stopping Powers and Ranges for Protons and Alpha Particles*. International Commission on Radiation Units and Measurements, Bethesda, MD.

Isabelle V., Franchet-Beuzit J., Sabattier R., Laine B., Spotheim-Maurizot M., and Charlier M. 1993. Radioprotection of DNA by a DNA-binding protein-MC1 chromosomal from the archaebacterium. Int. J. Radiat. Biol. **63**: 749-758.

Isabelle V., Prevost C., Spotheim-Maurizot M., Sabattier R., and Charlier M. 1995. Radiation-induced damages in single- and double-stranded DNA. Int.J.Rad.Biol. **67**: 169-176.

Joiner M. 1993. Models for radiation cell killing. In *Basic clinical radiobiology*, (G. Steel, Ed.), pp. 40-46. Edward Arnold Publishers, London.

Katz R., Zachariah R., Cucinotta F., and Zhang C. 1994. Survey of cellular radiosensitivity parameters. Rad. Res. **140**: 356-365.

Kiefer J. 1990. *Biological Radiation Effects*, Springer, Berlin.

Klassen N. 1987. Primary products in radiation chemistry. In *Radiation Chemistry, Principles and Applications* (Farathaziz, Rodgers M., eds.). pp. 29-64.VCH Publishers, New York, 1987.

Knoll G. 1989. *Radiation Detection and Measurement*. Wiley & Sons, New York.

Lahorte P., De Proft F., Vanhaelewyn G., Masschaele B., Cauwels P., Callens F., Geerlings P., and Mondelaers W. 1999a. Density functional calculations of hyperfine coupling constants in alanine-derived radicals. Journ. Phys. Chem. A **103**: 6650-6657.

Lahorte P., De Proft F., Callens F., Geerlings P., and Mondelaers W. 1999b. A density functional study of hyperfine coupling constants in steroid radicals. J. Phys. Chem. A 103: 11130-11135.

Lane D. 1992. p53, guardian of the genome. Nature **358**: 15-16.

Laverne J., Pimblott S. 1993. Yields of hydroxyl radical and hydrated electron scavenging reactions in aqueous-solutions of biological interest. Rad. Res. **135**: 16-23.
Lee J., Abrahamson J., Bernstein A. 1994. DNA-damage, oncogenesis and the p53 tumour-suppressor gene. Mutation Research **307**: 573-581.
Lee J., Bernstein A. 1995. Apoptosis, cancer and the p53 tumour suppressor gene. canc. Met. reviews **14**: 149-161.
Lehmann A. et al. 1992. Workshop on DNA-repair. Mutation Research **273**: 1-28.
Ma C.-H. 1999a. Monte Carlo modelling of electron beams from medical accelerators. Phys. Med. Biol. **44**: 157-189.
McLaughlin W. 1993. Dosimetry – new approaches. Radiat. Phys. Chem. **41**: 45-56.
McKenna W., Iliakis G., and Muschel R. 1992. Mechanism of radioresistance in oncogene transfected cell lines. In Radiation research: a twentieth century perspective (Dewey C. et al., Eds), pp. 392-397. Academic Press, San Diego.
McMillan T., Cassoni A., and Edwards S. 1990. The relation of DNA double-strand break induction to radiosensitivity in human tumour cell lines. Int. J. Radiat. Biol. **58**: 427-438.
McMillan T., Steel G. 1993. Molecular aspects of radiation biology. In *Basic clinical radiobiology*, (G. Steel, Ed.), pp. 211-224. Edward Arnold Publishers, London.
Mee L. 1987. Radiation chemistry of biopolymers. In *Radiation Chemistry, Principles and Applications* (Farathaziz, Rodgers M., eds.). pp. 477-499.VCH Publishers, New York.
Meek D. 1999. mechanisms of switching on p53: a role for covalent modification? Oncogene *13*: 7666-7675.
Michalik V., Maurizot M., and Charlier M. 1995. Calculation of hydroxyl radical attack on different forms of DNA. J. Biom. Struct. & Dyn. **13**: 565-575.
Michalik V., Spotheim-Maurizot M., and Charlier M. 1995. Calculated radiosensitivities of different forms of DNA in solutions. Nucl. Inst. Meth. **B105**: 328-331.
Murray A. 1992. Creative blocks: cell-cycle checkpoints and feedback controls. Nature **359**: 599-604.
Niemann E. 1982. Strahlenbiophysik. In *Biophysik* (Hoppe W., Lohmann W., Markl H., Eds.) pp. 300-312. Springer Verlag, Berlin.
Nikjoo H., Uehara S., Wilson W., Hoshi M., Goodhead D. 1998. Track structure in radiation biology: theory and applications. Int. J. Radiat. Biol. **73**: 355-364.
Nunez M., McMillan T., Valenzuela M., Dealmodovar J., Pedraza V. 1996. Relationship between DNA damage, rejoining and cell killing by radiation in mammalian cells. Radiotherapy and Oncology **39**: 155-165.
Obe G., Johannes C., and Schulte-Frohlinde D. 1992. DNA double-strand breaks induced by sparsely ionising radiation and endonucleases as critical lesions for cell death, chromosomal aberrations, mutations and oncogene transformation. Mutagenesis **7**: 3-12.
Oleinick N. 1990. Symposium summary – Ionising-radiation damage to DNA – molecular aspects. Radiat. Res. **124**: 1-6.
Olive P. 1999. DNA damage and repair in individual cells: applications of the comet assay in radiobiology. Int. J. Rad. Biol. **75**: 395-405.
Overgaard J., Horsman M. 1993. Overcoming hypoxic cell radioresistance. In *Basic clinical radiobiology*, (G. Steel, Ed.), pp. 163-172. Edward Arnold Publishers, London.
Overgaard J. 1994. Clinical evaluation of nitroimidazoles as modifiers of hypoxia in solid tumours. Oncol. Res. **6**: 509–518.
Platzer I., and Getoff N. 1998. Vitamin C act as radiation protecting agent. Radiat. Phys. Chem. **51**: 73-76.
Pollard E. 1983. Effect of radiation at the cellular and tissue level. In Preservation of food by ionising radiation. Vol. II. (Josephson E., Peterson M., Eds), pp. 219-242. CRC Press, Boca Raton, Florida.
Rathmell W., and Chu G. 1994. A DNA end-binding factor involved in double-strand repair and V(D)J recombination. Mol. Cell Biol. **14**: 4741-4748.
Rogers D. 1991. The role of Monte Carlo simulation of electron transport in radiation dosimetry. Int. J. Appl. Radiat. Isot. **42**: 965-974.
Rosenwald J. 1993. Physical background for protontherapy. Pathol. Biol. **41**: 115-116.
Rowley R., Phillips E., Schroeder A. 1999. The effects of ionising radiation on DNA synthesis in eukaryotic cells. Int. J. Rad. Biol. **75**: 267-283.
Sachs R., Hahnfeld P., Brenner D. 1997. The link between low-LET dose-response relations and the underlying kinetics of damage production/repair/misrepair. Int. J. Rad. Biol. **72**: 351-374.

Savitsky K. 1995. A single Ataxia Telangiectasia gene with a product similar to PI-3 kinase. Science **268**: 1749-1753.

Schulte-Frohlinde D. 1989. The effect of oxygen and thiols on the radiation-damage of DNA. Free Radical Res. Com. **6**: 181-183.

Seymour C., Mothersill C. 1989. Lethal mutations, the survival curve shoulder and split-dose recovery. Int. J. Radiat. Biol. **56**: 999–1010.

Shackleford R., Kaufmann W., Paules R. 1999. Cell cycle control, checkpoint mechanisms, and genotoxic stress. Env. Health Pers. **107**: 5-24.

Sharpatyi V. 1995. Aspects of radiation-chemistry of protein molecules – a review. High Energy Chemistry **29**: 77-90.

Sionov R., Haupt Y. 1999. The cellular response to p53: the decision between life and death. Oncogene **18**: 6145-6157.

Spotheim-Maurizot M., Franchet-Beuzit J., Isabelle V., Tartier L., and Carlier M. 1995. DNA radiolysis: mapping of gene regulations domains. Nucl. Inst. Meth. **B105**: 308-313.

Steel G. 1993. Response of normal and malignant tissues to radiation exposure. In *Basic clinical radiobiology*, (G. Steel, Ed.), pp. 211-224. Edward Arnold Publishers, London.

Steel G. 1999. From targets to genes: a brief history of radiosensitivity, Phys. Med. Biol. **41**: 205–222.

Steele R., Thompson A., Hall P., Lane D. 1998. The p53 tumour suppressor gene. Brit. J. Surg. **85**: 1460-1467.

Thames H., and Hendry J. 1987. *Fractionation in radiotherapy*. Taylor & Francis, London.

Thompson L., Brookman K., Jones N., Allen S., and Carrano A. 1990. Molecular cloning of human *XRCC1* gene, which corrects defective DNA strand break repair and sister chromatid exchange. Mol. Cell. Biol. **10**: 6160-6171.

Tomita H., Kai M., Kusama T., and Ito A. 1997. Monte Carlo simulation of physicochemical processes of liquid water radiolysis – The effects of dissolved oxygen and OH scavengers. Rad. Env. Biophys. **36**: 105-116.

Turner J. 1992. An introduction to Microdosimetry. Rad. Prot. Management **9(3)**: 25–58.

Turner J. 1995. *Atoms, Radiation and Radiation Protection*. John Wiley & Sons, New York.

Van Laere K., Onori S., Bartolotta A., and Callens F. 1993. Study of intrinsic energy dependence of alpha-alanine and dose intercomparison with ESR and ISE techniques. Appl. Radiat. Isot. **44**: 33-39.

von Sonntag C. 1987. *The Chemical Basis of Radiation Action*, Taylor and Francis, London.

von Sonntag C. 1991. The elucidation of peroxyl radical reactions in aqueous-solution with the help of radiation-chemical methods. Angew. Chem. (English) **30**: 1229-1253.

von Sonntag C. 1994. Radiation-chemistry in the 1990s – Pressing questions relating to the areas of radiation biology and environmental research. **65**: 19-26.

Ward J. 1990. The yield of DNA double-strand breaks produced intracellularly by ionising radiation: a review. Int. J. Radiat. Biol. **57**: 1141–1150.

Weichselbaum R., Hallahan D., Sukhatme V., Dritschilo A., Sherman M., Kufe D. 1991. Biological consequences of gene-regulation after ionising radiation exposure. J. Nat. Canc. Inst. **83**: 480-484.

Wetmore S., Boyd R., and Eriksson L. 1998a. Theoretical investigation of adenine radicals generated in irradiated DNA components. J. Phys. Chem. B **102**: 10602-10614.

Wetmore S., Boyd R., and Eriksson L. 1998b. Comparison of experimental and calculated hyperfine coupling constants. Which radicals are formed in guanine radicals? J. Phys. Chem. B **102**: 9332-9343.

WHO. 1984. *Codex General Standard for Irradiated Foods*. United Nations Food and Agriculture Organisation, WHO Codex Alimentarius Commission, Codex Alimentarius, XV.

Woods R., Pikaev A. 1994. *Applied Radiation Chemistry: Radiation Processing*. Wiley-Interscience, New York.

RADIATION-INDUCED BIORADICALS:
Technologies and research

Philippe Lahorte[*†] **and Wim Mondelaers**[*]
[*]*Laboratory of Subatomic and Radiation Physics, Radiation Physics group, Ghent University, Proeftuinstraat 86, B-9000 Ghent, BELGIUM. Member of IBITECH.*
[†]*Division of Nuclear Medicine, Ghent University Hospital, De Pintelaan 185, B-9000 Ghent, BELGIUM.*

Abstract

This chapter represents the second part of a review in which the production and application of radiation-induced radicals in biological matter are discussed. In part one the general aspects of the four stages (physical, physicochemical, chemical and biological) of interaction of radiation with matter in general and biological matter in particular, were discussed. Here an overview is presented of modern technologies and theoretical methods available for studying these radiation effects. The relevance is highlighted of electron paramagnetic resonance spectroscopy and quantum chemical calculations with respect to obtaining structural information on bioradicals, and a survey is given of the research studies in this field. We also discuss some basic aspects of modern accelerator technologies which can be used for creating radicals and we conclude with an overview of applications of radiation processing in biology and related fields such as biomedical and environmental engineering, food technology, medicine and pharmacy.

1. Introduction

In the previous chapter the basic aspects of ionising radiation and its interaction with (biological) matter have been explored. It has been shown that a complete description of all effects is indeed very complex and covers a range of physical, chemical and biological events. Therefore it should come as no surprise that a broad range of experimental techniques and theoretical methods have been developed of which the majority constitute individual research fields in their own right. Each of them allows us to explore certain aspects of radiation effects on biological matter. Often however, a

M. De Cuyper and J.W.M. Bulte (eds.), Physics and Chemistry Basis of Biotechnology, 277–303.

thorough understanding is only brought about as the result of the combined use of several techniques which thus requires an interdisciplinary approach.
The current chapter represents the second half of our overview. It consists of three major parts. First of all a survey is given of the experimental techniques and theoretical methods currently available for studying the effects of radiation on biological systems from the *physical* to the *biological* stage. In a second part we will elaborate on some important techniques (electron paramagnetic resonance spectroscopy and quantum chemical calculations) with which direct information concerning identity and structure can be obtained of the radicals involved in processes in the *chemical* stage. Also an overview will be given of the fundamental research in this field. This approach is partly inspired by our own research efforts in which the determination of radical identity and structure is often either a goal in itself or a necessary hurdle that has to be taken in the development of new applications. The final part will be devoted to technological aspects of the irradiation process. In specific, the basic principles of accelerator technologies will be elucidated and an overview will be given of the application-oriented irradiation research in biology and related fields such as biomedical and environmental engineering, food technology, medicine and pharmacy. Radiation research in the field of radiotherapy will not be treated as this is beyond the scope of the present review.

The target audience of this contribution being bio(techno)logical scientists, an attempt was made to describe technologies and methods from a qualitative point of view, focusing on conveying the overall ideas and limitations, and the biologic relevance of the data and information that can be obtained from them. For further exploration and a deeper understanding the reader will be referred to literature citations and reference works.

2. Experimental and theoretical methods for studying the effects of radiation

Figure 1 gives an overview of the broad spectrum of experimental and theoretical tools for studying physical, physico-chemical, chemical or biological aspects of the effects of radiation exposure on biological systems. Obviously, the delineation of the scope and the field of activity of each method is somehow subjective. The classification as shown in Fig. 1 is therefore to be interpreted as indicative rather then exactly corresponding to the boundaries of the four stages of interaction which eventually exist only by virtue of human intellect. In this section we will briefly discuss the experimental techniques and theoretical methods, available for the combined physical and physico-chemical stages, and the biological stage. The techniques for studying radicals in the chemical stage will be discussed in more detail in paragraph 3.

As has been extensively discussed in the first part of this overview the physical and physico-chemical stages on the time-scales of radiation effects are characterised by a distribution of the radiation energy among the irradiated specimen. Along the path of the primary ionising species, radicals and electrons are formed in tracks. In most environments these intermediates are highly reactive and, hence, can exist only for very short periods. In this case, these transient species diffuse into the medium and give rise to a set of chemical reactions. Although some information concerning the reaction intermediates can be inferred from the results of steady-state experiments, detailed characterisation and identification of these unstable species requires the ability to monitor kinetic behaviour, often of very low concentrations, over very short time intervals. The bulk of our understanding of radiation chemistry has resulted from the use of *pulse radiolysis* techniques. In principle this technique consists of delivering a very short pulse (nano- or picosecond time range) of ionising radiation to a chemical

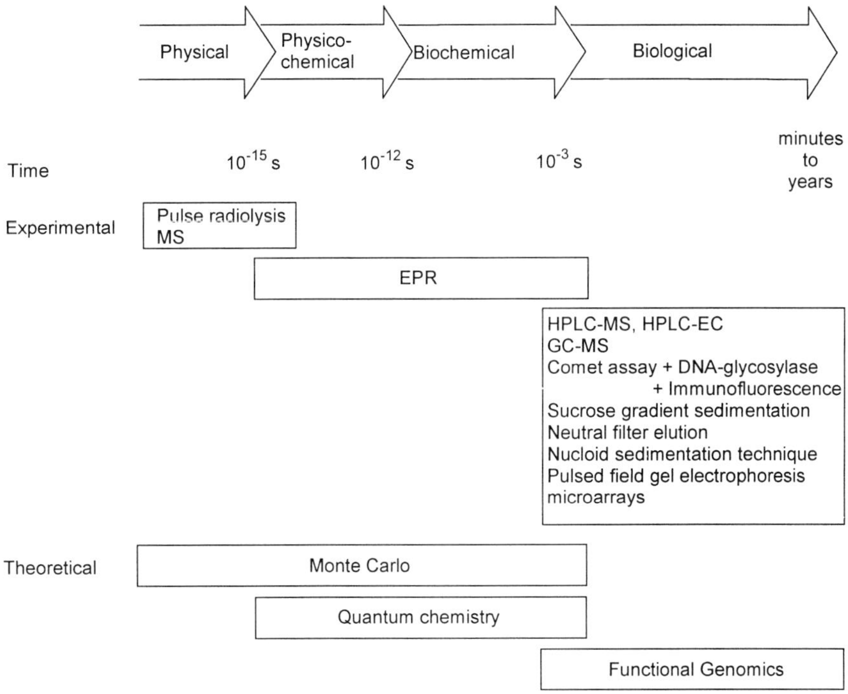

Figure 1. *Experimental and theoretical techniques for studying the effects of radiation on biological matter*

system so that a nonequilibrium system is produced containing significant concentrations of transient species. These can subsequently be monitored in time using an optical or electrochemical method such as *absorption spectroscopy* or *conductivity measurements*, respectively (Patterson 1987; Von Sonntag 1995). For the production of the ultrashort pulses electron accelerators are most often used. Some fundamental principles underlying accelerator technology are discussed in detail in paragraph 4. Wardman (1991, 1994) has reviewed the value and applications of pulse radiolysis measurements in free radical biology with respect to investigation of radical kinetic properties.

From a theoretical point of view, the Monte Carlo method is generally considered the most accurate method for calculating dose distributions. It is a statistical simulation method which allows one to calculate the tracks of individual particles by sampling appropriate quantities from the probability distributions governing the individual physical processes using machine generated (pseudo-)random numbers. Average values of macroscopic properties such as particle fluence, energy spectrum and absorbed dose can then be obtained by simulating a large number of particle histories. *Microdosimetric* Monte Carlo calculations allow for the modelling of the impact of radiation on an atomic level. In this way direct and indirect damage to DNA might be investigated by modelling the radiation formation of single and double strand breaks (Briden 1999, Hill 1999; Moiseenko 1998a-b). However, also macroscopic properties of chemical reactions such as rate constants, radiolysis end products and product yields are amenable to Monte Carlo simulations (Bolch 1998, Hamm 1998).

The methods described above may be used in a wide variety of systems but none of them yield substantial data that may be directly interpreted to give information on the *structure* of the transient species. By contrast, the techniques available for studying radicals in the *chemical* phase allow for the description of the chemical structure and the characterisation of radical processes. As already mentioned, these will be discussed in more detail in paragraph 3.

In the *biological* stage the consequences of irradiation are expressed in terms of altered biochemical functions in the living species which can emerge after periods of seconds (e.g. inactivation of enzymes) to years (e.g. development of cancers). Evidently, techniques developed for research in this stage probe for the detection of the radiation effects and the induced biochemical modifications. On the experimental side, high performance liquid chromatography (HPLC) and gas chromatography (GC) can be applied in combination with sensitive detection techniques such as electrochemistry (EC) or mass spectrometry (MS) to monitor the formation of oxidative base damage within cellular DNA (Cadet 1998, 1999). Alternatively, the single cell gel electrophoresis (comet) assay can be used (Fairbairn 1995; Koppen 1999). The alkaline type version of this assay can be applied to isolated cells and allows the measurement of DNA strand breaks and alkali-labile lesions. The comet assay offers high sensitivity at the cost of reduced specificity. The latter can be increased by using the assay in combination with either DNA-glycosylases (in order to convert base lesions in additional DNA strand breaks) (Pouget 1999a, b) or immunofluorescence detection (Sauvaigo 1998). Because of the importance of strand breaks as critical lesions produced by ionising radiation, a lot of effort has gone into developing methods (currently routine methods in molecular biology and biotechnology) with which strand breaks can be measured. Of these we mention the sucrose gradient sedimentation, the non-denaturing filter elution and the nucleoid sedimentation technique (Prise 1998). Of special importance has been the development of pulsed field gel electrophoresis (Iliakis

1991; Shao 1999) and derived variants which have revolutionised the separation of large molecules.
A very rapidly emerging field with huge potential for biotechnology and medicine is *functional genomics* (Claverie 1999; Dyer 1999; Langer-Safer 1997). With this term a platform of technologies are described, comprising among others differential display, proteomics and bioinformatics, which aim to establish a functional relationship between a particular genotype and a given disease state. Advances in microarrays and gene chips form the core of *differential display* technology which enables researchers to understand differences in gene expression in normal and diseased cells and tissues. This approach is very useful in identifying gene patterns which are up- or down-regulated in a given disease state, in a specific environment or in response to external stimuli. *Proteomics* is a complementary approach which aims to produce high-resolution protein maps for a given organism. It enables the analysis of changes in protein abundance and post-translation modification in both in the normal and pathologic state. Finally, molecular *bioinformatics* comprises the development and application of computational algorithms for the purpose of analysis, interpretation and prediction of the vast amount of molecular biologic data currently generated with modern biotechnology techniques such as the ones mentioned above. Main applications of these sophisticated informatics tools are primarily DNA sequence analysis, protein structure prediction and structure-function relationships in DNA and proteins. While the overall majority of efforts in genomics have concentrated on uncovering genetic and biochemical pathways and mechanisms for the development of new drugs and therapies, the same ensemble of techniques holds the potential for studying the effects of ionising radiation on living organisms. As an example one might investigate, using micro-array technology, the up- or down-regulation of genes coding, for instance, for repair enzymes in normal or tumour cells exposed to irradiation. Research in this area is however still in its infancy and major developments remain to be expected.

3. Studying radiation-induced radicals

3.1. EXPERIMENTAL AND THEORETICAL METHODS FOR DETECTING AND STUDYING RADICALS

Now we will focus on the most important experimental technique and theoretical method (*electron paramagnetic resonance* spectroscopy and *quantum chemistry*, respectively) currently available for gathering information on radiation-induced bioradicals in the *chemical stage*. The most prominent questions in this respect are often related to chemical reactivity and thus concern the *identity* of the radicals involved as well as their *geometry* and *electronic structure*. The basic aspects of the technique in question will be discussed together with the physical and chemical information that can be obtained from it. In this respect; attention will also be paid to the interaction between experiment and theory.

3.1.1. Electron Paramagnetic Resonance

The unique feature of electron paramagnetic resonance (EPR) or electron spin resonance (ESR) is that it is a spectroscopic technique that is applicable to paramagnetic systems. These are species containing one or more unpaired electrons resulting in a net electron spin angular momentum. It is an example of a *magnetic-dipole* spectroscopic technique in which magnetic fields are used to induce transitions between various energy levels in the atomic or molecular species by interaction with the electronic dipole moment.

An electron possesses a spin magnetic moment, and in the presence of an applied magnetic field its two permitted spin states α and β have different energies as can be

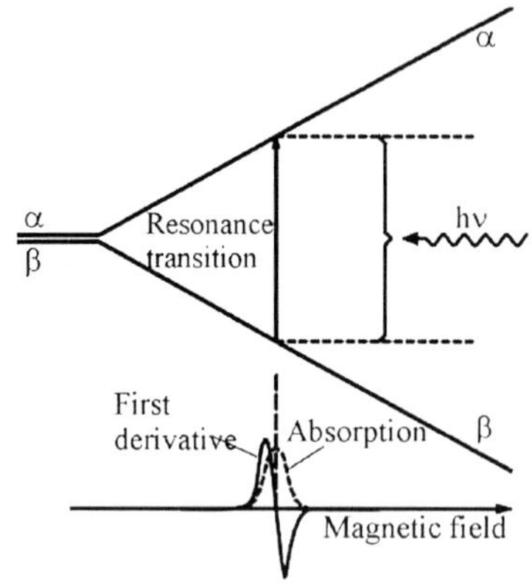

Figure 2a. *Energy-level diagram and resonant absorption for a particle of spin ½ in a magnetic field.*

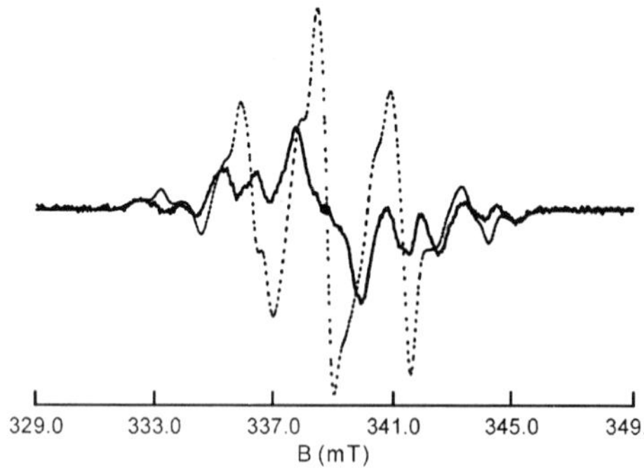

Figure 2b. *EPR spectrum of L-alanine X-irradiated at room temperature (dotted line) or at high temperature (solid line) (Vanhaelewyn 1999)*

seen from Fig. 2a. The transition from the lower to the higher energy state is made most efficiently when the energy-level separation equals hν, where h is Planck's constant and ν is the frequency of a microwave electromagnetic field. Then the sample and the electromagnetic field are in resonance. Thus an EPR spectrometer will basically consist of a source of microwaves (most frequently a klystron), a magnet capable of providing a tuneable but stable and homogeneous magnetic field and a device to detect the absorption.

In practice the absorption is monitored upon performing a scan of the static field B while the frequency ν of the microwave radiation at which the spectrometer is operated, is fixed. EPR spectra are the first-derivative of the energy absorption with respect to the magnetic field B (Fig. 2a).

EPR spectra may convey a remarkable wealth of significant information both concerning the identity and the electronic structure of the contributing radicals. This can readily be understood upon inspection of the spin hamiltonian

$$H_{spin} = g \cdot \beta_e \cdot \mathbf{S} \cdot \mathbf{B} - g_N \cdot \beta_N \cdot \mathbf{I} \cdot \mathbf{B} + \mathbf{S} \cdot \mathbf{A} \cdot \mathbf{I}$$

which is the quantum mechanical operator describing the spin-related energy changes in radical species. The first two contributions are the electronic and the nuclear Zeeman terms, respectively, caused by the interaction of the magnetic field **B** and the magnetic moments of the electrons (**S**) or nuclei (**I**) in the system. g and g_N are the electron and nuclear magnetogyric ratios, and β_e, β_N the Bohr and nuclear magnetons. The remaining term is the *hyperfine interaction* term. It arises from the interaction of the unpaired electron(s) with nearby nuclei in the molecule with a net spin magnetic moment (e.g. ^1H, ^{13}C and ^{14}N).

The principal, chemically relevant information that can be obtained from an EPR spectrum is contained by the g-factor and the hyperfine tensor **A**. From these quantities valuable information can be gained about the structure and orientation of free radicals in single crystals as well as the distribution of the unpaired electron over the magnetic nuclei in the radical species. However, the unequivocal determination of these parameters can be seriously hampered due to the potential complex character of real-life spectra such as the ones illustrated in Fig. 2b. In principle, *quantum chemical calculations* allow for the theoretical determination of the main EPR spectroscopic parameters. Accurate calculations of this type are however computationally demanding. Therefore they are starting to be performed on a routine basis only since the last few years, facilitated by the ever-increasing available computer power. In this respect, the calculation of the matrix elements of the hyperfine tensor A (the so-called hyperfine coupling constants) in radical species of biological interest is developing into a powerful tool to the experimental scientist for the elucidation and interpretation of EPR spectra. This is illustrated, for instance, in the case of the intensively investigated radicals of the X-irradiated amino acid alanine (Lahorte 1999a). A treatment of the basic concepts of quantum chemical methods will be the topic of the next paragraph.

The atoms and molecules amenable to study by EPR can either exist in a paramagnetic ground state or be (temporarily) excited into a paramagnetic state (e.g. by irradiation). Typical systems that can been studied include free radicals in the solid, liquid or gaseous phase, transitions ions, point defects in solids, systems with more than one unpaired electron (triplet-state systems or biradicals) and systems with conducting electrons. In biological systems paramagnetic species are mostly found as transition metal complexes (e.g. Cu in copper proteins and Fe in haem proteins), as intermediary products in electron transfer processes or as radiation degradation products of biomolecules. In the present review, we will concentrate on the latter category. For a compilation of the literature in the former fields, the reader is referred to separate overviews (Atherton 1996).

If in the same system two or more nuclei of the same spin are present, ambiguity can arise in the assignment of hyperfine couplings. Furthermore, if the spacing within a set of hyperfine lines do not exceed their individual widths, only a broadening might be noted and the splitting will not be detected. These shortcomings of EPR are in many cases resolved by an important elaboration of EPR in the form of electron-nuclear double-resonance (ENDOR) experiments. The major advantage of this technique is the simplification of spectra, which facilitates the measurement and interpretation of spectral parameters for all unpaired-electron systems in which nuclear spins are present.

Excellent reference works are available covering both ENDOR theory and applications as well as the underlying mathematical and physical concepts and technological aspects of EPR (Atherton 1993; Gordy 1980; Weil 1994). The reader is also referred to the literature for an introduction to some of the many new EPR developments in progress such as time-resolved spectroscopy (which involves the acquisition of EPR-data for short-lived species), EPR-imaging, *in vivo* spin-trapping and applications in food and medical sciences (Eaton 1996; Goldfarb 1996; Mader 1998; Stehlik 1997; Turro 1999).

3.1.2. Quantum chemistry

The term quantum chemistry denotes the application of the principles of quantum mechanics to topics of chemical interest. It includes the calculation of molecular structures and properties, the analysis of spectroscopic data and the description of chemical reactions in terms of individual molecular events.
At the heart of the field lies the solution of the of the nonrelativistic, time-independent electronic Schrödinger equation

$$H\Psi = E\Psi$$

Here H is the notation for the hamiltonian operator corresponding to the total energy of the system which operates on the wave function Ψ, with E being the total energy of the system.

There are two main approaches to the solution of this equation. In *ab initio* calculations, a model is chosen for the electronic wave function and the equation is solved using as input only the values of the fundamental constants (in specific, Planck's constant and the electronic mass and charge) and the atomic numbers of the atoms present. The accuracy of this approach is mainly determined by the model chosen for the wave function. For large molecules however, accurate ab initio calculations impose a substantial computational burden and various *semi-empirical* methods have been developed that are computationally less expensive. In a semi-empirical method a simplified form for the hamiltonian is used in combination with adjustable parameters obtained from experimental data.

The solution of the Schrödinger equation is a very challenging task. Normally a one-particle, or orbital, basis set of atom-centred Gaussian functions is used as a model for constructing the molecular wave function. In traditional approaches one starts with the *Hartree-Fock* approximation where the molecular orbitals are solved for in terms of the chosen orbital basis set assuming that the electrons move in the average field of all others. While this approximation allows a great many molecular properties to be determined with a (near) experimental accuracy, erroneous results can be produced in those cases where the incorporation of electron correlation, the instantaneous interaction of the electrons, is quintessential. This is particularly true for the calculation of magnetic properties of radicals such as hyperfine coupling constants. The electron correlation can be accounted for by so-called *post-Hartree-Fock* approaches that are however characterised by a dramatic increase of the computational work involved. Consequently, their use is limited to small molecular systems, even with the formidable computer power currently available. Therefore, resort has to be sought to other methods if one is interested in routinely calculating properties to a high degree of accuracy in systems that are of true interest to a biologist or chemist such as radicals derived from biomolecules.

Density functional calculations have proved valuable in this respect over the last few years. The basis for density functional theory (DFT) is the proof by Hohenberg and Kohn (1964) that the ground state electronic energy of an atomic or molecular species is completely determined by the electron density ρ. This means that there exists a one-to-one correspondence between the electron density of the system and the energy. The problem is that although it has been proven that each different density yields a different ground-state energy, the *functional*[2] connecting these two quantities is unknown. The goals of DFT methods is to design functionals that connect the electron density with the energy.

In analogy with the wave function approach, the energy functional can be divided into three parts, kinetic energy, $T[\rho]$, attraction between the nuclei and electrons, $E_{ne}[\rho]$, and electron-electron repulsion, $E_{ee}[\rho]$. The $E_{ee}[\rho]$ term may be further divided into a Coulomb and exchange part, $J[\rho]$ and $K[\rho]$, respectively. The exchange energy is a correction that should be made to the Coulomb energy to take into account the effect of spin correlation.[3] Thus the energy functional is written as

$$E[\rho] = T[\rho] + E_{ne}[\rho] + J[\rho] + K[\rho].$$

The key for the practical use of DFT methods in computational chemistry lies in the Kohn-Sham theory. Central in this theory is the calculation of the kinetic energy $T_s[\rho]$ assuming that the electrons are not interacting (in the same way as non-interacting electrons are described in wave mechanics by the HF method). In reality, the electrons are interacting, so the Kohn-Sham kinetic energy does not provide the total kinetic energy. However, the difference between the exact kinetic energy and that calculated

[2] A functional is a prescription for producing a number from a function which in turn depends on variables. The electron density ρ is a function (of spatial and spin co-ordinates) whereas an energy depending on the density is a functional, denoted with brackets, $E[\rho]$.

[3] Spin correlation refers to the quantum mechanical effect that gives electrons of opposite spin an increased probability of being found near to each other, whereas those of the same spin are preferentially kept apart.

assuming non-interacting electron orbitals is small. The remaining kinetic energy is absorbed into an *exchange-correlation* term, and thus the general DFT energy can be written as

$$E_{DFT}[\rho] = T_S[\rho] + E_{ne}[\rho] + J[\rho] + E_{xc}[\rho].$$

By equating E_{DFT} to the expression for the exact total energy E, it can be seen the exchange-correlation energy is defined as

$$E_{xc}[\rho] = (T[\rho] - T_S[\rho]) + (E_{ee}[\rho] - J[\rho]).$$

From this expression it can readily be seen that the exchange-correlation functional accounts for the residual energy differences between the true energy and the one obtained when using an approximate model in which correlation between electron orbitals is neglected. Indeed, by construction both $T_S[\rho]$ and $J[\rho]$ are readily calculated.

The major problem in DFT however consists of deriving suitable formulas for the exchange-correlation term. Although it is possible to prove that the exchange-correlation potential is a unique functional, valid for all systems, an explicit functional form for this potential has been elusive. The difference between DFT methods is the choice of the functional form of the exchange-correlation energy and the development of new and improved functionals represents an active field in DFT research.

The power of DFT is that only the total density needs to be considered. Whereas in the wave function approach the complexity increases with the number of electrons, the electron density has the same number of variables independent of the system size. Several excellent reference works, treating quantum chemical techniques in general (Jenssen 1999, Levine 1991) as well as DFT theory (Baerends 1997; Kohn 1996; Seminario 1995) and applications toward radical magnetic properties (Barone 1995; Malkin 1995, Ventura 1997) are available.

3.2. FUNDAMENTAL STUDIES OF RADIATION EFFECTS ON BIOMOLECULES

In Table 1 an overview is presented of the multitude of available studies of radiation-induced radicals in biomolecules, partially or completely conducted with the use of the experimental or theoretical methods described in paragraph 3.1.

Most of the studies on biological systems employing EPR spectroscopy have focused extensively and successfully on the detection and identification of low-molecular-weight reactive radicals. To a lesser extent the technique has also been used to investigate high-molecular-weight species generated as a result of radical-induced damage to biological macromolecules, such as DNA, RNA, proteins and carbohydrates (Davies 1993).

Complementary to the available experimental techniques, theoretical quantum chemical calculations of hyperfine coupling constants (hfcc's) have proven very valuable in the investigation of the identity and the electronic structure of the radicals involved. As mentioned before, DFT calculations of hyperfine couplings in biologically relevant radicals are steadily being recognised as a powerful method to the

experimental scientist for verifying and predicting experimental values thus contributing substantially to a better understanding of EPR spectra.
One of the main goals in the study of radiation effects on DNA has been the understanding of the processes leading from the primary excitations and ionisations of the DNA constituents to the stable damage of biological relevance. Although substantial progress has been made (e.g. an almost complete description is now possible of the radiation-induced decomposition of the guanine moiety in DNA (Cadet 1999)), in general relatively little detailed knowledge exists regarding the exact reaction mechanisms and the identity of the radical species involved. This can at least partly be explained by the relative complexity of DNA and its molecular environment.

The numerous EPR and ENDOR studies on DNA model systems indicate that in addition to the various bases, sugar radicals may be a site of significant radiation damage although their detection in full DNA is not straightforward.
The available theoretical studies concern geometries, relative energies, spin density distributions and hyperfine coupling constants of base and sugar radicals generated in irradiated DNA. In most cases, the calculated hyperfine coupling constants directly support the assignment of experimentally observed couplings to the specific radicals.

Many of the research efforts on amino acids (in particular alanine) and sugars are inspired by their potential dosimetric applications. L-a-alanine is currently used as a reference dosimeter suitable over a wide dose range (McLaughlin 1993) whereas the development of a dosimetric protocol for various sugars (e.g. glucose, fructose, sucrose) plays an important role in the detection of some irradiated foodstuff (Fattibene 1996).

Table 1. *Experimental and theoretical studies of radiation-induced radicals in biologically relevant molecules*

DNA	
Experimental	Becker 1993, 1994; Close 1993, 1997; Cullis 1990; Davies 1993; Lin 1998; Sanderud 1996; Spalletta 1993; Weiland 1996; Yan 1992;
Calculations	Colson 1995b
DNA bases	
Experimental	Ambroz 1998; Barnes 1991, 1994, 1995a-b; Becker 1996; Bernhard 1989, 1994a-b; Close 1988, 1989, 1993, 1994, 1998; Cullis 1990, 1992a-b, 1995, 1996; Davies 1995; Herak 1994, 1997, 1999; Hole 1989, 1991a-b, 1992a-c, 1995, 1998; Kabiljo 1990; Kim 1989; Malone 1995; Mroczka 1995, 1997; Nelson 1988, 1989, 1992a-b, 1998; Podmore 1991; Sagstuen 1988, 1989a-b, 1992a-b, 1996, 1998; Sankovic 1991, 1996; Sevilla 1989, 1991; Swarts 1992, 1996; Wang 1994a-b, 1997.
Calculations	Colson 1992a-b, 1993a-b, 1995c, 1996; Sevilla 1995; Wetmore 1998 a-d.

Table 1. Experimental and theoretical studies of radiation-induced radicals in biologically relevant molecules

DNA sugars	
Experimental	Box 1990; Close 1997.
Theoretical	Colson 1995a; Luo 1999; Miaskiewicz 1994; Wetmore 1998e.
Amino acids	
Experimental	Berthomieu 1996; Davies 1996; Hawkins 1998; Huang 1998; Lassmann 1999; Sagstuen 1997; Sanderud 1998.
Calculations	Ban 1999; Lahorte 1999a;
Steroids	
Experimental	Krzyminiewski 1990, 1995; Szyczewski 1994, 1996, 1998.
Calculations	Kubli-Garfias 1998a-e, 1999; Lahorte 1999b.
Sugars	
Experimental	Raffi 1993; Sagstuen 1986; Triolet 1990, 1991, 1992a-b.

4. Applications of irradiation of biomolecules and biomaterials

4.1. RADIATION SOURCES FOR THE PRODUCTION OF BIORADICALS

During the early days of radiation chemistry, investigations were carried out with the aid of radioactive sources producing α, β and γ-radiation. Also neutrons from nuclear reactors were used. Nowadays bioradicals are preferentially created with electron beams from electron accelerators, and with photon beams produced either by radioactive γ-sources or by electron accelerators via the bremsstrahlung process. Here we have to notice that in the literature there is some confusion when the terms 'X-rays' and γ-rays' are used. Both stand for photon beams. From the physical point of view they are indistinguishable. Actually there is a general consensus to use 'X-rays' when these photons are produced by an accelerator and 'γ-rays' when the photons are emitted by a radioactive source. We will follow this convention. The γ-source of choice is ^{60}Co. ^{60}Co produces γ-rays with energies of 1.173 and 1.333 MeV, with a half-life of about 5 years.

An electron accelerator imparts high kinetic energy to electrons. Two types of electron accelerators are used most often for the production of electron or photon beams in the required energy range from 1 to 10 MeV: the DC (direct current) accelerator and the linear electron accelerator (also called linac). In both types electrons are accelerated to a high kinetic energy in an evacuated tube (typically 1 – 3 m long).

In **DC accelerators** this is achieved by applying a high voltage across the terminals of this vacuum tube. For optimum insulation the vacuum tube and the high-voltage generator are contained in a pressurised tank filled with SF_6 gas. Electrons emitted from an electron source at the negative terminal are pushed away and attracted to the positive end which is placed at earth potential. The potential difference between the terminals defines the energy of the electrons. The different types of DC accelerators differ primarily only in the way the accelerating voltage of several million volts is produced. There are four major types: the Van de Graaff, the Cockroft-Walton, the Dynamitron and the ICT (Insulating Core Transformer) accelerator.

It is very difficult to handle accelerating potentials higher than 5 MV in DC accelerators, while **linear electron accelerators** are in principle unlimited in energy. In an electron linac bunches of electrons are accelerated in an evacuated tube by the electric field of a high-frequency electromagnetic wave travelling down the tube. The electrons ride on the top of this travelling wave (in a simplified picture comparable with a surfer riding on the crest of a water wave). The travelling waves with a typical frequency of several GHz and peak powers of several MW are produced in a klystron. By repeating this process in consecutive accelerator tubes, the electron energy can be increased up to the required energies.

At the end of the accelerator the electron beam can leave the vacuum through a window to be used directly, or it can be sent to a bremstrahlung converter target to be transformed in a photon beam. The optimum power conversion efficiency for 5 MeV electrons into a photon beam is about 10 %. The remaining 90 % of the power is dissipated as heat. The thermal design of the bremsstrahlung target is one of the critical design parameters of an electron irradiator for industrial radiation applications. In contrast to the applications in medicine where typical irradiation doses are in the 2 – 4 Gy range, in radiation processing typical doses are situated in the kGy to MGy-range. Due to the high doses required, radioactive sources of 1 million Curie are typical in an industrial environment. To produce the equivalent photon beam power with a 5 MeV accelerator an electron beam power of about 150 kW is required. Only in recent years compact, reliable electron accelerators became available at these extremely high beam power levels (Van Lancker 1999). That is one of the main reasons why Cobalt sources were predominant in the past. As an example: a 100 kW electron accelerator with double-sided irradiation can sterilise about 10 tons/h.

The penetration of radiation energy into a material can be represented by depth-dose curves in which the relative absorbed dose at a particular point is plotted against the distance (i.e. the depth) of that point from the irradiated face of the sample. The shape of these depth-dose curves depends on the nature of the radiation (electrons or photons), their energy, the density of the irradiated material and the irradiation geometry.

A sample with a thickness of 20 cm can be irradiated reasonably uniformly throughout its depth with a Co-source or with 4 MV X-rays. In contrast electrons are less penetrating than photons of the same energy. The useful range of electrons, that is the depth of the material at which the entrance and exit doses are the same, varies proportional with the energy. For 10 MeV electrons, it is about 3.3 cm in water. The useful range for double-sided irradiation is about 2.5 times higher. The reduced penetration depth of electrons is one of the limitations of the industrial applications of electron beams, because the maximum energy is limited to 10 MeV. For biological applications, such as sterilisation of disposables, or irradiation of foodstuffs, dose variations ± 30 % are acceptable, provided all parts of the sample receive at least the specified minimum dose. However smaller dose variations of ± 10 % are needed in polymer processing, such as in the production of heat-shrinkable and foamed products, where the homogeneity of the product is important. For the synthesis of biomaterials or the sterilisation of pharmaceuticals, even better homogeneities are required. Because an electron beam has a characteristic diameter of a few mm, the whole power of the beam would be concentrated on a small area of the irradiated specimen. In a magnet attached to the end of the accelerator tube, the electron can be scanned so that a large specimen receives a homogeneous radiation dose. A detailed description of the various available types of accelerators can be found in Scharf (1994). Various overviews of both research and industrial applications of electron accelerators are available (Mehnert 1996, Mondelaers 1998, Lahorte 1999c).

4.2. BIO(TECHNOL)OGICAL IRRADIATION APPLICATIONS

Table 2 gives an overview of applied irradiation research in biology and related fields. A number of studies have focused on enhancing macroscopic plant characteristics in flowers, vegetables and fruits. The majority of irradiation research efforts in biology have focused on understanding at a fundamental level biological effects of ionising radiation on cells and living (micro-)organisms. Important applications of this research can be situated in the fields of radiobiology and radiotherapy but as these are beyond the scope of the present work and have already been extensively reviewed (Lett 1994; Steele 1993; Ward 1990), they will not be incorporated here.

Biomedical engineering is one the fields that have substantially benefited from the development of innovative irradiation applications. One of the most exciting research areas in this respect has been the radiation-prepared hydrogels[4] of which various applications have been achieved. Of these can be mentioned wound dressings, supports for immobilisation of bioactive substances, artificial skin, soft contact lenses and artificial organs. Furthermore a new area of applications has been opened by the development of stimuli-sensitive and responsive gels. These can reversibly undergo volume changes between shrinkage and expansion in response to the on-off switching of stimuli such as changes in pH, temperature, solvent and solute concentration or electric field.

As a consequence of the rapid development of agriculture and industry, in combination with the growth of large population centres, the management and purification of both municipal and industrial wastes is becoming an important problem. Radiation technology offers interesting opportunities in this respect. More specifically, radiation has been used to treat and purify water (natural and polluted), industrial liquid wastes, sewage sludge and flue gases.

---

[4] A hydrogel can be defined as crosslinked hydrophilic polymers in a water-swollen state.

Sterilisation of food products, one of the earliest applications of radiation processing in food technology, is now a well-established technology. Together with the radiation sterilisation of biomedical devices (e.g. surgical supplies such as orthopaedic implants, bandages, syringes) and pharmaceuticals it represents a substantial industrial activity which is quickly developing due to the availability in recent years of powerful and compact linear electron accelerators. Complementary to the contributions mentioned in Table 2, a comprehensive overview can be found in Woods (1994) of irradiation processing applications such as sterilisation of medical and pharmaceutical products, treatment of food and waste management. More recent applications include the inhibition of sprouts (e.g. in potatoes, onions, garlic), the delay of maturation of fresh fruits and vegetables (e.g. bananas, oranges, mushrooms) and the extension of shelf life (e.g. strawberries, fresh meat and fish). The study of the interaction of food products with irradiated packaging materials is one of the current topics under investigation.

A few applications can also be noted in the field of medicine. The irradiation of blood has been investigated for the prevention of graft-versus-host disease on immunodeficient patients by the abrogation of T-lymphocytes. Furthermore the technique of high-dose extracorporeal irradiation and subsequent re-implantation has been developed for the treatment of malignant primary bone tumour and cartilage tumours. A long-term follow-up survey of the clinical data of the first 50 patients, treated during the beginning period, revealed that long-term bone stability complications were detected for knee and hip grafts, which are complication patterns very similar to those encountered by the treatment with allografts. The technique remains very useful, however, for other tumour localisations.

Finally, quite a number of studies have focused on the investigation of the degradation of and radical mechanisms in irradiated pharmaceutical products (e.g. cephalosporin antibiotics) using techniques such as EPR, HPLC and pulse radiolysis.

Table 2. Applications of X-irradiation in biology and related fields

Biology	
Bud regeneration in plants after irradiation	Ibrahim 1998.
Seed radiation resistance	Kumagai 2000
Pollen irradiation	Dewitte 1994; Dore 1993; Falque 1994; Lardon 1999; Musial 1998, 1999; Naess 1998.
Biomedical engineering	
Biomaterials	
Responsive gels	Carenza 1994; Kaetsu 1996, 1999; Mathur 1996; Rosiak 1995, 1999;
Sterilisation	Burg 1996a; Shalaby 1996;.
Environmental Engineering	
Waste management	Binks 1996; Capizzi 1999; IAEA 1997, 1998;

Table 2. Applications of X-irradiation in biology and related fields

Food technology	
Applications of food irradiation	Burg 1996b; Monk 1995; Sendra 1996; Thakur 1994
Interaction with food packaging materials	Buchalla 1993, 1999; Demertzis 1999; Garde 1998; Goulas 1996; Riganakos 1999.
Detection of irradiated food	Dodd 1995; Haire 1997.
Medicine	
Blood and blood components	Jacobs 1998; Moroff 1997
Extracorporeal bone irradiation	Bohm 1998; Mondelaers 1993; Sabo 1999.
Pharmacy	
Degradation and detection of pharmaceuticals.	Barbarin 1996; Basly 1996a-c, 1997a-h, 1998a-g; Crucq 1995, 1996; Duroux 1996, 1997; Fauconnet 1996; Gibella 1993, 1994; Miyazaki 1994a-b; Onori 1996, Signoretti 1993, 1994; Zeegers 1993, 1997.
Irradiation of alkaloids, morphine derivatives, antibiotics	Boess 1996
Sterilisation of pharmaceutical products	Gopal 1988; Jacobs 1988, 1995; Talrose 1995, Tilquin 1999.

5. Conclusions

In the first part of this review a survey was given of the broad range of both experimental techniques and theoretical methods currently available for studying radiation effects on biological material from the initial physical phase of energy deposition through the biological time scale. Consequently, the most important techniques which yield information with regard to the intermediate chemical phase, were treated in more detail. It was shown that electron paramagnetic spectroscopy and quantum chemical calculations yield valuable direct information on important radical properties. An overview was given of studies in this respect on biologically relevant molecular species such as DNA, amino acids and sugars.

Finally, attention was paid to modern accelerator technologies available for creating radicals and the application of X-irradiation in biology-related fields was reviewed.

Acknowledgements

This work has been performed within the framework of a Concerted Research Action (GOA-12050100) with financial support from the Research Board of the Ghent University. W. Mondelaers acknowledges the F.W.O. Flanders (Belgium). The assistance of Mr. R. Verspille in producing the figures is highly appreciated.

References

Ambroz, H. B., Kemp, T. J., Kornacka, E. M., and Przybytniak, G. K. 1998. The role of copper and iron ions in the gamma-radiolysis of DNA. Part 1. EPR studies at cryogenic temperatures. *Radiat. Phys. Chem.* **53**:491-499.

Atherton, N. M.1993. *Principles of electron spin resonance*, Ellis Horwood Limited, Chichester

Atherton, N. M., Davies, M. J., and Gilbert, B. C. 1996. *Electron spin resonance, Volume 15*. The Royal Society of Chemistry, Cambridge.

Baerends, E. J., and Gritsenko, O. V. 1997. A quantum chemical view of density functional theory. *J. Phys. Chem. A* **101**:5383-5403.

Ban, F. Q., Wetmore, S. D., and Boyd, R. J. 1999. A density-functional theory investigation of the radiation products of L-alpha-alanine. *J. Phys. Chem. A* **103**:4303-4308.

Barbarin, N., Crucq, A. S., and Tilquin, B. 1996. Study of volatile compounds from the radiosterilisation of solid cephalosporins. *Radiat. Phys. Chem.* **48**:787-794.

Barnes, J., Bernhard, W. A., and Mercer, K. R. 1991. Distribution of electron trapping in DNA - protonation of one-electron reduced cytosine. *Radiat. Res.* **126**:104-107.

Barnes, J., and Bernhard, W. A. 1994. Irreversible protonation sites of one-electron-reduced adenine - comparisons between the C5 and the C2 or C8 protonation sites. *J. Phys. Chem.* **98**:10969-10977.

Barnes, J., and Bernhard, W. A. 1995a. The distribution of electron-trapping in DNA - one-electron-reduced oligodeoxynuclotides of adenine and thymine. *Radiat. Res.* **143**:85-92.

Barnes, J., and Bernhard, W. A. 1995b. The electron scavenging ability of the DNA bases in glassy matrices X-irradiated at 4 K. *J. Phys. Chem.* **99**:11248-11254.

Barone, V. 1995. Structure, magnetic properties and reactivities of open-shell species from density functional and self-consistent hybrid methods. In *Recent advances in density functional methods, part 1* (D. P. Chong, Eds.), pp. 287-334. World scientific, Singapore.

Basly, J. P., Duroux, J. L., and Bernard, M. 1996a. Gamma radiation induced effects on metronidazole. *Int. J. Pharm.* **139**:219-221.

Basly, J. P., Duroux, J. L., and Bernard, M. 1996b. Gamma irradiation sterilisation of orciprenaline and fenoterol. *Int. J. Pharm.* **142**:125-128.

Basly, J. P., Duroux, J. L., Bernard, M., and Penicaut, B. 1996c. Gamma radiolysis of three antiparasitic agents: Metronidazole, ornidazole, and ternidazole. *J. Chim. Phys. Phys-Chim. Biol.* **93**:1-6.

Basly, J. P., Longy, I., and Bernard, M. 1997a. ESR identification of radiosterilised pharmaceuticals: latamoxef and ceftriaxone. *Int. J. Pharm.* **158**:241-245.

Basly, J. P., Longy, I., and Bernard, M. 1997b. Influence of radiation treatment on two antibacterial agents and four antiprotozoal agents: ESR dosimetry. *Int. J. Pharm.* **154**:109-113.

Basly, J. P., Longy, I., and Bernard, M. 1997c. Influence of radiation treatment on theodrenaline: ESR and HPLC study. *Int. J. Pharm.* **152**:201-206.

Basly, J. P., Longy, I., and Bernard, M. 1997d. Radiation effects on dopamine and norepinephrine. *Pharmaceut. Res.* **14**:1192-1196.

Basly, J. P., Duroux, J. L., and Bernard, M. 1997e. The effect of gamma radiation on the degradation of Salbutamol. *J. Pharmaceut. Biomed. Anal.* **15**:1137-1141.

Basly, J. P., and Bernard, M. 1997f. radiosterilisation dosimetry by ESR spectroscopy: Ritodrine hydrochloride and comparison with other sympathomimetics. *Int. J. Pharm.* **149**:85-91.

Basly, J. P., Longy, I., and Bernard, M. 1997g. Radiation sterilisation of formoterol. *Pharmaceut. Res.* **14**:810-814.

Basly, J. P., Longy, I., and Bernard, M. 1997h. ESR dosimetry of irradiated ascorbic acid. *Pharmaceut. Res.* **14**:1186-1191.

Basly, J. P., Basly, I., and Bernard, M. 1998a. Radiosterilisation dosimetry of vitamins: an ESR study. *Int. J. Radiat. Biol.* **74**:521-528.

Basly, J. P., Basly, I., and Bernard, M. 1998b. Radiation-induced effects on cefotaxime: ESR study. *Free Radical Res.* **29**:67-73.

Basly, J. P., Basly, I., and Bernard, M. 1998c. ESR spectroscopy applied to the study of pharmaceuticals radiosterilisation: cefoperazone. *J. Pharmaceut. Biomed. Anal.* **17**:871-875.

Basly, J. P., Longy, I., and Bernard, M. 1998d. Radiosterilisation dosimetry by electron-spin resonance spectroscopy: Cefotetan. *Anal. Chim. Acta* **359:**107-113.

Basly, J. P., Basly, I., and Bernard, M. 1998e. Electron spin resonance identification of irradiated ascorbic acid: Dosimetry and influence of powder fineness. *Anal. Chim. Acta* **372:**373-378.

Basly, J. P., Basly, I., and Bernard, M. 1998f. Electron spin resonance detection of radiosterilisation of pharmaceuticals: application to four nitrofurans. *Analyst* **123:**1753-1756.

Basly, J. P., Basly, I., and Bernard, M. 1998g. Influence of radiation treatment on dobutamine. *Int. J. Pharm.* **170:**265-269.

Basly, J. P., Basly, I., and Bernard, M. 1999. Radiation induced effects on cephalosporins: an ESR study. *Int. J. Radiat. Biol.* **75:**259-263.

Becker, D., and Sevilla, M. 1993. The chemical consequences of radiation-damage to DNA. *Advan. Radiat. Biol.* **17:**121-180.

Becker, D., Lavere, T., and Sevilla, M. 1994. ESR detection at 77 K of the hydroxyl radical in the hydration layer of gamma-irradiated DNA. *Radiat. Res.* **140:**123-129.

Becker, D., Razskazovskii, Y., Callaghan, M. U., and Sevilla, M. 1996. Electron spin resonance of DNA irradiated with a heavy-ion beam (O-16(8+)): Evidence for damage to the deoxyribose phosphate backbone. *Radiat. Res.* **146:**361-368.

Bernhard, W. A. 1989. Sites of electron trapping in DNA as determined by ESR of one-electron reduced oligonucleotides. *J. Phys. Chem.* **93:**2187-2189.

Bernhard, W. A., Barnes, J., Mercer, K. R., and Mroczka, N. 1994. The influence of packing on free-radical yields in crystalline nucleic-acids - the pyrimidine bases. *Radiat. Res.* **140:**199-214.

Bernhard, W. A., Mroczka, N., and Barnes, J. 1994. Combination is the dominant free-radical process initiated in DNA by ionising-radiation - An overview based on solid-state EPR studies. *Int. J. Radiat. Biol.* **66:**491-497.

Berthomieu, C., and Boussac, A. 1995. FTIR and EPR study of radicals of aromatic-amino-acids 4-Methylimidazole and phenol generated by UV irradiation. *Biospectroscopy* **1:**187-206.

Binks, P. R. 1996. Radioresistant bacteria: Have they got industrial uses? *J. Chem. Technol. Biotechnol.* **67:**319-322.

Boess, C., and Bogl, K. W. 1996. Influence of radiation treatment on pharmaceuticals - A review: Alkaloids, morphine derivatives, and antibiotics. *Drug Develop. Ind. Pharm.* **22:** 495-529.

Bohm, P. 1998. The re-implantation of extracorporeally devitalised bone segments for defect reconstruction in tumour orthopaedics - A review of the literature. *Z. Orthop. Grenzgeb.* **136:**197-204.

Bolch, W. E., Turner, J. E., Yoshida, H., Jacobson, K. B., Hamm, R. N., and Crawford, O. H. 1998. Product yields from irradiated glycylglycine in oxygen-free solutions: Monte Carlo simulations and comparison with experiments. *Radiat. Environ. Biophys.* **37:**157-166.

Box, H. C., Budzinski, E. E., and Freund, H. G. 1990. Electrons trapped in single crystals of sucrose - Induced spin-densities. *J. Chem. Phys.* **93:**55-57.

Briden, P. E., Holt, P. D., and Simmons, J. A. 1999. The track structures of ionising particles and their application to radiation biophysics - I. A new analytical method for investigating two biophysical models. *Radiat. Environ. Biophys.* **38:**175-184.

Buchalla, R., Schuttler, C., and Bogl, K. W. 1993. Effects of ionising radiation on plastic food-packaging materials - A review .2. Global migration, sensory changes and the fate of additives. *J. Food. Protect.* **56:**998-1005.

Buchalla, R., Boess, C., and Bogl, K. W. 1999. Characterisation of volatile radiolysis products in radiation-sterilised plastics by thermal desorption-gas chromatography- mass spectrometry: screening of six medical polymers. *Radiat. Phys. Chem.* **56:**353-367.

Burg, K. J. L., and Shalaby, S. W.1996.Advances in food irradiation research, 254-261.

Cadet, J., D'Ham, C., Douki, T., Pouget, J. P., Ravanat, J. L., and Sauvaigo, S. 1998. Facts and artefacts in the measurement of oxidative base damage to DNA. *Free Radical Res.* **29:**541-550.

Cadet, J., Douki, T., Gasparutto, D., Gromova, M., Pouget, J. P., Ravanat, J. L., Romieu, A., and Sauvaigo, S. 1999. Radiation-induced damage to DNA: mechanistic aspects and measurement of base lesions. *Nucl. Instrum. Meth. Phys. Res. B* **151:**1-7.

Capizzi, S., Chevallier, A., and Schwartzbrod, J. 1999. Destruction of *Ascaris* ova by accelerated electron. *Radiat. Phys. Chem.* **56:**591-595.

Carenza, M., and Veronese, F. M. 1994. Entrapment of biomolecules into hydrogels obtained by radiation-induced polymerisation. *J. Control. Release* **29:**187-193.
Chalak, L., and Legave, J. M. 1997. Effects of pollination by irradiated pollen in Hayward kiwifruit and spontaneous doubling of induced parthenogenetic trihaploids. *Sci. Hort.* **68:**83-93.
Claverie, J. M. 1999. Computational methods for the identification of differential and co-ordinated gene expression. *Hum. Mol. Genet.* **8:**1821-1832.
Close, D. M., Sagstuen, E., and Nelson, W. H. 1988. Radical formation in X-irradiated single-crystals of guanine hydrochloride monohydrate .3. Secondary radicals and reaction mechanisms. *Radiat. Res.* **116:**379-392.
Close, D. M., Nelson, W. H., Sagstuen, E., and Hole, E. O. 1989. ESR and ENDOR studies of X-irradiated single-crystals of guanine derivatives. *Free Radical Res. Commun.* **6:**83-85.
Close, D. M. 1993. Radical ions and their reactions in DNA constituents - ESR/ENDOR studies of radiation-damage in the solid state. *Radiat. Res.* **135:**1-15.
Close, D. M., Nelson, W. H., Sagstuen, E., and Hole, E. O. 1994. ESR and ENDOR study of single-crystals of deoxyadenosine monohydrate X-irradiated at 10 K. *Radiat. Res.* **137:**300-309.
Close, D. M. 1997. Where are the sugar radicals in irradiated DNA? *Radiat. Res.* **147:**663-673.
Close, D. M., Hole, E. O., Sagstuen, E., and Nelson, W. H. 1998. EPR and ENDOR studies of X-irradiated single crystals of deoxycytidine 5 '-phosphate monohydrate at 10 and 77 K. *J. Phys. Chem. A* **102:**6737-6744.
Colson, A. O., Besler, B., and Sevilla, M. 1992a. Ab-initio molecular-orbital calculations on DNA-base pair radical ions - effects of base pairing on proton-transfer energies, electron-affinities, and ionisation-potentials. *J. Phys. Chem.* **96:**9787-9794.
Colson, A. O., Besler, B., Close, D. M., and Sevilla, M. 1992b. Ab initio molecular-orbital calculations of DNA bases and their radical ions in various protonation states - evidence for proton-transfer in GC base pair radical-anions. *J. Phys. Chem.* **96:**661-668.
Colson, A. O., Besler, B., and Sevilla, M. 1993a. Ab-initio molecular-orbital calculations on DNA-radicals .4. effect of hydration and electron-affinities and ionisation-potentials of base pairs. *J. Phys. Chem.* **97:**13852-13859.
Colson, A. O., Besler, B., and Sevilla, M. 1993b. Ab-initio molecular-orbital calculations on DNA radical ions .3. ionisation-potentials and ionisation sites in components of the DNA sugar-phosphate backbone. *J. Phys. Chem.* **97:**8092-8097.
Colson, A. O., and Sevilla, M. 1995a. Structure and relative stability of deoxyribose radicals in a model DNA backbone - Ab-initio molecular-orbital calculations. *J. Phys. Chem.* **99:**3867-3874.
Colson, A. O., and Sevilla, M. 1995b. Elucidation of primary radiation-damage in DNA through application of ab-initio molecular-orbital theory. *Int. J. Radiat. Biol.* **67:**627-645.
Colson, A. O., and Sevilla, M. 1995c. Ab-initio molecular orbital calculations of radicals formed by $H^{\bullet}$ and $^{\bullet}OH$ addition to the DNA bases - electron affinities and ionisation potentials. *J. Phys. Chem.* **99:**13033-13037.
Colson, A. O., and Sevilla, M. 1996. Ab initio molecular orbital study of the structures of purine hydrates. *J. Phys. Chem.* **100:**4420-4423.
Crucq, A. S., Tilquin, B. L., and Hickel, B. 1995. Radical mechanisms of cephalosporins - A pulse-radiolysis study. *Free Radical Biol. Med.* **18:**841-847.
Crucq, A. S., and Tilquin, B. L. 1996. Attack of cefotaxime by different radicals: Comparison of the effects. *Free Radical Biol. Med.* **21:**827-832.
Cullis, P. M., Davis, A. S., Malone, M. E., Podmore, I. D., and Symons, M. C. R. 1992a. Electron-paramagnetic resonance studies of the effects of 1/1 electrolytes on the action of ionising-radiation on aqueous DNA. *J. Chem. Soc. Perkin Trans. 2* 1409-1412.
Cullis, P. M., McClymont, J. D., Malone, M. E., Mather, A. N., Podmore, I. D., Sweeney, M. C., and Symons, M. C. R. 1992b. Effects of ionising-radiation on deoxyribonucleic-acid .7. electron-capture at cytosine and thymine. *J. Chem. Soc. Perkin Trans. 2* 1695-1702.
Cullis, P. M., Malone, M. E., Podmore, I. D., and Symons, M. C. R. 1995. Site of protonation of one-electron-reduced cytosine and its derivatives in aqueous-methanol glasses. *J. Phys. Chem.* **99:**9293-9298.

Cullis, P. M., Malone, M. E., and MersonDavies, L. A. 1996. Guanine radical cations are precursors of 7,8-dihydro-8-oxo-2'- deoxyguanosine but are not precursors of immediate strand breaks in DNA. *J. Amer. Chem. Soc.* **118:**2775-2781.
Davies, M. J. 1993. Detection and identification of macromolecule-derived radicals by EPR-spin trapping. *Res. Chem. Intermediates* **19:**669-679.
Davies, M. J., Gilbert, B. C., Hazlewood, C., and Polack, N. P. 1995. EPR spin-trapping studies of radical-damage to DNA. *J. Chem. Soc. Perkin Trans. 2* 13-21.
Davies, M. J. 1996. Protein and peptide alkoxyl radicals can give rise to C- terminal decarboxylation and backbone cleavage. *Arch. Biochem. Biophys.* **336:**163-172.
Demertzis, P. G., Franz, R., and Welle, F. 1999. The effects of gamma-irradiation on compositional changes in plastic packaging films. *Packag. Technol. Sci.* **12:**119-130.
Dewitte, K., and Keulemans, J. 1994. Restriction of the efficiency of haploid plant-production in apple cultivar Idared, through parthenogenesis in-situ. *Euphytica* **77:**141-146.
Dodd, N. J. F. 1995. Free radicals and food irradiation. *Free Radicals and Oxidative Stress: Environment, Drugs and Food Additives* 247-258.
Dore, C., and Marie, F. 1993. Production of gynogenetic plants of onion (Allium-Cepra L) after crossing with irradiated pollen. *Plant Preed.* **111:**142-147.
Duroux, J. L., Basly, J. P., Penicaut, B., and Bernard, M. 1996. ESR spectroscopy applied to the study of drugs radiosterilisation: Case of three nitroimidazoles. *Appl. Radiat. Isotopes* **47:**1565-1568.
Duroux, J. L., Basly, J. P., and Bernard, M. 1997. Drugs radiosterilisation. Importance of electron spin resonance in dosimetry. *J. Chim. Phys. Phys-Chim. Biol.* **94:**405-409.
Dyer, M. R., Cohen, D., and Herrling, P. L. 1999. Functional genomics: from genes to new therapies. *Drug Discov. Today* **4:**109-114.
Eaton, S. S., and Eaton, G. R. 1996. EPR imaging. In *Electron spin resonance, Volume 15*, pp. 169-185. The Royal Society of Chemistry, Cambridge.
Fairbairn, D. W., Olive, P. L., and ONeill, K. L. 1995. The comet assay - A comprehensive review. *Mutat. Res. -Rev. Genet. Toxicol.* **339:**37-59.
Falque, M. 1994. Pod and seed development and phenotype of the M1 plants after pollination and fertilisation with irradiated pollen in cacao (*Theobroma cacao L*). *Euphytica* **75:**19-25.
Fattibene, P., Duckworth, T. L., and Desrosiers, M. F. 1996. Critical evaluation of the sugar-EPR dosimetry system. *Appl. Radiat. Isotopes* **47:**1375-1379.
Fauconnet, A. L., Basly, J. P., and Bernard, M. 1996. Gamma radiation induced effects on isoproterenol. *Int. J. Pharm.* **144:**123-125.
Garde, J. A., Catala, R., and Gavara, R. 1998. Global and specific migration of antioxidants from polypropylene films into food simulates. *J. Food. Protect.* **61:**1000-1006.
Gibella, M., Crucq, A. S., and Tilquin, B. 1993. ESR measurements and the detection of radiosterilisation of drugs. *J. Chim. Phys. Phys-Chim. Biol.* **90:**1041-1053.
Gibella, M., Pronce, T., and Tilquin, B. 1994. ESR study of irradiated and photolysed drugs. *J. Chim. Phys. Phys-Chim. Biol.* **91:**1868-1872.
Goldfarb, D. 1996. Time domain EPR. In *Electron spin resonance, Volume 15*, pp. 186-243. The Royal Society of Chemistry, Cambridge.
Gopal, N. G. S., Patel, K. M., Sharma, G., Bhalla, H. L., Wills, P. A., and Hilmy, N. 1988. Guide for radiation sterilisation of pharmaceuticals and decontamination of raw materials. *Radiat. Phys. Chem.* **32:**619-622.
Gordy, W.1980.*Theory and applications of electron spin resonance*, John Wiley & Sons, Inc., New York
Goulas, A. E., and Kontominas, M. G. 1996. Migration of dioctyladipate plasticiser from food-grade WC film into chicken meat products: Effect of gamma-radiation. *Z. Lebensmittel-Untersuch. Fors.* **202:**250-255.
Haire, D. L., Chen, G. M., Janzen, E. G., Fraser, L., and Lynch, J. A. 1997. Identification of irradiated foodstuffs: a review of the recent literature. *Food. Res. Int.* **30:**249-264.
Hamm, R. N., Stabin, M. G., and Turner, J. E. 1998. Investigation of a Monte Carlo model for chemical reactions. *Radiat. Environ. Biophys.* **37:**151-156.
Hawkins, C. L., and Davies, M. J. 1998. Reaction of HOCl with amino acids and peptides: EPR evidence for rapid rearrangement and fragmentation, reactions of nitrogen-centred radicals. *J. Chem. Soc. Perkin Trans. 2* 1937-1945.

Herak, J. N., Sankovic, K., and Hutterman, J. 1994. Thiocytosine as radiation energy trap in a single-crystal of cytosine hydrochloride. *Int. J. Radiat. Biol.* **66**:3-9.

Herak, J. N., Sankovic, K., Krilov, D., Jaksic, M., and Huttermann, J. 1997. Radiation energy transfer and trapping in single crystals of semihydrate and hydrochloride of 5-methylcytosine doped with 5-methylthiocytosine - An EPR study. *Radiat. Phys. Chem.* **50**:141-148.

Herak, J. N., Sankovic, K., Krilov, D., and Huttermann, J. 1999. An EPR study of the transfer and trapping of holes produced by radiation in guanine(thioguanine) hydrochloride single crystals. *Radiat. Res.* **151**:319-324.

Hill, M. A. 1999. Radiation damage to DNA: The importance of track structure. *Radiat. Meas.* **31**: 15-23.

Hohenberg, P., and Kohn, W. 1964. XXX. *Phys. Rev. B* **136**:864.

Hole, E. O., Sagstuen, E., Nelson, W. H., and Close, D. M. 1989. Free-radical formation in single-crystals of 2'-deoxyguanosine 5'-monophosphate, and guanine hydrobromide monohydrate after X-irradiation at 10 K and 65 K - An ESR, ENDOR and FSE study. *Free Radical Res. Commun.* **6**:87-90.

Hole, E. O., Sagstuen, E., Nelson, W. H., and Close, D. M. 1991a. Primary reduction and oxidation of thymine derivatives - ESR/ENDOR of thymidine and 1-methylthymidine X-irradiated at 10 K. *J. Phys. Chem.* **95**:1494-1503.

Hole, E. O., Sagstuen, E., Nelson, W. H., and Close, D. M. 1991b. Environmental effects on primary radical formation in guanine - solid-state ESR and ENDOR of guanine hydrobromide monohydrate. *Radiat. Res.* **125**:119-128.

Hole, E. O., Nelson, W. H., Sagstuen, E., and Close, D. M. 1992a. Free-radical formation in X-irradiated anhydrous crystals of inosine studied by EPR and ENDOR spectroscopy. *Radiat. Res.* **130**:148-159.

Hole, E. O., Sagstuen, E., Nelson, W. H., and Close, D. M. 1992b. The structure of the guanine cation - ESR/ENDOR of cyclic guanosine-monophosphate single-crystals after X-irradiation at 10 K. *Radiat. Res.* **129**:1-10.

Hole, E. O., Nelson, W. H., Sagstuen, E., and Close, D. M. 1992c. Free-radical formation in single-crystals of 2'-deoxyguanosine 5'-monophosphate tetrahydrate disodium salt - An EPR ENDOR study. *Radiat. Res.* **129**:119-138.

Hole, E. O., Sagstuen, E., Nelson, W. H., and Close, D. M. 1995. Free-radical formation in single-crystals of 9-methyladenine X-irradiated at 10K - An electron-paramagnetic-resonance and electron-nuclear double-resonance study. *Radiat. Res.* **144**:258-265.

Hole, E. O., Nelson, W. H., Sagstuen, E., and Close, D. M. 1998. Electron paramagnetic resonance and electron nuclear double resonance studies of X-irradiated crystals of cytosine hydrochloride. Part I: Free radical formation at 10 K after high radiation doses. *Radiat. Res.* **149**:109-119.

Huang, W. D., Han, J. W., Wang, X. Q., Yu, Z. L., and Zhang, Y. H. 1998. keV ion irradiation of solid glycine: an EPR study. *Nucl. Instrum. Meth. Phys. Res. B* **140**:137-142.

IAEA-Tecdoc-971. 1997. *Sewage sludge and wastewater for use in agriculture*

IAEA -Tecdoc-1023. 1998. *Radiation technology for conservation of the environment*

Ibrahim, R., Mondelaers, W., and Debergh, P. C. 1998. Effects of X-irradiation on adventitious bud regeneration from in vitro leaf explants of *Rosa hybrida. Plant Cell Tissue Organ Cult.* **54**:37-44.

Iliakis, G. 1991. The role of DNA double strand breaks in ionising radiation-induced killing of eukaryotic cells. *Bioessays* **13**:641-648.

Jacobs, G. P., and Wills, P. A. 1988. Recent developments in the radiation sterilisation of pharmaceuticals. *Radiat. Phys. Chem.* **31**:685-691.

Jacobs, G. P. 1995. A review of the effects of gamma-radiation on pharmaceutical materials. *J. Biomater. Appl.* **10**:59-96.

Jacobs, G. P. 1998. A review on the effects of ionising radiation on blood and blood components. *Radiat. Phys. Chem.* **53**:511-523.

Jensen, F.1999.*Introduction to computational chemistry*, John Wiley & Sons, Ltd.

Kabiljo, Z., Sankovic, K., and Herak, J. N. 1990. ESR study of the thymine anion radical in a single-crystal of thymine monohydrate. *Int. J. Radiat. Biol.* **58**:439-447.

Kaetsu, I. 1996. Biomedical materials, devices and drug delivery systems by radiation techniques. *Radiat. Phys. Chem.* **47**:419-424.

Kaetsu, I., Uchida, K., Shindo, H., Gomi, S., and Sutani, K. 1999. Intelligent type controlled release systems by radiation techniques. *Radiat. Phys. Chem.* **55**:193-201.

Kim, H., Budzinski, E. E., and Box, H. C. 1989. The radiation-induced oxidation and reduction of guanine - Electron-spin-resonance electron nuclear double-resonance studies of irradiated guanosine cyclic monophosphate. *J. Chem. Phys.* **90:**1448-1451.

Kohn, W., and Sham, L. J. 1965. Self-consistent equations including exchange and correlation effects. *Phys. Rev. A* **140:**1133-1138.

Kohn, W., Becke, A. D., and Parr, R. G. 1996. Density functional theory of electronic structure. *J. Phys. Chem.* **100:**12974-12980.

Koppen, G., Toncelli, L. M., Triest, L., and Verschaeve, L. 1999. The comet assay: a tool to study alteration of DNA integrity in developing plant leaves. *Mech. Age. Dev.* **110:** 13-24.

Krzyminiewski, R., Pietrzak, J., and Konopka, R. 1990. An ESR study of the stable radical in a gamma-irradiated single crystal of 17-alpha-hydroxy-progesterone. *J. Mol. Struct.* **240:**133-140.

Krzyminiewski, R., Bernhard, W., and Mercer, K. 1995. Conversion of free-radicals upon annealing of X-irradiated single-crystal of cholets-4-ene-3-one. *Radiat. Phys. Chem.* **45:** 883-888.

Kubli-Garfias, C. 1998a. Ab initio comparative study of the electronic structure of testosterone, epitestosterone and androstenedione. *Theochem-J. Mol. Struct.* **422:**167-177.

Kubli-Garfias, C., and Vazquez-Ramirez, R. 1998b. Ab initio calculations of the electronic structure of glucocorticoids. *Theochem-J. Mol. Struct.* **454:**267-275.

Kubli-Garfias, C. 1998c. Comparative study of the electronic structure of estradiol, epiestradiol and estrone by ab initio theory. *Theochem-J. Mol. Struct.* **452:**175-183.

Kubli-Garfias, C., Vazquez, R., and Mendieta, J. 1998d. Austin Model 1 study of the effect of carbonyl and hydroxyl functional groups on the electronic structure of androstane. *Theochem-J. Mol. Struct.* **428:**189-194.

Kubli-Garfias, C. 1998e. Ab initio study of the electronic structure of progesterone and related progestins. *Theochem-J. Mol. Struct.* **425:**171-179.

Kubli-Garfias, C. 1999. Comparative study of the electronic structure of pregnanolones by ab initio theory. *Int. J. Quantum Chem.* **71:**433-440.

Kumagai, J., Katoh, H., Kumada, T., Tanaka, A., Tano, S., and Miyazaki, T. 2000. Strong resistance of *Arabidopsis thaliana* and *Raphanus sativus* seeds for ionising radiation as studied by ESR, ENDOR, ESE spectroscopy and germination measurement: Effect of long-lived and super-long-lived radicals. *Rad. Phys. Chem.* **57:**75-83.

Lahorte, P., De Proft, F., Vanhaelewyn, G., Masschaele, B., Cauwels, P., Callens, F., Geerlings, P., and Mondelaers, W. 1999a. Density functional calculations of hyperfine coupling constants in alanine-derived radicals. *J. Phys. Chem. A* **103:**6650-6657.

Lahorte, P., De Proft, F., Callens, F., Geerlings, P., and Mondelaers, W. 1999b. A density functional study of hyperfine coupling constants in steroid radicals. *J. Phys. Chem. A* **103:**11130-11135.

Lahorte, P., Mondelaers, W., De Frenne, D., Callens, F., Vanhaelewyn, G., Schacht, E., Van Calenberg, S., Van Cleemput, O., and Huyghebaert, A. 1999c. Applied radiation research around a 15 MeV high-average-power linac. *Radiat. Phys. Chem.* **55:**761-765.

Langer-Safer, P. R., Fitz, L. J., Whitley, M. Z., Wood, C. R., and Beier, D. R. 1997. Strategies for the application of functional genomics technology to biopharmaceutical drug discovery. *Drug Develop. Res.* **41:**173-179.

Lardon, A., Georgiev, S., Aghmir, A., Le Merrer, G., and Negrutiu, I. 1999. Sexual dimorphism in white campion: Complex control of carpel number is revealed by Y chromosome deletions. *Genetics* **151:**1173-1185.

Lassmann, G., Eriksson, L. A., Himo, F., Lendzian, F., and Lubitz, W. 1999. Electronic structure of a transient histidine radical in liquid aqueous solution: EPR continuous-flow studies and density functional calculations. *J. Phys. Chem. A* **103:**1283-1290.

Lett, J. T.1994.The renaissance in basic cellular radiobiology and its significance for radiation therapy181-223.

Levine, I. N.1991.Quantum chemistry4th:

Lin, W. Z., Tu, T. C., Dong, J. R., Zhang, J. S., and Lin, N. Y. 1998. ESR studies of gamma-irradiated histone octamer and the histone H3. *Radiat. Phys. Chem.* **53:**651-655.

Luo, N., Litvin, A., and Osman, R. 1999. Theoretical studies of ribose and its radicals produced by hydrogen abstraction from ring carbons. *J. Phys. Chem. A* **103:**592-600.

Mader, K. 1998. Pharmaceutical applications of in vivo EPR. *Phys. Med. Biol.* **43:**1931-1935.

Malkin, V. G., Malkina, O. L., Eriksson, L. A., and Salahub, D. R. 1995. The calculation of NMR and EPR spectroscopy parameters using density functional theory. In *Modern density functional theory, a tool for chemistry* (J. M. Seminario and P. Politzer, Eds.), pp. 273-347. Elsevier, Amsterdam.
Malone, M. E., Cullis, P. M., Symons, M. C. R., and Parker, A. W. 1995. Biphotonic photoionisation of cytosine and its derivatives with UV-radiation at 248 nm - An EPR study in low-temperature perchlorate glasses. *J. Phys. Chem.* **99**:9299-9308.
Mathur, A. M., Moorjani, S. K., and Scranton, A. B. 1996. Methods for synthesis of hydrogel networks: A review. *J. Macromol. Sci. -Rev. Macromol.* **36**:405-430.
McLaughlin, W. L. 1993. ESR dosimetry. *Radiat. Prot. Dosim.* **47**:255-262.
Mehnert, R. 1996. Review of industrial applications of electron accelerators. *Nucl. Instrum. Meth. Phys. Res. B* **113**:81-87.
Miaskiewicz, K., and Osman, R. 1994. Theoretical study of the deoxyribose radicals formed by hydrogen abstraction. *J. Amer. Chem. Soc.* **116**:232-238.
Miyazaki, T., Arai, J., Kaneko, T., Yamamoto, K., Gibella, M., and Tilquin, B. 1994a. Estimation of irradiation dose of radiosterilised antibiotics by electron-spin-resonance - Ampicillin. *J. Pharm. Sci.* **83**:1643-1644.
Miyazaki, T., Kaneko, T., Yoshimura, T., Crucq, A. S., and Tilquin, B. 1994b. Electron-spin-resonance study of radiosterilisation of antibiotics - Ceftazidime. *J. Pharm. Sci.* **83**:68-71.
Moiseenko, V. V., Hamm, R. N., Waker, A. J., and Prestwich, W. V. 1998a. Modelling DNA damage induced by different energy photons and tritium beta-particles. *Int. J. Radiat. Biol.* **74**:533-550.
Moiseenko, V. V., Hamm, R. N., Waker, A. J., and Prestwich, W. V. 1998b. The cellular environment in computer simulations of radiation induced damage to DNA. *Radiat. Environ. Biophys.* **37**:167-172.
Mondelaers, W., Van Laere, K., and Uyttendaele, D. 1993. Treatment of primary tumours of bone and cartilage by extracorporeal irradiation with a low-energy high-power linac. *Nucl. Instrum. Meth. Phys. Res. B* **79**:898-900.
Mondelaers, W. 1998. Low-energy electron accelerators in industry and applied research. *Nucl. Instrum. Meth. Phys. Res. B* **139**:43-50.
Monk, J. D., Beuchat, L. R., and Doyle, M. P. 1995. Irradiation inactivation of food-borne microorganisms. *J. Food. Protect.* **58**:197-208.
Moroff, G., Leitman, S. F., and Luban, N. L. C. 1997. Principles of blood irradiation, dose validation, and quality control. *Transfusion* **37**:1084-1092.
Mroczka, N. E., and Bernhard, W. A. 1995. Electron-paramagnetic-resonance investigation of X-irradiated poly(U), poly(A) and poly(A)-poly(U) - Influence of hydration, packing and conformation on radical yield at 4 K. *Radiat. Res.* **144**:251-257.
Mroczka, N. E., Mercer, K. R., and Bernhard, W. A. 1997. The effects of lattice water on free radical yields in x- irradiated crystalline pyrimidines and purines: A low- temperature electron paramagnetic resonance investigation. *Radiat. Res.* **147**:560-568.
Musial, K., and Przywara, L. 1998. Influence of irradiated pollen on embryo and endosperm development in kiwifruit. *Ann. Bot.* **82**:747-756.
Musial, K., and Przywara, L. 1999. Endosperm response to pollen irradiation in kiwifruit. *Sex. Plant. Reprod.* **12**:110-117.
Naess, S. K., Swartz, H. J., and Bauchan, G. R. 1998. Ploidy reduction in blackberry. *Euphytica* **99**:57-73.
Nelson, W. H., Hole, E. O., Sagstuen, E., and Close, D. M. 1988. ESR/ENDOR study of guanine.HCl.2H2O X-irradiated at 20 K. *Int. J. Radiat. Biol.* **54**:963-986.
Nelson, W. H., Close, D. M., Sagstuen, E., and Hole, E. O. 1989. Radiation-chemistry of adenine-derivatives following direct ionisation in solids - ESR and ENDOR investigations. *Free Radical Res. Commun.* **6**:81-82.
Nelson, W. H., Sagstuen, E., Hole, E. O., and Close, D. M. 1992a. On the proton transfer behaviour of the primary oxidation-product in irradiated DNA. *Radiat. Res.* **131**:10-17.
Nelson, W. H., Sagstuen, E., Hole, E. O., and Close, D. M. 1992b. Ionisation of adenine-derivatives - EPR and ENDOR studies of X-irradiated adenine.HCl.1/2H2O and adenosine.HCl. *Radiat. Res.* **131**:272-284.
Nelson, W. H., Sagstuen, E., Hole, E. O., and Close, D. M. 1998. Electron spin resonance and electron nuclear double resonance study of X-irradiated deoxyadenosine: Proton transfer behaviour of primary ionic radicals. *Radiat. Res.* **149**:75-86.

Onori, S., Pantaloni, M., Fattibene, P., Signoretti, E. C., Valvo, L., and Santucci, M. 1996. ESR identification of irradiated antibiotics: Cephalosporins. *Appl. Radiat. Isotopes* **47:**1569-1572.

Patterson, L. K. 1987. Instruments for measurement of transient behaviour in radiation chemistry. In *Radiation chemistry: principles and applications* (Farhataziz and M. Rodgers, Eds.), pp. 65-96. VCH Publishers, Inc., New York.

Podmore, I. D., Malone, M. E., Symons, M. C. R., Cullis, P. M., and Dalgarno, B. G. 1991. Factors controlling the site of protonation of the one-electron adduct of cytosine and its derivatives. *J. Chem. Soc. Faraday Trans.* **87:**3647-3652.

Pouget, J. P., Ravanat, J. L., Douki, T., Richard, M. J., and Cadet, J. 1999a. Use of the comet assay to measure DNA damage in cells exposed to photosensitisers and gamma radiation. *J. Chim. Phys. Phys-Chim. Biol.* **96:**143-146.

Pouget, J. P., Ravanat, J. L., Douki, T., Richard, M. J., and Cadet, J. 1999b. Measurement of DNA base damage in cells exposed to low doses of gamma-radiation: comparison between the HPLC-EC and comet assays. *Int. J. Radiat. Biol.* **75:**51-58.

Prise, K. M., Ahnstrom, G., Belli, M., Carlsson, J., Frankenberg, D., Kiefer, J., Lobrich, M., Michael, B. D., Nygren, J., Simone, G., and Stenerlow, B. 1998. A review of dsb induction data for varying quality radiations. *Int. J. Radiat. Biol.* **74:**173-184.

Raffi, J., Thiery, C., Battesti, C., Agnel, J. P., Triolet, J., and Vincent, P. 1993. Electron-spin resonance studies of gamma-irradiated saccharides. *J. Chim. Phys. Phys-Chim. Biol.* **90:**1009-1019.

Riganakos, K. A., Koller, W. D., Ehlermann, D. A. E., Bauer, B., and Kontominas, M. G. 1999. Effects of ionising radiation on properties of monolayer and multilayer flexible food packaging materials. *Radiat. Phys. Chem.* **54:**527-540.

Rosiak, J. M., Ulanski, P., Pajewski, L. A., Yoshii, F., and Makuuchi, K. 1995. Radiation formation of hydrogels for biomedical purposes - Some remarks and comments. *Radiat. Phys. Chem.* **46:**161-168.

Rosiak, J. M., and Yoshii, F. 1999. Hydrogels and their medical applications. *Nucl. Instrum. Meth. Phys. Res. B* **151:**56-64.

Sabo, D., Bernd, L., Ewerbeck, V., Eble, M., Wannenmacher, M., and Schulte, M. 1999. Intraoperative extracorporeal irradiation and replantation (IEIR) in the treatment of primary malignant bone tumours. *Unfallchirurg* **102:**580-588.

Sagstuen, E., Lund, A., Awadelkarim, O., Lindgren, M., and Westerling, J. 1986. Free radicals in X-irradiated single crystals of sucrose: a reexamination. *J. Phys. Chem.* **90:**5584-5588.

Sagstuen, E., Hole, E. O., Nelson, W. H., and Close, D. M. 1988. Electron-spin-resonance ENDOR study of guanosine 5'-monophosphate (free acid) single-crystals X-irradiated at 10 K. *Radiat. Res.* **116:**196-209.

Sagstuen, E., Hole, E. O., Nelson, W. H., and Close, D. M. 1989a. Structure of the primary reduction product of thymidine after X-irradiation at 10 K. *J. Phys. Chem.* **93:**5974-5977.

Sagstuen, E., Hole, E. O., Nelson, W. H., and Close, D. M. 1989b. Free-radical formation in nucleosides and nucleotides of guanine - ESR and ENDOR of guanosine 5'-monophosphate and guanosine-dimethylformamide X-irradiated at 10 K. *Free Radical Res. Commun.* **6:**91-92.

Sagstuen, E., Hole, E. O., Nelson, W. H., and Close, D. M. 1992a. Radiation-induced free-radical formation in thymine derivatives - EPR/ENDOR of anhydrous thymine single-crystals X-irradiated at 10 K. *J. Phys. Chem.* **96:**1121-1126.

Sagstuen, E., Hole, E. O., Nelson, W. H., and Close, D. M. 1992b. Protonation state of radiation-produced cytosine anions and cations in the solid-state - EPR ENDOR of cytosine monohydrate single-crystals X-irradiated at 10 K. *J. Phys. Chem.* **96:**8269-8276.

Sagstuen, E., Hole, E. O., Nelson, W. H., and Close, D. M. 1996. Radiation damage to DNA base pairs .1. Electron paramagnetic resonance and electron nuclear double resonance study of single crystals of the complex 1-methylthymine-9-methyladenine X- irradiated at 10 K. *Radiat. Res.* **146:**425-435.

Sagstuen, E., Hole, E. O., Haugedal, S. R., and Nelson, W. H. 1997. Alanine radicals: Structure determination by EPR and ENDOR of single crystals X-irradiated at 295 K. *J. Phys. Chem. A* **101:**9763-9772.

Sagstuen, E., Hole, E. O., Nelson, W. H., and Close, D. M. 1998. Radiation damage to DNA base pairs. II. Paramagnetic resonance studies of 1-methyluracil center dot 9-ethyladenine complex crystals X-irradiated at 10 K. *Radiat. Res.* **149:**120-127.

Sanderud, A., and Sagstuen, E. 1998. EPR and ENDOR studies of single crystals of alpha-glycine X-ray irradiated at 295 K. *J. Phys. Chem. B* **102:**9353-9361.

Sankovic, K., Krilov, D., and Herak, J. N. 1991. Postirradiation long-range energy-transfer in a single crystal of cytosine monohydrate - An EPR study. *Radiat. Res.* **128:**119-124.

Sankovic, K., Krilov, D., PranjicPetrovic, T., Huttermann, J., and Herak, J. N. 1996. Nature of the chlorine-centred paramagnetic species in irradiated crystals of cytosine hydrochloride doped with thiocytosine. *Int. J. Radiat. Biol.* **70:**603-608.

Sauvaigo, S., Serres, C., Signorini, N., Emonet, N., Richard, M. J., and Cadet, J. 1998. Use of the single-cell gel electrophoresis assay for the immunofluorescent detection of specific DNA damage. *Anal. Biochem.* **259:**1-7.

Scharf, W. H. 1994. Methods of accelerating charged particles. In *Biomedical particle accelerators*, pp. 45-134. American Institute of Physics, New York.

Seminario, J. M. 1995. An introduction to density functional theory in chemistry. In *Modern density functional theory: a tool for chemistry* (J. M. Seminario and P. Politzer, Eds.), pp. 1-27. Elsevier, Amsterdam.

Sendra, E., Capellas, M., Guamis, B., Felipe, X., MorMur, M., and Pla, R. 1996. Review: Food irradiation. General aspects. *Food Sci. Technol. Int.* **2:**1-11.

Sevilla, M., Yan, M., Becker, D., and Gillich, S. 1989. ESR investigations of the reactions of radiation-produced thiyl and DNA peroxyl radicals - formation of sulphoxyl radicals. *Free Radical Res. Commun.* **6:**99-102.

Sevilla, M., Becker, D., Yan, M. Y., and Summerfield, S. R. 1991. Relative abundance of primary ion radicals in gamma-irradiated DNA-cytosine vs. thymine anions and guanine vs. adenine cations. *J. Phys. Chem.* **95:**3409-3415.

Sevilla, M., Besler, B., and Colson, A. O. 1995. Ab-initio molecular-orbital calculations of DNA radical ions .5. scaling of calculated electron-affinities and ionisation-potentials to experimental values. *J. Phys. Chem.* **99:**1060-1063.

Shalaby, S. W. 1996. Radiochemical sterilisation: a new approach to medical device processing. In *Irradiation of polymers, fundamentals and technological applications* (R. L. Clough and S. W. Shalaby, Eds.), pp. 246-253. American Chemical Society, Washington, D.C.

Shao, C., Saito, M., and Yu, Z. 1999. Radiation induced DNA strand breaks measured by a modified method of gel scanning. *Radiat. Phys. Chem.* **56:**547-551.

Signoretti, E. C., Onori, S., Valvo, L., Fattibene, P., Savella, A. L., Desena, C., and Alimonti, S. 1993. Ionising-radiation induced effects on cephradine - influence of sample moisture-content, irradiation dose and storage-conditions. *Drug Develop. Ind. Pharm.* **19:**1693-1708.

Signoretti, E. C., Valvo, L., Fattibene, P., Onori, S., and Pantaloni, M. 1994. Gamma-radiation induced effects on cefuroxime and cefotaxime - investigation on degradation and syn-anti isomerisation. *Drug Develop. Ind. Pharm.* **20:**2493-2508.

Spalletta, R. A., and Bernhard, W. A. 1993. Influence of primary structure on initial free-radical products trapped in A-T polydeoxynucleotides X-irradiated at 4 K. *Radiat. Res.* **133:**143-150.

Steel, G. G. E.1993.Basic clinical radiobiology

Stehlik, D., and Mobius, K. 1997. New EPR methods for investigating photoprocesses with paramagnetic intermediates. *Annu. Rev. Phys. Chem.* **48:**745-784.

Swarts, S. G., Sevilla, M., Becker, D., Tokar, C. J., and Wheeler, K. T. 1992. Radiation-induced DNA damage as a function of hydration .1. release of unaltered bases. *Radiat. Res.* **129:**333-344.

Swarts, S. G., Becker, D., Sevilla, M., and Wheeler, K. T. 1996. Radiation-induced DNA damage as a function of hydration .2. Base damage from electron-loss centers. *Radiat. Res.* **145:**304-314.

Szyczewski, A., and Mobius, K. 1994. An ENDOR study of the radicals in a gamma-irradiated single-crystal of 17-alpha,21-dihydroxyprogesterone. *J. Mol. Struct.* **318:**87-93.

Szyczewski, A. 1996. EPR/ENDOR investigations of gamma-irradiated steroid hormone single crystals. *Appl. Radiat. Isotopes* **47:**1675-1681.

Szyczewski, A., Endeward, B., and Mobius, K. 1998. ENDOR study of gamma-irradiated hydrated testosterone orthorhombic single crystals. *Appl. Radiat. Isotopes* **49:**59-65.

Talrose, V. L., and Trofimov, V. I. 1995. Cryoradiation sterilisation - contemporary state and outlook. *Radiat. Phys. Chem.* **46:**633-637.

Thakur, B. R., and Singh, R. K. 1994. Food irradiation - Chemistry and applications. *Food. Rev. Int.* **10:**437-473.

Tilquin, B., and Crucq, A. S. 1999. The chemistry of the radiosterilisation of solid pharmaceuticals. *J. Chim. Phys. Phys-Chim. Biol.* **96:**167-173.

Triolet, J., Thiery, C., Agnel, J. P., Battesti, C., Raffi, J., and Vincent, P. 1990. ESR spin trapping analysis of gamma-induced radicals in sucrose. *Free Radical Res. Commun.* **10:**57-61.

Triolet, J., Thiery, C., Battesti, C., Agnel, J. P., Raffi, J., and Vincent, P. 1991. Spin trapping study of gamma-radiolysis of sucrose. *J. Chim. Phys. Phys-Chim. Biol.* **88:**1237-1244.

Triolet, J., Raffi, J., Agnel, J. P., Battesti, C., Thiery, C., and Vincent, P. 1992a. Electron-spin resonance study of spin-trapped radicals from gamma-irradiation of fructans. *Magn. Reson. Chem.* **30:**1051-1053.

Triolet, J., Thiery, C., Agnel, J. P., Battesti, C., Raffi, J., and Vincent, P. 1992b. ESR spin trapping analysis of gamma-induced radicals in sucrose .2. *Free Radical Res. Commun.* **16:**183-196.

Turro, N. J., and Khudyakov, I. V. 1999. Applications of chemically induced dynamic electron polarisation to mechanistic photochemistry. *Res. Chem. Intermediates* **25:**505-529.

Vanhaelewyn, G., Mondelaers, W., and Callens, F. 1999. Effect of temperature on the electron paramagnetic resonance spectrum of irradiated alanine. *Radiat. Res.* **151:**590-594.

Van Lancker, M., Herer, A., Cleland, M. R., Jongen, Y., and Abs, M. 1999. The IBA Rhodotron: an industrial high-voltage high-powered electron beam accelerator for polymers radiation processing. *Nucl. Instrum. Meth. Phys. Res. B* **151:**242-246.

Ventura, O. N., Kieninger, M., and Irving, K. 1997. Density functional theory: A useful tool for the study of free radicals. *Advan. Quantum. Chem.* **28:**293-309.

Von Sonntag, C., Bothe, E., Ulanski, P., and Deeble, D. J. 1995. Pulse-radiolysis in model studies toward radiation processing. *Radiat. Phys. Chem.* **46:**527-532.

Wang, W., and Sevilla, M. 1994a. Reaction of cysteamine with individual DNA-base radicals in gamma-irradiated nucleotides at low-temperature. *Int. J. Radiat. Biol.* **66:**683-695.

Wang, W., Razskazovskii, Y., and Sevilla, M. 1997. Secondary radical attack on DNA nucleotides: Reaction by addition to DNA bases and abstraction from sugars. *Int. J. Radiat. Biol.* **71:**387-399.

Wang, W. D., Yan, M. Y., Becker, D., and Sevilla, M. 1994b. The influence of hydration on the absolute yields of primary free-radicals in gamma-irradiated DNA at 77 K .2. Individual radical yields. *Radiat. Res.* **137:**2-10.

Ward, J. F. 1990. The yield of DNA double-strand breaks produced intracellularly by ionising radiation - A review. *Int. J. Radiat. Biol.* **57:**1141-1150.

Wardman, P., and Ross, A. B. 1991. Radiation-chemistry literature compilations - Their wider value in free-radical research. *Free Radical Biol. Med.* **10:**243-247.

Wardman, P., Candeias, L. P., Everett, S. A., and Tracy, M. 1994. Radiation-chemistry applied to drug design. *Int. J. Radiat. Biol.* **65:**35-41.

Weil, J., Bolton J., and Wertz, E.1994. *Electron paramagnetic resonance: elementary theory and practical applications.* John Wiley & Sons, Inc. New York.

Weiland, B., Huttermann, J., Malone, M. E., and Cullis, P. M. 1996. Formation of C1' located sugar radicals from X-irradiated cytosine nucleosides and -tides in BeF2 glasses and frozen aqueous solutions. *Int. J. Radiat. Biol.* **70:**327-336.

Wetmore, S. D., Boyd, R. J., and Eriksson, L. A. 1998a. Theoretical investigation of adenine radicals generated in irradiated DNA components. *J. Phys. Chem. B* **102:**10602-10614.

Wetmore, S. D., Himo, F., Boyd, R. J., and Eriksson, L. A. 1998b. Effects of ionising radiation on crystalline cytosine monohydrate. *J. Phys. Chem. B* **102:**7484-7491.

Wetmore, S. D., Boyd, R. J., and Eriksson, L. A. 1998c. Radiation products of thymine, 1-methylthymine, and uracil investigated by density functional theory. *J. Phys. Chem. B* **102:**5369-5377.

Wetmore, S. D., Boyd, R. J., and Eriksson, L. A. 1998e. A comprehensive study of sugar radicals in irradiated DNA. *J. Phys. Chem. B* **102:**7674-7686.

Wetmore, S. D., Boyd, R. J., and Eriksson, L. A. 1998d. Comparison of experimental and calculated hyperfine coupling constants. Which radicals are formed in irradiated guanine? *J. Phys. Chem. B* **102:**9332-9343.

Woods, R., and Pikaev, A.1994. Applied radiation chemistry: radiation processing. John Wiley & Sons, Inc. New York.

Yan, M. Y., Becker, D., Summerfield, S., Renke, P., and Sevilla, M. 1992. Relative abundance and reactivity of primary ion radicals in gamma-irradiated DNA at low-temperatures .2. single-stranded vs. double-stranded DNA. *J. Phys. Chem.* **96:**1983-1989.

AROMA MEASUREMENT:
Recent developments in isolation and characterisation

SASKIA M. VAN RUTH
University College Cork, Department of Food Science and Technology, Nutritional Sciences, Western Road, Cork, Ireland

Abstract

This chapter gives an overview of techniques for isolation and characterisation of aroma compounds. Isolation methods of volatile compounds, such as extraction, distillation, headspace techniques, mouth analogues and in-mouth measurements are described. Furthermore, a section is focused on applications of analytical and sensory-instrumental characterisation of aroma compounds. Finally, the relevance of aroma isolation and characterisation techniques with respect to flavour compounds generated by biotechnology are discussed.

1. Introduction

Most scientists agree that aroma, taste, texture and mouthfeel account for the major stimuli that make up flavour. The stimuli occur when chemicals from the food come into contact with sensory cells in the nose (odour/aroma) and mouth (taste), interact with mucous membranes, or when food structures such as emulsions or rigid cell walls affect the chewing process (texture) or interact with the mouth lining [mouthfeel) [1]. Mouthfeel responses consider the heat of spices and the cooling of menthol. Taste is concerned with the sensations of sweet, sour, salty, bitter and umami, which are associated with receptors on the tongue. Aroma is a much broader sensation, encompassing an estimated 10,000 or more different odours [2].

Foods have highly complex chemical compositions. They contain both volatile and non-volatile substances, which attribute to the flavour of a food. Human taste-buds can differentiate four stimuli, whereas the nose is capable of discerning thousands of different odours. Therefore, it is not surprising that a major part of flavour research has dealt with analysis of volatile compounds [3]. When flavour isolation of flavour chemistry is discussed, most often only the volatile component of flavour –aroma- is considered, ignoring the other components of flavour. While very often it is the aroma

M. De Cuyper and J.W.M. Bulte (eds.), Physics and Chemistry Basis of Biotechnology, 305–328.

component of flavour that most effectively characterises a flavour, one should not forget the importance of the other flavour components in giving the complete view of a certain flavour [2].

Flavour chemistry developed as a specialised application of organic chemistry. Organic chemistry started in the 19th century when it was shown that chemical transformations took place without supernatural influences. The first era consisting of the theory of vitalism died, and the second period of organic chemistry began with the birth of structural theory, to explain how atoms were organised to form molecules, the network of carbon to carbon bonds [4]. In the third era of organic chemistry, the present modern era, there are three major developments [5]:

- Electronic theories of valences and bonds.
- Understanding of mechanisms of organic reactions.
- Development of instruments for separating, analysing and identifying organic compounds.

Chemical industries utilising these developments have developed considerably to produce dyestuffs, explosives, medicines, plastics, elastomers, fibres, films, perfumes, flavours, etc. Studies of natural products, volatiles from food materials, have been the backbone of the flavour industry [4].

In the early stages of flavour research, most emphasis was on development of methods to establish chemical identity of constituents found only in trace quantities. Because flavour chemistry can be considered a special application of organic chemistry, the methods worked out in flavour chemistry can be applied in other fields in which small amounts of organic compounds can have profound biological effects and vice versa. Examples are found in nutrition, air or water pollution, plant and animal hormones, insect attractants, etc. [4]. The analytical task in aroma chemistry is rather complicated, as a relatively simple flavour may have 50-200 constituents, which give a combined effect to yield the characteristic aroma of a food. The large number of volatile compounds that make up the total amount of aroma chemicals (>6000, [6]) complicates the analytical task even further. Now that many constituents have been identified, the task remains to determine biological activities; that is, which constituent, or constituents, is/are contributing to the characteristic sensory properties of the food products being investigated [4].

1.1. OVERVIEW

An instrumental approach to aroma characterisation can be regarded as a two-phase arrangement, when focusing on those compounds relevant for sensory perception. The first phase comprises representative isolation of volatile compounds [7]. It should be considered that no analytical method is valid unless the isolates represents the material being studied and no logical sensory conclusions can be made if isolates do not have the characteristic sensory properties of the food product being studied. The importance of the isolation step cannot be overemphasised. Sample size requirements of analytical instruments have become smaller and smaller with new developments. Nevertheless, great care must be taken in concentration and isolation so that isolates have sensory properties of the food being studied. Care must be taken so that heat labile compounds are not destroyed by harsh conditions, highly volatile compounds not lost in concentration by distillation, or low solubility compounds lost in extractions [4].

This review paper is focused on several analytical techniques which were designed to measure two types of volatile profiles: (1) the total volatile content of a food and (2) the profile of volatile compounds present in the air around a food. The first group incorporates several extraction methods and is presented in Section 2. This total volatile content is of importance in the production of aroma compounds, but no attempt is made to reflect the aroma composition in the mouth during eating. The second group of isolation techniques includes headspace techniques (Section 3), mouth analogues (Section 4) and in-mouth measurements (Section 5). The aroma patterns determined by these techniques are considered to be more closely related to the composition of volatiles available for perception during eating.

The second phase of an instrumental aroma characterisation involves selection of volatile compounds relevant to the flavour, i.e. those compounds, which can be perceived by humans. This implicates correct determination of the relevant aroma compounds from the whole range of volatiles present in a product. In the present paper an analytical characterisation and sensory-instrumental characterisation section is included (Section 6).

Finally, implementation of the reviewed isolation and analysis techniques in bioprocess production of flavours in general will be discussed in Section 7.

2. Isolation techniques for measurement of total volatile content

Most isolation techniques use some form of extraction or distillation, or a combination of both. These methods utilise differences in solubility in different solvents (extraction and chromatography) and differences in vapour pressures (distillation). The present discussion will concentrate on the following techniques:

- Extraction
- Distillation, including fractional and steam distillation
- Distillation-extraction combinations

2.1. EXTRACTION

Extraction is a technique, which has been used very frequently in the sixties, seventies and eighties. For certain foods it is possible to extract the volatile compounds directly from the food material. This is particularly true with thin materials or relatively dry powders. More often it is necessary to grind the material [8]. The most common used solvents are organic compounds that have a low boiling point (pentane, hexane, methanol, ethanol, dichloromethane, chloroform, ethyl acetate, ethyl ether and acetone [9]. Supercritical gases, such as ammonia, carbon dioxide, nitrous oxide, ethylene, and some fluorocarbons, have also been used to extract thermolabile natural products [10]. Some researchers used a Soxhlet apparatus for the extraction of aroma compounds from foods [11-13]. The procedure varied but often the food material was ground and placed under ether or dichloromethane overnight, followed by Soxhlet extraction. The advantage of direct extraction is that volatiles of both low and high water solubility are isolated in one operation. In direct solvent extraction highly water soluble compounds can be isolated effectively in comparison with other techniques.

One of the problems with direct extraction is that non-volatiles, such as waxes and fats are extracted as well. After direct extraction in most cases separation of volatiles from non-volatiles is necessary. A transfer of the volatiles to another container under high vacuum is usually possible [12-13], although viscous material can retard the movement of the volatiles to the surface. If the extract is spread in a thin film or if it is stirred, the molecules do not have as far to diffuse to the surface, and vaporisation is more efficient [14].

Extraction has long been a popular isolation technique, but it suffers from several limitations. The biases imposed on the aroma profile by solvent extraction relate to the relative solubility of various aroma compounds in the organic/aqueous phases. Leahy and Reineccius [15] as well as Cobb and Bursey [16] showed low and variable recoveries (0-100 %) of aroma compounds, which depended on the solvent chosen and the aroma compound extracted. Attempts to remove appreciable amounts of residual solvent from the extract invariably lead to quantitative changes in the sample. Furthermore, the solvent might interfere with the volatile compounds during the gas chromatographic analysis of the extract. Solvent odours complicate the sensory analysis of extracted materials and the toxicity of many common solvents poses stringent limitations on the use of their extracts [17].

2.2. DISTILLATION

Distillation has also been a very important method for isolating and concentrating volatiles from foods over the last decades. All liquid substances have a vapour pressure that is constant at a given temperature. When the temperature is raised so that the vapour pressure of the substance equals that of the external pressure, the substance boils, and this temperature is defined as the boiling point of the substance.

While distillation techniques can provide useful information, it is now realised that certain procedures can cause the formation of artefacts, especially when fresh foods are subjected to distillation techniques. A classic example is provided in the analysis of marjoram [18] where direct sampling of the oils from leaves followed by gas chromatography showed four major compounds. Distillation gave rise to a complex chromatogram that included a large number of degradation products formed due to oxidation or thermal processes [1]. Formation of artefacts is a major drawback of the technique.

Distillation is divided into two categories: fractional and steam distillation. In both cases, there must be sufficient starting material and the desired aroma compounds must be stable to the conditions of distillation [10].

2.2.1. Fractional distillation

Distillation is a powerful method for obtaining samples of very narrow boiling point ranges. Factors such as theoretical plates, throughput, column hold-up, ease of approaching equilibrium conditions, pressure drop across column, and boil-up rate must be considered for satisfactory operation. The type of column is of critical importance. Packed columns are mostly used for purification of solvents. Spinning band columns are good for preparing samples of relatively broad boiling point ranges. The concentric tube column, in which the vapour is going up in close proximity to the liquid going down, can be used more effectively at very low pressures and with smaller starting mixtures. Each of type of column has its own advantage and usefulness [10]. In fractional distillation of a homogeneous mixture, consisting of substances soluble in each other, Raoult's law prevails. This is expressed by

$$P_{AS} = P_{AP} * N_A$$

Where P_{AS} is the partial pressure of compound A from the solution, P_{AP} is the vapour pressure of the pure compound A and N_A is the mole fraction of A in the liquid.

2.2.2. Steam distillation

In the distillation of heterogeneous mixtures, consisting of substance not soluble in each other, if one of the substances is water, the term 'steam distillation' is used. Water and volatile organic matter not soluble in water distil together when the total vapour pressure adds up to the external pressure. Steam distillation is used to isolate volatile, water-insoluble organic materials from non-volatile materials such as carbohydrates and proteins. This is a crude separation, because the vapour pressure of the mixture is entirely independent of the concentration. Water cannot be separated from the organic compounds distilled by this method. Isolation can sometimes be accomplished by separating the organic material floating on top of the aqueous layer [10].

Methods using some type of atmospheric or reduced pressure steam distillation have been historically the most commonly used, and essential oils were and are still frequently isolated this way on a commercial scale [2].

2.3. DISTILLATION-EXTRACTION COMBINATIONS

As stated above, the organic compounds isolated by distillation have to be separated from the water, which is frequently accomplished by solvent extraction or by using a combination of a steam distillation-continuous extraction apparatus.

Solvent extraction after distillation often results in rather large quantities of steam distillate to be handled. The classical method does have some advantages using steam distillation under reduced pressure. As the extraction occurs later at atmospheric pressure, a lower boiling solvent can be used [8]. The aroma profile ultimately obtained is influenced by volatility of the aroma compounds (initial isolation), solubility during solvent extraction of the distillate, and, finally, volatility again during the concentration of the solvent extract [2].

The combination of steam distillation and continuous extraction has been known to organic chemists for many years. The method became popular with flavour chemists after Nickerson and Likens [19] used it for isolating beer volatiles. The Simultaneous

Distillation-Extraction (SDE) or Likens-Nickerson extraction method was an extension of the Essential Oil industry's technique of "cohobation" where the condensed steam is separated from the volatile oil using a simple trap and returned to the still in order to minimise loss of water soluble volatiles. SDE had actually been described earlier in practical organic chemistry textbooks [20]. The Likens Nickerson head though was simpler and easier to construct and control than the earlier organic chemists version. More recently other versions of the head have been designed [21].

The advantages of SDE are that desired volatile compounds can be concentrated about 1:10000 from a dilute mixture in a single operation within 1-3 hrs. Furthermore, small volumes of water and solvent are used to minimise artefact introduction from solvent and water. Thirdly, thermal degradation and artefact formation is reduced because the SDE system can be used at reduced pressure and temperature [10].

While it is obvious that even a simple solvent extraction introduces substantial bias on the volatile content of a product, combining solvent extraction with another technique (in order to separate volatile compounds from non-volatiles) adds more bias [2]. Leahy and Reineccius [15] showed that the aroma isolate prepared by SDE contained nearly all the volatiles in a food, but their proportions only poorly represented the true volatile content of the food. It was also shown that SDE is less efficient under vacuum [22].

3. Headspace techniques

The total volatile content represents the volatile composition in a food, however, it is often difficult to relate this profile to the volatile profile expressed when foods are consumed. In simple aqueous systems, it is more likely that a direct relationship between total volatile content and the expressed profile exists on the basis of partition between the air and water phases at mouth temperature. However, even in this simple system, interaction between the volatiles and between the volatiles and the mouth should not be ignored [1].

An important characteristic of aroma compounds is that they must exhibit sufficient vapour pressure to be present in the gas phase at a concentration detectable by the olfactory system. This basis of aroma isolation appears most reasonable. It is understandable that many aroma isolation techniques are based on volatility, such as static headspace, dynamic headspace, direct injection techniques (where food is placed in the gas chromatograph itself and heated to volatilise aroma compounds), distillation, mouth analogues and even in-mouth analysis. These techniques are focused on the volatility of the compounds in the system to be studied and not on the concentration of the compounds in the food. Volatility is dependent upon the volatility of the aroma compound in the food itself. The amounts of aroma compounds in the headspace do not follow in order of volatility (vapour pressure) of the pure compounds but, instead, depends upon their vapour pressure over the food system. An example of this difference in volatility due to the food matrix is shown in Figure 1, which presents the volatility of five aldehydes in sunflower oil (SFO) and in a 40 % sunflower oil-in-water emulsion (SFO-E). The author published recently more studies on release and volatility of aroma compounds in oils and emulsions [23].

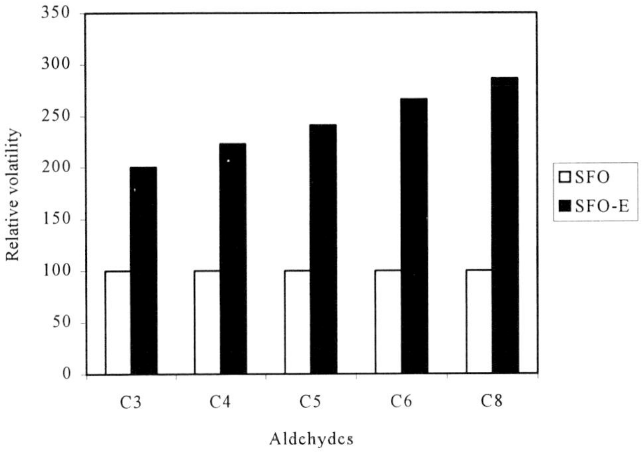

Figure 1. Relative volatility of propanal (C3), butanal (C4), pentanal (C5), hexanal (C6) and octanal (C8) in sunflower oil (SFO) and in a 40 % sunflower oil-in-water emulsion (SFO=100).

3.1. STATIC HEADSPACE

Many compounds exist as gases at the temperature at which they are being sampled or have sufficiently high vapour pressure to evaporate and produce a gas phase solution. In these cases, the gas itself may be injected into a gas chromatograph, either by syringe or by transferring a known volume of vapour from a sample loop attached to a valve. The amount of gas that can be injected is limited by the capacity of the injection port and the analytical column. In practical terms, injections are almost always in the low millilitre range, with sizes of 0.1-2.0 ml being typical [24].

If a complex material, such as a food, is placed into a sealed vial and allowed to stand, the volatile compounds in the sample matrix will leave the sample and distribute over the headspace around it. If the concentration of such a compound reaches about 1 ppm in the headspace, then it may be assayed by a simple injection of an aliquot of the headspace in the vial. The concentration of the compounds in the headspace depends on several factors, including the concentration in the original sample, the volatility of the compound, the solubility of that compound in the sample matrix, the temperature of the sample, and combination of the size of the vial and the time the sample has been inside the vial [24]. At equilibrium, the relationship between the concentration of the volatile compound in the product phase and in the vapour phase can be expressed by Henry's law. This law states that the mass of vapour dissolved in a certain volume of solvent is directly proportional to the partial pressure of the vapour that is in equilibrium with the solution [25]. This law is only valid if the volatile compound is present in infinitely diluted concentrations. Volatile compounds are generally present in foods at relatively low concentrations (ppm-ppb range), and therefore obey Henry's law at the concentrations found in foods. For a linear relationship between product and vapour phase concentrations, also Raoult's law needs to be ideal. This law defines the behaviour of the solvent, and under ideal conditions there is no interaction between volatile compound and solvent [26]. There are exceptions to Henry's law besides those linked to concentration, e.g. volatile compounds (such as acids) should not dissociate [27].

Static headspace isolation normally involves taking a sample of the equilibrium headspace immediately above a sample. This can be directly injected onto a gas chromatographic column. Static headspace is useful for the analysis of very volatile compounds. However, detection limits require a certain amount of material into the headspace and therefore the sample sometimes has to be heated to 60-100 °C. These elevated temperatures could give an unrealistic picture of the volatiles of the sample, as a result of aroma formation or distortion of the quantitative volatile composition. Static headspace is a very rapid method, but it does not give a comprehensive analysis of the volatiles [8].

When interested in the volatile content of a food, a disadvantage of static headspace methods is that it is difficult to do quantitative studies. The analytical data concerns amounts of volatiles in the headspace of the sample. The relationship between vapour pressure and concentration in the sample can be very complex and must be determined experimentally [2]. In classical static headspace analysis, volatiles are removed without any attempt to simulate the conditions in the mouth during eating. Therefore, amounts determined do not necessarily represent the compounds and in the quantities available for perception during eating. Classic headspace is probably more closely related to the odour orthonasally perceived as food approaches the mouth prior to eating [1].

3.2. DYNAMIC HEADSPACE

In the dynamic mode of headspace isolation, a flow of gas is passed over the food to strip off volatiles, which results in a greater yield of material for analysis than in static headspace isolation. In a separate part of the apparatus, volatiles are trapped from the gas stream. The headspace of the sample is continually renewed. Often the sample is stirred or otherwise agitated to increase mass transfer from the sample into the headspace [28]. Normally a solid absorbent is used to trap the volatiles. Absorbents which ignore the water are the most useful. General methods using solid absorbents have been compared by various authors [29-32]. Volatiles have been concentrated on absorbents, such as charcoal, Porapak Q, Chromosorb 101-105 or Tenax, and they are usually thermally desorbed prior to gas chromatographic analysis.

Dynamic headspace, although not as fast as static headspace, gives isolation which approaches being comprehensive. Dynamic headspace methods permit analyses with the minimum of introduction of artefacts developed or introduced during sampling. Thus, headspace analyses permit analyses of volatiles emitted by plants or food products to provide data representing fresh flowers, fresh fruits and vegetables, of products as they are, not data containing artefacts formed during isolation [4]. The thermal desorption of the compounds from the absorbent, although convenient and rapid, has the disadvantage of causing molecular change in some important unstable aroma compounds, such as (Z)-3-hexenal, alkadienals and certain sulphur compounds. This may be due to the elevated temperatures applied during desorption and the metal parts frequently used in thermal desorption units [8,33].

In the actual eating process, equilibrium in not likely, if ever attained in the mouth, the dynamic mode seems, therefore, more closely aligned with what happens in the mouth during eating than static headspace isolation. It should be noted that the times for headspace collection are substantially greater (minutes or hours) compared with the time that food remains in the mouth (seconds).

4. Model mouth systems

More recently, mouth analogues have been developed to mimic aroma release in the mouth more precisely and to consider changes in aroma release during eating. In actual eating situations, aroma concentrations are determined kinetically rather than thermodynamically [34]. Only a few instrumental methods of aroma release have incorporated the crushing, mixing, dilution, and temperature conditions required to simulate aroma release in the mouth from solid foods. Lee [35] reported an instrumental technique for measuring dynamic flavour release. A mass spectrometer was coupled with a dynamic headspace system, i.e. a vial with several small metal balls. The vial was shaken and the balls simulated chewing, while the headspace was flushed with helium gas in order to displace volatile compounds, one of which was analysed directly by mass spectrometry. Roberts and Acree [34] reported a "retronasal aroma simulator", a purge-and-trap device made from a blender. It simulated mouth conditions by regulating temperature to 37 °C, adding artificial saliva, and using mechanical forces. Naβl et al. [36] described a "mouth imitation chamber", which consisted of a thermostated 800 ml vessel with a stirrer, while artificial saliva was added to the system. Withers et al. [37] reported a simple mouth analogue for drinks, which consisted of a glass flask filled with liquid sample and artificial saliva with glass beads. Hydrated air passed over the headspace of the flask, while it was situated in a shaking waterbath. Volatile compounds were trapped on an adsorbent.

The author of the present work presented the model mouth system developed in their group for the first time in 1994 [38]. The model mouth system (Figure 2) has a volume of 70 ml, which is similar to the volume of the human mouth. Temperature is kept constant at 37 °C by water circling through a double wall of the sample flask. Artificial saliva is usually added, which consists of water, salts, mucin and α-amylase [39]. During isolation of the volatiles a plunger makes up and down screwing movements in order to simulate mastication. Nitrogen gas flushes the headspace of the sample at 20-250 ml/min and volatile compounds are swept towards a small cooler (-10 °C, cooled with ethanol) to freeze out the water, before they are being trapped at Tenax TA. The Tenax is placed in a glass tube.

Release of volatiles from rehydrated vegetables in this model mouth system was compared with dynamic headspace (nitrogen gas flushed the headspace of the sample), with purge-and-trap (nitrogen gas purged the vegetable-saliva mixture) and release from the vegetables in the mouths of twelve volunteers (details are described in paragraph 5). Results are presented in Figure 3. Purge-and-trap gave the greatest recovery of volatiles, while dynamic headspace gave lower amounts. The amounts released over 12 minutes did not differ significantly for the model mouth system and the mouths of assessors. The other techniques all showed significant differences [40]. This is one of the few cases that a model mouth system has been validated by comparison with the real situation that exists in-mouth. However, because volatiles were collected over periods of 12 minutes, it should be noted that the volatile profiles obtained represents the average profile that exists over the collection period and does not provide information on the temporal profile during eating.

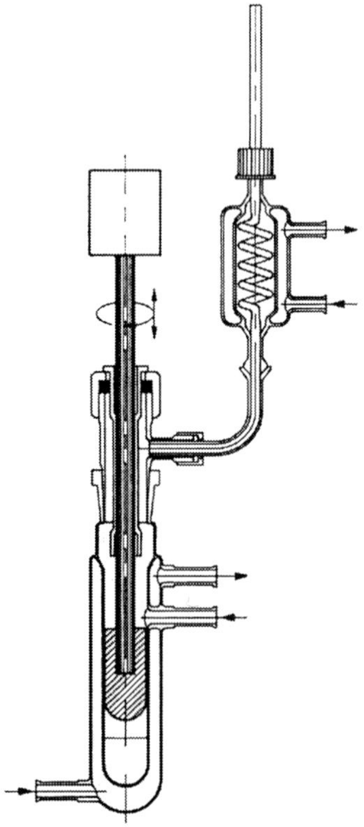

***Figure** 2. Model mouth system developed by van Ruth et al. modified from ref. 40 Copyright 2000 with permission from Elsevier Science.*

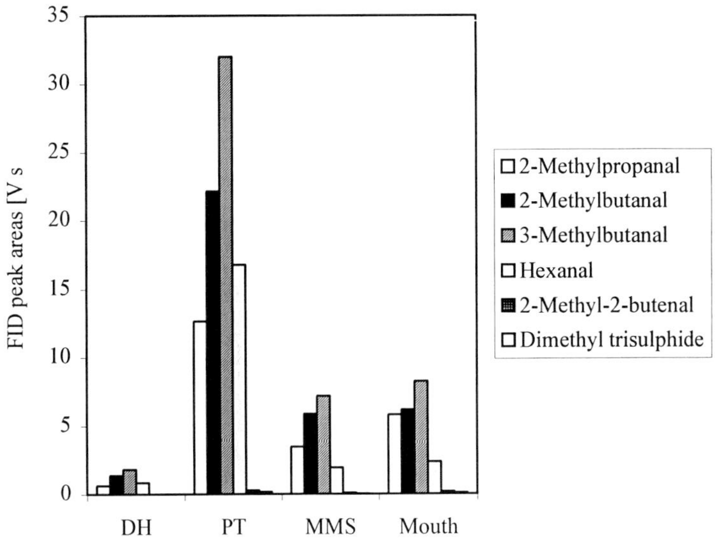

***Figure** 3. Comparison of techniques for isolation of aroma compounds of rehydrated French beans: dynamic headspace (DH), purge-and-trap (PT), model mouth system (MMS) and the mouths of twelve assessors (Mouth) drafted after prinary data from ref. 40 Copyright 2000 with permission from Elsevier Science.*

5. In-mouth measurements

Over the last decade, techniques were developed to measure the release of volatile compounds in the mouth and nose of people. The advantage over model systems is that the results are real, however, the disadvantage is that the results are highly variable, as conditions can not be controlled as in a model system.

A simple way of measuring volatile release from foods is to analyse compounds in the expired air drawn from the noses of people eating foods. Unfortunately, many aroma compounds have extremely low thresholds. Although they contribute to the aroma they are present in very low amounts, with detection limits causing analysis problems. Philips and Greenberg [41] reported techniques for collecting expired air, but these methods were mainly concerned with obtaining a suitable sample of breath for analysis rather than monitoring changes in volatiles over the short times associated with eating. A membrane separator method using direct mass spectrometry was developed by Soeting and Heidema [42] for determining aroma profiles in the expired air of assessors as a function of time. Haring [43] used this method and found a fairly consistent pattern of aroma release when one individual was sampled, but when six different people were given the same solution (2-butanone in water), a considerable variation in the aroma release profiles was observed.

In the beginning of the nineties, attention turned to analyses that concentrate the volatiles and remove air and water from the sample prior to analysis. Linforth and Taylor [44] compared expired air collected from the nose and mouth of people eating foods onto lengths of capillary tubes, as well as cryotrapping, and trapping on solid carbon dioxide and Tenax. Furthermore, Roozen and Legger-Huysman [45] developed a similar technique for measuring aroma release in the mouth at various time intervals. Delahunty et al. [46] have also reported a Tenax-based system for in-mouth measurements of volatile release from cheese. The work above described aroma release from foods in terms of a time-averaged aroma profile.

Time-intensity sensory analysis has suggested that the aroma profile changes temporally, and attempts have been made especially to measure the release of individual aroma compounds over short periods of time using instrumental means. Real time analysis requires a detector capable of continuously monitoring compounds in air. Several different systems have been developed to allow real time gas phase analysis [42,47-49]. With atmospheric pressure chemical ionisation compounds can be ionised in air containing water vapour at atmospheric pressure and exhibits high sensitivity and rapid response times. Taylor and co-workers modified an atmospheric pressure chemical ionisation source of a mass spectrometer to allow the introduction of gas phase samples. It is a soft ionisation technique, which adds a proton to the compound of interest and does not normally induce fragmentation. Consequently compounds present in the breath are monitored in selected ion mode, further enhancing sensitivity. Aroma release from mints, strawberry-flavoured sweets [50], biscuits, sausages [51], gelatine and pectin gels [52], as well as aqueous model food systems [53] have been studied. Grab and Gfeller [54] adapted the technique for studies on strawberries.

6. Analitical techniques

6.1. INSTRUMENTAL CHARACTERISATION

6.1.1. Gas chromatography

Gas chromatography is a very powerful separation method in flavour science. Since the early days of aroma research, more stable columns with greater separation resolution are now commercially available. Gas chromatography became even more powerful and useful with high resolution capillary columns used in combination with mass spectrometers. Most of the extracts, distillates and headspace samples are nowadays analysed by gas chromatography. For analysis of aroma compounds, gas chromatographs are often combined to "chemical/physical" detectors, such as thermal conductivity, flame ionisation, electron capture, flame photometric and mass spectrometric detectors [55].

Located at the exit of the separation column, the detector senses the presence of the individual components as they leave the column. The detector output, after suitable amplification, is usually acquired and processed by a computerised data system. The result is a chromatogram of concentration versus time [56]. Detectors can be divided in detectors that respond well to a wide variety of compounds and specific detectors. General detectors are the thermal conductivity detector and the flame ionisation detector. The thermal conductivity detector responds to any compound with a thermal conductivity significantly different from that of the carrier gas, whereas the flame ionisation detector responds to most carbon-containing compounds. Specific detectors respond only to compounds containing specific elements or element classes. Examples of specific detectors include the electron capture detector (halogens, nitrogen or sulphur), flame photometric detector (sulphur and phosphorus) and the nitrogen-phosphorus detector [57-58].

6.1.1.1. General detectors. The thermal conductivity detector (TCD) does not interact with solutes chemically but instead measures a bulk physical property, the thermal conductivity of the solute and carrier-gas mixture passing through the detector. The TCD responds in proportion to the concentration of any solute that has a thermal conductivity that is significantly different from the carrier-gas conductivity. The minimum detectable concentration is thus related to the conductivities of the carrier and the solute. Solutes are unaffected by passage through a TCD because it does not chemically interact with them. A second specific detector can therefore be connected in series after the TCD to provide additional information. The TCD is also widely used in preparative gas chromatography by connecting a fraction collector to the detector outlet and in aroma research by humidifying the effluent before olfactory evaluation. Because of its simplicity, the TCD often is preferred for survey work and for moderate sensitivity work [56,59].

The flame ionisation detector (FID) is currently one of the most popular detectors because of its high sensitivity, wide range and great reliability. For the FID, as ions are formed inside the detector, they are accelerated by an electric potential toward an electrode, producing a minute current on the order of pico-amperes (10^{-12} A). This current is converted to a voltage. When $-CH_2-$ groups are introduced into the flame, a complex process takes place in which positively charged carbon species and electrons are formed. The current is greatly increased. The FID responds only to substances that produce charged ions when burned in a hydrogen/air flame. In an organic compound the response is proportional to the number of oxidisable carbon atoms. For example, butane has twice as many carbons as an equivalent volume of ethane. Within narrow limits, response of the FID to 1 mol of butane will be twice that to ethane. There is no response from fully oxidised carbons such as carbonyl or carboxyl groups (and thio analogs) and ethers, and response diminishes with increasing substitution of halogens, amines and hydroxyl groups. The detector does not respond to inorganic compounds apart from those easily ionised in a hydrogen/air flame at 2100 °C. The FID is a mass flow detector. It thus depends directly on flow rate of carrier gas. It also varies in response with the applied voltage (until a plateau is reached), and with the temperature of the flame, which is a function of the hydrogen/air mixture ratio, as distinguished from the temperature of the detector housing. Detection limits are about 5 ng/s for light hydrocarbon gases. Response is linear over seven orders of magnitude. The FID provides high sensitivity to a wide range of compounds as well as reliable routine operation [56,60].

6.1.1.2. Specific detectors. Many specific detectors are ionisation detectors, e.g. electron capture, nitrogen-phosphorus, photoionisation detectors and others. They interact with the solute eluting from the gas chromatographic column to produce a current that varies in proportion to the amount of solute present. These detectors rely on substance-specific ionisation mechanisms and respond only to certain hetero-atoms (halogens, nitrogen, or sulphur for the electron capture detector; nitrogen and phosphorus for the nitrogen-phosphorus detector) or chemical structures (aromatics for the photoionisation detector with a 10.2 eV photon source) [57].

The electron capture detector (ECD) has two electrodes with the column effluent passing between. One of the electrodes is treated with a radioisotope that emits high-energy electrons as it decays. These emitted electrons produce copious amounts of low-energy (thermal) secondary electrons in the carrier gas, all of which are collected by the other positively polarised electrode. Molecules that have an affinity for thermal electrons capture electrons as they pass between the electrodes and reduce this steady-state current, thus providing an electrical reproduction of the gas chromatographic peak. The ECD responds to electrophilic species which gives the detector its specificity. Polyhalogenated compounds give excellent response; detection is at the picogram level. The order of increasing response $F<Cl<Br<I$. Other groups exhibiting good selectivity include anhydrides, peroxides, conjugated carbonyls, nitriles and nitrates, plus ozone and organometallics and sulphur-containing compounds.

In the photoionisation detector (PID), the sensor contains a sealed interchangeable ultraviolet lamp that emits a selected energy line. The absorption of ultraviolet light by a molecule leads to ionisation. A chamber adjacent to the ultraviolet source contains a pair of electrodes. A positive potential applied to the accelerating electrode creates a field which drives ions formed by absorption of lamp energy to the collecting electrode where the current, proportional to concentration, is measured. The PID has a dynamic range of seven orders of magnitude, extending from 2 pg through 30 μg. Compounds whose ionisation potentials are lower than the lamp ionising energy will respond. This allows detection of aliphatics (except CH_4), aromatics, ketones, aldehydes, esters, heterocyclics, amines, organic sulphur compounds and some organometallics [56].

The flame photometric detector (FPD) is essentially a flame emission photometer. The eluted compounds pass into a flame, usually a hydrogen-enriched, low-temperature plasma, inside a shielded jet which supplies sufficient energy first to produce atoms and simple molecular species, and then to excite them to a higher electronic state. The excited atoms and molecules subsequently return to their ground states with emission of characteristic atomic line or molecular band spectra which are measured by the photomultiplier tube, A narrow bandpass filter isolates the appropriate analytical wavelength range. The most highly developed FPDs are selective for phosphorus and sulphur, These elements are detected by monitoring band emissions from the molecular species HPO at 526 nm and S_2 at 394 nm. Detector response to phosphorus compounds is linear, whereas the response to compounds containing a single atom of sulphur is proportional to the square of the compound concentration. The high sensitivity and selectivity of the FPD for compounds containing sulphur and phosphorus has given it superiority over FID or ECD detectors for such analyses [56].

6.1.1.3. Identification: mass spectrometry. Mass spectrometry has been used by petroleum chemists long before flavour chemists used it. Possible applications of mass spectrometry in flavour research were recognised by James and Martin [61] even before the advent of gas chromatography. Direct introduction of samples separated by temperature programmed gas chromatography capillary columns into fast-scan mass spectrometers solved the initial problems. Compounds analysed by mass spectrometry range from acetals, furans and furanones and other compounds from the Maillard browning reaction, to non-volatile bio-active and flavour-active compounds such as phenols, flavonoids, glycosoids, saponins, etc. Mass spectrometry is very widely used for identification of flavour compounds. New techniques, such as capillary gas chromatography coupled to on-line isotope ratio mass spectrometers are used to determine the origin of various flavours, especially peppermint, vanilla, etc. Atmospheric pressure chemical ionisation mass spectrometry was already mentioned in the previous section because of its value for direct in-mouth analysis. Atmospheric pressure chemical ionisation-tandem mass spectrometry and inductively coupled plasma-mass spectrometry has been reported for analysis of organosulphur compounds. Recent reviews and books with sections on novel methods of flavour analysis provide more detailed information on these techniques [62-65].

6.1.2. Liquid chromatography

Liquid partition chromatography, especially countercurrent liquid chromatography is used in studies of flavour precursors, which are usually water-soluble low volatility compounds containing sugar moieties. This separation method is done at ambient temperatures, and it is therefore a very powerful tool for isolating labile compounds [4].

6.2. SENSORY-INSTRUMENTAL CHARACTERISATION

Progress in analysis techniques has led to long lists of volatiles [66]. Unfortunately an identified volatile compound, even analysed at concentrations as present in the human mouth, does not necessarily has to be an aroma compound. This implicates a need for correct determination of the relevant aroma compounds from the whole range of volatiles present in a particular food product samples. An interesting 'sensory' approach is sniffing the gas chromatographic effluent (gas chromatography olfactometry), in order to associate odour activity with the eluting compounds. This method was proposed by Fuller and co-workers as early as 1964 [66].

Many of the "chemical/physical" detectors are not as sensitive as the human nose for many odourants [55]. The interest in determining the individual contribution of volatile compounds present in foods, has led to a new generation of gas chromatography olfactometry techniques, which can be classified in four categories.

- Dilution analysis methods for producing titer or potency values based on stepwise dilution to threshold, e.g. CharmAnalysis [67] and Aroma Extraction Dilution Analysis (AEDA;) [68].
- Detection frequency methods for recording time duration of perceived odours over a group of assessors. The number of assessors perceiving an odour also estimates a titer or potency [69].
- Time-intensity methods for producing subjective estimates of perceived intensity recorded simultaneously with the elution of the chromatographic peak, e.g. Osme [70].
- Posterior intensity methods for producing subjective estimates of perceived intensity, which are recorded after a peak has eluted [71].

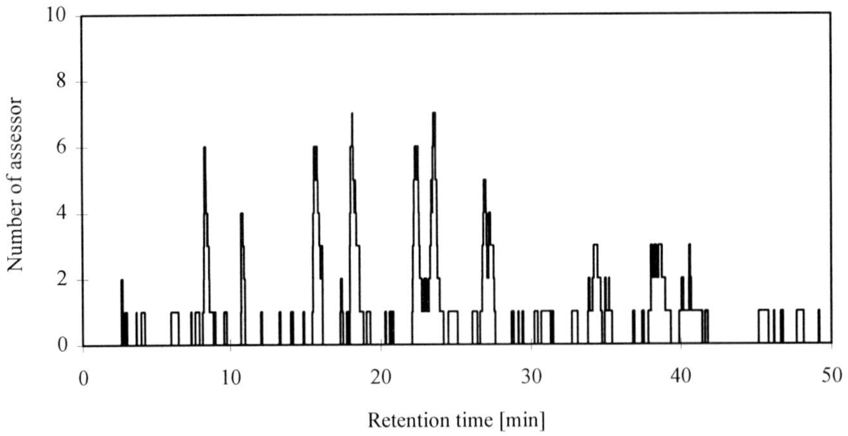

Figure 4. *Aromagram of a vegetable oil-in-water emulsion obtained by gas chromatography olfactometry showing the detection frequency of assessors perceiving a compound*

Both CharmAnalysis and AEDA are based on the odour-detection threshold. The dilution value obtained for each compound is proportional to its odour activity value in air, i.e. its concentration. Several injections are required to reach a dilution of the aroma extract in which odorous regions are no longer detected. More recently, Guth and Grosch [72] reported a new concept of AEDA using static headspace instead of extracts. Dilution steps are made by injecting decreasing headspace volumes to evaluate the relative odour conditions, so that identification can be performed on the basis of odour qualities and retention indices. The aromagrams obtained by sniffing successive dilutions of the same extract give valuable information on the number of odour units of each aroma compound eluted from the column. This protocol allows the scientist to obtain reliable results because of the simplicity of the task asked of the assessors and because of a validation of the final result obtained from the multiple detection of the same odour in the various dilutions of the same extract. Nevertheless, the major drawbacks of the dilution approach are, the difficulty of using more than one assessor, as is advisable in sensory analysis, because the method is very time-consuming. Secondly, the results are based on detection thresholds and not on real intensities.

Two other types of methods presented above (detection frequency and time intensity) overcome these serious limitations. The detection frequency method as proposed by Roozen and co-workers [69] overcomes the problem of the limited number of assessors. The method uses a panel of ten or twelve assessors. The number of assessors have been related to the posterior intensity and it was shown that the number of assessors perceiving an odour is a sufficient measure for odour intensity [73].

This method of gas chromatography olfactometry was later evaluated for various quantities of aroma compounds. Different sampling times (1-12 min) in a model mouth system) resulted in various quantities of compounds but resulted in an identical selection of odour active compounds by the panel, showing the robustness of the method [74]. An example of a sniffing chromatogram of a sunflower oil-in-water emulsion is shown in Figure 4. This aroma analysis technique is described in detail for analysis of oils and emulsions by van Ruth et al. [39].

The second attempt to overcome the problems mentioned at the previous page was reported by Sanchez et al. [70] and da Silva et al. [75]. Their time-intensity method is based on magnitude estimation of the odour intensity. The assessors use a variable resistor with a pointer moving along a category scale. A simultaneous computerised graphical feedback of the settled position of the cursor helps the assessor to adjust this position to the perceived intensity. The authors demonstrated that the estimation of the intensity of odours detected in the chromatographic effluent can be well correlated for each assessor with the concentration of the corresponding compounds. Etiévant et al. [76] reported a cross-modality matching with the finger span based on the same principle. They described a prototype for the precise measurement and acquisition of the distance between the thumb and another finger during analysis. Their four-member panel was able to determine most characteristics of the solutions with reference compounds and to create a finger span multidimensional space highly correlated with the theoretical intensity space.

7. Relevance of techniques for biotechnology

Microorganisms and enzymes have been employed for centuries in traditional biotechnological methods for preservation of foods. Modern biotechnology makes use of the increasing knowledge of the underlying scientific principles and is starting to exploit the advantages offered by biocatalysts in a more specific way. Due to the advances in microbial fermentation and enzyme technology and consumer demands for natural products, production of individual flavour compounds or complex flavour mixtures, are increasingly becoming targets for production on an industrial scale [77-78]. The exploitation of such techniques is especially attractive, because flavour compounds obtained via biotechnology are considered to be natural by regulatory authorities in many countries [79]. The biotechnological processes leading to flavour production can be divided into two major groups: (1) microbiologically mediated syntheses and (2) plant production methods [80]. The latter methods will not be discussed any further as they are beyond the scope of this review.

Flavour-producing microorganisms are primarily fungi, as they are able to carry out secondary metabolism. Factors, such as media composition, seed culture generation, pH, fermentation time and temperature, determine the amount and type of flavour substances produced. The group of flavour compounds produced by fermentation methods cover a broad flavour area and include acids (dairy flavours), alcohols (mushroom), lactones (dairy, peach), esters (fruity), aldehydes (fruity), ketones (dairy, cheese), pyrazines (nutty, roasted green) and terpenoids (citrus, mint flavours) [81]. Currently, the creation of so-called pure aromatic chemicals is one of the major applications of biotechnology to flavour production. However, simple or complex mixtures of aroma compounds are also useful. Cheese flavours, dairy-, fruit-, mushroom- and fish flavours, as well as mixtures of flavour enhancers are produced by the action of microorganisms or enzymes on substrate materials and are commercialised as flavours.

Research related to biotechnological flavour synthesis can be divided into three areas: non-volatile precursors, biotransformations and enzymes.

7.1. BIOTECHNOLOGICAL FLAVOUR SYNTHESIS

7.1.1. Non-volatile precursors

An essential prerequisite for optimum flavour generation by microorganisms and enzymes is detailed knowledge of the composition and availability of precursors needed. Many intensive studies have been focused on precursors. Some examples are: lipids and fatty acids [82], strawberry aroma precursors [83] and passion fruit aroma precursors [84]. Several studies of other precursors have been published in the book edited by Teranishi et al. [85]. This type of studies and how enzymatic action releases characteristic volatiles from precursor compounds has progressed because of the advance in methods of handling glycosides [65]. Countercurrent chromatographic techniques have been successfully employed in the isolation of flavour precursors, such as glycosides. Development of cyclodextrin columns and advances in multidimensional gas chromatography combined with mass spectrometry has led to progress in analysis of these desirable flavour compounds. Atmospheric pressure chemical ionisation and electrospray ionisation, liquid chromatography coupled with mass spectrometry and tandem mass spectrometry are other techniques, which are tools to provide elucidation of structures of flavour precursors [65]. This is very important because such data are necessary in biotechnological development of more flavourful foods. Also, knowledge of precursors permit processing parameters to be adjusted for maximum yield of flavour [4].

7.1.2. Biotransformations

Biotransformations include transformations (single step reactions) and conversions (multi-step reactions) of suitable flavour precursors either by microorganisms or by enzymes. Natural sources containing the required precursors in high amounts can be used directly as substrates. Alternatively, single compounds isolated from abundant natural sources can be subjected to highly specific microbiologically catalysed reaction sequences [79].

7.1.3. Enzymes

The use of enzymes is an integral part of many important processes in the food industry. The high substrate specificity in complex matrices, high reaction specificity, mild reaction conditions and reduction in waste product formation are of importance in the synthesis of flavour compounds. Furthermore, the enzymes, e.g. lipases and proteases, are stabile in organic solvents which allows the catalysis of reactions which are not feasible in aqueous media.

7.2. FLAVOUR ANALYSIS RESEARCH

There are two research areas to be covered from a flavour analysis point of view with respect to biotransformations and enzymes. Firstly, the progress of the formation of the flavour compounds and the yield is important in relation to flavour production. Secondly, the sensorial relevance of the aroma compounds and their application in foods is another point of concern.

The progress of formation can be followed off-line by extraction of the compounds from the medium [86-88]. Formation can also be easily monitored by static headspace analysis [89-90]. For on-line volatile measurements are gas sensors available. Pattern recognition is feasible with the use of an array of sensors. Systems called "electronic noses", which are obviously electronic but not noses (in the meaning of mammalian noses) generally use headspace techniques and can be used for flavour production monitoring with some adaptations [91-93]. Drawbacks of these sensors are that they respond to groups of compounds and not to specific compounds, they are more or less sensitive to water and they exhibit considerable drifts over time.

Currently, there is much interest in flavour compounds, which are responsible for characteristic flavours. Several novel techniques for isolation and characterisation of these important compounds are extensively described in Section 4 (Model mouth systems), Section 5 (In-mouth measurements) and Section 6 (Characterisation techniques).

8. Conclusions

The fundamental concepts of several isolation and characterisation methods for analysis of aroma compounds have been discussed in this paper. The presently available methodology can be divided into two groups. One group of methods is concerned with the volatile content of the products. The other group of techniques is focused on those aroma compounds released in the mouth during eating and those compounds contributing to the sensory perception of a food or flavour. Both groups of techniques are valuable for specific tasks. Eventually, it is the flavour scientist who has to define the problem and consider which isolation methods suits best. He/she has the difficult task to wisely choose among the techniques available to yield a solution.

References

1 Taylor, A.J. (1996) Volatile flavour release from foods during eating, *Critical Reviews in Food Science and Nutrition 36*, 765-784.

2 Reineccius, G. (1993) Biases in analytical flavour profiles introduced by isolation method, in C.-T. Ho and C.H. Manley (eds), *Flavour Measurement*, Marcel Dekker: New York, pp. 61-76.

3 Hoff, J.T., Chicoye, E., Herwig, W.C. and Helbert, J.R. (1978) Flavour profiling of beer using statistical treatments of G.L.C. headspace data, in G. Charalambous (ed.), *Analysis of Foods and Beverages. Headspace Techniques*, Academic Press: New York, pp. 187-201.

4 Teranishi, R. (1998) Challenges in flavour chemistry: an overview, in C.J. Mussinan, M.J. Morello (eds), *Flavour Analysis. Developments in Isolation and Characterisation*, American Chemical Society: Washington, DC, pp. 1-6.

5 Cram, D.J. and Hammond, G.S. (1964) *Organic Chemistry*, Mc Graw-Hill, New York.

6 Maarse, H. and Visscher, C.A. (1991) *Volatile Compounds in Foods*, TNO-Voeding: Zeist.

7 Dirinck, P. and De Winne, A. (1994) Advantages of instrumental procedures for measurement of flavour characters, in H. Maarse and D.G. van der Heij (eds), *Trends in Flavour Research*, Elsevier: Amsterdam, pp. 259-265.

8 Buttery, R.G. and Ling, L.C. (1996) Methods for isolating food and plant volatiles, in G.R. Takeoka, R. Teranishi, P.J. Williams and A. Kobayashi (eds), *Biotechnology for Improved Foods and Flavors*, American Chemical Society, Washington, pp. 240-248.

9 Shibamoto, T. (1984) Gas chromatography, in G. Charalambous (ed.), *Analysis of Foods and Beverages*, Academic Press: New York, pp. 93-115.

10 Teranishi, R. (1984) Sample preparation, in Charalambous, G. (ed.), *Analysis of Foods and Beverages*, Academic Press: New York, pp. 1-12.

11 Ropkins, K. and Taylor, A.J. (1996) The isolation of flavour compounds from foods by enhanced solvent extraction methods, in A.J. Taylor and D.S. Mottram (eds), *Flavour Science. Recent Developments*, Royal Society of Chemistry: Cambridge, pp. 297-300.

12 Sen, A., Laskawy, G., Schieberle, P. and Grosch, W. (1991) Quantitative determination of beta-damascenone in foods using a stable isotope dilution essay, *Journal of Agricultural and Food Chemistry 39*, 757-759.

13 Guth, H. and Grosch, W. (1989) 3-Methylnonane-2,4-dione: an intense odour compound formed during flavour reversion of soybean oil, *Fat Science and Technology 91*, 225-230.

14 Buttery, R.G., Takeoka, G.R., Krammer, G.E. and Ling, L.C. (1994) Identification of 2,5-dimethyl-4-hydroxy-3(2H)-furanone in fresh and processed tomato, *Lebensmittel-Wissenschaft und –Technologie 27*, 592-594.

15 Leahy, M.M. and Reineccius, G.A. (1984) Comparison of methods for the analysis of volatile compounds from aqueous model systems, in P. Schreier (ed.), *Analysis of Volatiles: Methods and Application*, Walter de Gruyter: Berlin, pp. 19-47.

16 Cobb, C.S. and Bursey, M.M. (1978) Comparison of extracting solvents for typical volatile components of Eastern wines in model aqueous alcoholic systems, *Journal of Agricultural and Food Chemistry 26*, 197-199.

17 Takeoka, G.R., Ebeler, S. and Jennings, W. (1985) Capillary gas chromatographic analysis of volatile flavour compounds, in D.D. Bills and C.J. Mussinan (eds), Characterisation and Measurement of Flavour Compounds, American Chemical Society: Washington, pp. 95-108.

18 Fischer, N., Nitz, S. and Drawert, F. (1988) Original flavour compounds of marjoram flavour and its changes during processing, *Journal of Agricultural and Food Chemistry 36*, 996-1003.

19 Nickerson, G.B. and Likens, S.T. (1966) Gas chromatographic evidence for the occurrence of hop oil components in beer, *Journal of Chromatography 21*, 1-4.

20 Wiberg, K.B. (1960) *Laboratory Techniques of Organic Chemistry*, McGraw-Hill: New York.

21 Teranishi, R and Kint, S. (1993) Sample preparation, in T.E. Acree, R. Teranishi (eds), *Flavour Science: Sensible Principles and Techniques*, American Chemical Society: Washington, pp. 137-167.

22 Buttery, R.G., Ling, L.C. and Bean, M.M. (1978) Coumarin off-odor in wheat flour due to sweetclover (*Melilotus*) seeds, *Journal of Agricultural and Food Chemistry 26*, 179-180.

23 van Ruth, S.M., Roozen, J.P.and Jansen, F.J.H.M. (2000) Release of odour active compounds from oxidised sunflower oil and its oil-in-water emulsion, in P. Schieberle and K.-H. Engel (eds), *Frontiers of Flavour Science*, Deutsche Forschungsanstalt für Lebensmittelchemie: Garching, pp. 292-299.

24 Wampler, T.P. (1997), Analysis of food volatiles using headspace-gas chromatographic techniques, in R. Marsili (ed.), *Techniques for Analysing Food Aroma*, Marcel Dekker: New York, pp. 27-58.

25 Morris, J.G. (1968) *A Biologist's Physical Chemistry*, London: Edward Arnold.

26 Taylor, A.J. (1998) Physical chemistry of flavour, *International Journal of Food Science and Technology 33*, 53-62.

27 de Roos, K.B. and Sarelse, J.A. (1996) Volatile acids and nitrogen compounds in prawn powder, in A.J. Taylor and D.S. Mottram (eds), *Flavour Science, Recent Developments*, Royal Society of Chemistry: Cambridge, pp. 13-18.

28 Sucan, M.K., Fritz-Jung, C. and Ballam, J. (1998) Evaluation of purge-and-trap parameters optimisation using a statistical design, in C.J. Mussinan and M.J. Morello (eds), *Flavour Analysis. Developments in Isolation and Characterisation*, American Chemical Society: Washington, pp. 22-37.

29 Laye, I., Karleskind, D. and Morr, C.V. (1995) Dynamic headspace analysis of accelerated storage commercial whey protein concentrate using four different adsorbent traps, *Milchwissenschaft 50*, 268-271.

30 Nunez, A., Gonzalez, L.F. and Janak, J. (1984) Pre-concentration of headspace volatiles for trace organic analysis by gas chromatography, *Journal of Chromatography 300*, 127-162.

31 Schaefer, J. (1981) Comparison of adsorbents in headspace sampling, in P. Schreier (ed.), *Flavour '81*, Walter de Gruyter: Berlin, pp. 301-313.

32 Wyllie, S.G., Alves, S., Filsoof, M. and Jennings, W.G. (1978) Headspace sampling: use and abuse, in G. Charalambous, *Analysis of Foods and Beverages*, Wiley: New York, pp. 1-16.

33 Park, P.S.W. (1993) Loss of volatile lipid oxidation products during thermal desorption in dynamic headspace-capillary gas chromatography, *Journal of Food Science 58*, 220-222.

34 Roberts, D.D.; Acree, T.E. (1995) Simulation of retronasal aroma using a modified headspace technique: investigating the effects of saliva, temperature, shearing, and oil on flavour release, *Journal of Agricultural Food Chemistry 43*, 2179-2186.

35 Lee III, W.E. (1986) A suggested instrumental technique for studying dynamic flavour release from food products, *Journal of Food Science 51*, 249-250.

36 Naβl, K.; Kropf, F. and Klostermeyer, H. (1995) A method to mimic and to study the release of flavour compounds from chewed food, *Zeitschrift für Lebensmittel-Untersuchung und Forschung 201*, 62-68.

37 Withers, S.J., Conner, J.M., Piggott, J.R. and Paterson, A. (1998) A simulated mouth to study flavour release from alcoholic beverages, in P. Schieberle (ed.), *Food Science and Technology Cost 96, Interaction of Food Matrix with Small Ligands Influencing Flavour and Texture, volume 3*, Office for Official Publications of the European Communities: Luxembourg, pp. 13-18.

38 van Ruth, S.M., Roozen, J.P. and Cozijnsen, J.L. (1994) Comparison of dynamic headspace mouth model systems for flavour release from rehydrated bell pepper cuttings, in H. Maarse and D.G. van der Heij (eds), *Trends in Flavour Research*, Elsevier: Amsterdam, pp. 59-64.

39 van Ruth, S.M., Roozen, J.P., Posthumus, M.A. and Jansen, F.J.H.M. (1999) Influence of ascorbic acid and ascorbyl palmitate on the aroma composition of an oxidised vegetable oil and its emulsion, *Journal of the American Oil Chemists' Society 76*, 1375-1381.

40 van Ruth, S.M. and Roozen, J.P. (2000) Influence of mastication and artificial saliva on aroma release in a model mouth system, *Food Chemistry 71*, 393-399.

41 Philips, M. and Greenberg, J. (1991) Methods for the collection and analysis of volatile compounds in the breath, *Journal of Chromatography 564*, 242.

42 Soeting, W.J. and Heidema, J. (1988). A mass spectrometric method for measuring flavour concentration/time profiles in human breath, *Chemical Senses 13*, 607-617.

43 Haring, P.G.M. (1990) Flavour release: from product to perception, in Y. Bessière and A.F. Thomas (eds), *Flavour Science and Technology*, Wiley: Chichester, pp. 351-354.

44 Linforth, R.S.T. and Taylor, A.J. (1993) Measurement of volatile release in the mouth, *Food Chemistry 48*, 115-120.

45 Roozen, J.P. and Legger-Huysman (1995) Sensory analysis and oral vapour gas chromatography of chocolate flakes, in M. Rothe and H.-P. Kruse (eds), *Aroma. Perception, Formation, Evaluation*, Eigenverlag Deutsches Institut für Ernährungsforschung: Potsdam-Rehbrücke, pp. 627-632.

46 Delahunty, C.M., Piggott, J.R., Conner, J.M. and Paterson, A. (1994) Low-fat Cheddar cheese flavour: flavour release in the mouth, in H. Maarse and D.G. van der Heij (eds), *Trends in Flavour Research*, Elsevier: Amsterdam, pp. 47-52.

47 Benoit, F.M., Davidson, W.R., Lowett, A.M., Nacson, S. and Ngo, A. (1983) Breath analysis by atmospheric pressure ionisation mass spectrometry, *Analytical Chemistry 55*, 805-807.

48 Hansel, A., Jordan, A., Holzinger, R., Prazeller, P., Vogel, W. and Lindinger, W. (1995) Proton transfer reaction mass spectrometry: on-line trace gas analysis at the ppb level. *International Journal of Mass Spectrometry and Ion Processes 149/150*, 609-619.

49 Smith, D. and Spanel, P. (1996) The novel selected-ion flow tube approach to trace gas analysis of air and breath, *Rapid Communications in Mass Spectrometry 10*, 1183-1198.

50 Linforth, R.S.T., Ingham, K.E. and Taylor, A.J. (1996) Time course profiling of volatile release from foods during the eating process, in A.J. Taylor and D.S. Mottram (eds), *Flavour Science, Recent Developments*, Royal Society of Chemistry: Cambridge, pp. 361-368.

51 Ingham, K.E., Taylor, A.J., Chevance, F.F.V. and Farmer, L.J. (1996). Effect of fat content on volatile release from foods, in A.J. Taylor and D.S. Mottram (eds), *Flavour Science, Recent Developments*, The Royal Society of Chemistry: Cambridge, pp. 386-391.

52 Linforth, R.S.T., Baek, I. and Taylor, A.J. (1999) Simultaneous instrumental and sensory analysis of volatile release from gelatine and pectin/gelatine gels, *Food Chemistry 65*, 77-83.

53 Marin, M., Baek, I., Taylor, A.J. (1999). Volatile release from aqueous solutions under dynamic headspace dilution conditions, *Journal of Agricultural and Food Chemistry 47*, 4750-4755.

54 Grab, W. and Gfeller, H. (1998). Fast dynamic flavour changes and flavour release during eating, in P. Schieberle (ed.), *Food Science and Technology Cost 96, Interaction of Food Matrix with Small Ligands Influencing Flavour and Texture, volume 3*, Office for Official Publications of the European Communities: Luxembourg, pp. 132-135.

55 Acree, T.E. and Barnard, J. (1994) Gas chromatography-olfactometry and CharmAnalysis, in H. Maarse and D.G. van der Heij (eds), *Trends in Flavour Research*, Elsevier: Amsterdam, pp. 211-220.

56 Willard, H.H., Merritt, L.L., Dean, J. and Settle, F.A. (1981) *Instrumental Methods of Analysis*, Wadsworth Publishing Company: Belmont.

57 Hinshaw, J.V. (1990) Use and care of electrolytic conductivity detectors, *LC-GC International 3* (3), 22-26.

58 Perkins, E.G. (1989) Gas chromatography and gas chromatography mass spectrometry of odor/flavour components in lipid foods, in D.B. Min and T.H. Smouse (eds), *Flavour Chemistry of Lipid Foods*, American Oil Chemists' Society: Champaign, IL, pp. 35-56.

59 Hinshaw, J.V. (1990) Thermal conductivity detectors, *LC-GC International 3* (5), 26-31.

60 Hinshaw, J.V. (1990) Flame ionisation detectors, *LC-GC International 3* (4), 34-39.

61 James, A.T. and Martin, A.J.P. (1952) Gas-liquid partition chromatography: the separation and micro-estimation of volatile fatty acids from formic acid to dodecanoic acid, *Biochemical Journal 50*, 679-690.

62 Mussinan, C.J. and Morella, M.J. (eds) (1998) *Flavour Analysis: Developments in Isolation and Characterisation*, American Chemical Society: Washington.
63 Marsili, R. (1997) *Techniques for Analysing Food Aroma*, Marcel Dekker: New York.
64 Taylor, A.J. and Mottram, D.S. (eds) (1996) *Flavour Science, Recent Developments*, The Royal Society of Chemistry: Cambridge.
65 Takeoka, G.R., Kobayashi, A. and Teranishi, R. (1996) Analytical methodology in biotechnology: an overview, in G.R. Takeoka, R. Teranishi, P.J. Williams and A. Kobayashi (eds), *Biotechnology for Improved Foods and Flavors*, American Chemical Society: Washington.
66 Fuller, G.H., Steltenkamp, R. and Tisserand, G.A. (1964) The gas chromatograph with human sensor: perfumer model, *Annual NY Acad. Sci. 116*, 711-724.
67 Acree, T.A., Barnard, J. and Cummingham, D. (1984) A procedure for the sensory analysis of gas chromatographic effluents, *Food Chemistry 14*, 273-286.
68 Ullrich, F. and Grosch, W. (1987) Identification of the most intense volatile flavour compounds formed during autoxidation of linoleic acid, *Zeitschrift für Lebensmittel-Untersuchung und Forschung 184*, 277-282.
69 Linssen, J.P.H., Janssens, J.L.G.M., Roozen, J.P. and Posthumus, M.A. (1993) Combined gas chromatography and sniffing port analysis of volatile compounds of mineral water packed in laminated packages, *Food Chemistry 8*, 1-7.
70 Sanchez, N.B., Lederer, C.L., Nickerson, G.B., Libbey, L.M. and McDaniel, M.R. (1992) Sensory and analytical evaluation of beers brewed with three varieties of hops and an unhopped beer, in G. Charalambous (ed.), *Proceedings of the 6th International Flavour conference, Rethymnon, Crete*, Elsevier: Amsterdam, pp. 403-426.
71 Casimir, D.J. and Whitfield, F.B. (1978) Flavour impact values: a new concept for assigning numerical values for potency of individual flavour components and their contribution to overall flavour profile, *Berichte der Internationalen Fruchtsaftunion 15*, 325-345.
72 Guth, H. and Grosch, W. (1995) Comparison of the juices of stewed beef and stewed pork by instrumental analyses of the odourants and by sensory studies, in P. Étiévant and P. Schreier, *Bioflavour 95*, Institut National de la Recherche Agronomique: Dijon, pp. 201-205.
73 van Ruth, S.M., Roozen, J.P., Hollmann, M.E. and Posthumus, M.A. (1996) Instrumental and sensory analysis of the flavour of French beans (*Phaseolus vulgaris*) after different rehydration conditions. *Zeitschrift für Lebensmittel-Untersuchung und Forschung 203*, 7-13.
74 van Ruth, S.M., Roozen, J.P. and Cozijnsen, J.L. (1996) Gas chromatography/sniffing port analysis evaluated for aroma release from rehydrated French beans (*Phaseolus vulgaris*), *Food Chemistry 56*, 343-346.
75 da Silva, M.A.A.P., Lundahl, D.S. and McDaniel, M.R. (1994) The capability and psychophysics of Osme: a new GC-olfactometry technique, in H. Maarse and D.G. van der Heij (eds), *Trends in Flavour Research*, Elsevier: Amsterdam, pp. 191-209.
76 Étiévant, P.X., Callement, G., Langlois, D., Issanchou, S. and Coquibus, N. (1999) Odor intensity evaluation in gas chromatography-olfactometry by finger span method, *Journal of Agricultural and Food Chemistry 47*, 1673-1680.
77 Benz, L. and Muheim, A. (1996) Biotechnological production of vanillin, in A.J. Taylor and D.S. Mottram (eds), *Flavour Science, Recent Developments*, Royal Society of Chemistry: Cambridge, pp. 111-117.
78 Stam, H., Boog, A.L.G.M. and Hoogland, M. (1996) the production of natural flavours by fermentation, in A.J. Taylor and D.S. Mottram (eds), *Flavour Science. Recent Developments*, Royal Society of Chemistry: Cambridge, pp. 122-125.
79 Engel, K.-H and Roling, I. (1996) Generation of flavours by microorganisms and enzymes: an overview, in G.R. Takeoka, R. Teranishi, P.J. Williams and A. Kobayashi (eds), *Biotechnology for Improved Foods and Flavors,* American Chemical Society: Washington, pp. 120-123.
80 Welsh, F.W. (1994) Overview of bioprocess flavour and fragrance production, in A. Gabelman (ed.), *Bioprocess, Production of Flavour, Fragrance, and Colour Ingredients*, Wiley: New York, pp. 1-18.
81 Manley, C.H. (1994) The development and regulation of flavour, fragrance, and colour ingredients produced by biotechnology, in A. Gabelman (ed.), *Bioprocess, Production of Flavour, Fragrance, and Colour Ingredients*, Wiley: New York, pp. 19-40.

82 Ratledge, C. and Dickinson, F.M. (1995) Lipids and fatty acids as potential flavour components using microbial systems, in P. Étiévant and P. Schreier (eds), *Bioflavour 95*, Institut National de la Recherche Agronomique: Dijon, pp. 153-166.

83 Zabetakis, I. and Holden, M.A. (1995) A study of strawberry flavour biosynthesis, in P. Étiévant and P. Schreier, *Bioflavour 95*, Institut National de la Recherche Agronomique: Dijon, pp. 211-216.

84 Chassagne, D., Bayonove, C., Crouzet, J. and Baumes, R. (1995) Formation of aroma by enzymatic hydrolysis of glycosidically bound components of passion fruit, in P. Étiévant and P. Schreier (eds), *Bioflavour 95*, Institut National de la Recherche Agronomique: Dijon, pp. 217-222.

85 Teranishi, R., Takeoka, G.R. and Guenters, M. (eds) (1992) *Flavour Precursors: Thermal and Enzymatic Conversions*, American Chemical Society: Washington.

86 Demyttenaere, J.C.R., Konickx, I.E.I. and Meersman, A. (1996) Microbial production of bioflavours by fungal spores, in A.J. Taylor and D.S. Mottram (eds), *Flavour Science, Recent Developments*, Royal Society of Chemistry: Cambridge, pp. 105-110.

87 Latrasse, A., Guichard, E., Fournier, N., Le Quéré, L., Dufossé, L. and Spinnler, H.E. (1994) Biosynthesis and flavour properties of the lactones formed by *Fusarium poase*, in H. Maarse and D.G. van der Heij (eds), *Trends in Flavour Research*, Elsevier: Amsterdam, pp. 493-498.

88 van der Schaft, P., van Geel, I., de Jong, G. and ter Burg, N. (1994) Microbial production of natural furfurylthiol, in H. Maarse and D.G. van der Heij (eds), *Trends in Flavour Research*, Elsevier: Amsterdam, pp. 59-64.

89 Stahnke, L.H. (1996) Volatile produced by Staphylococcus xylosus growing in meat/fat systems simulating fermented sausage minced with different ingredient levels, in A.J. Taylor and D.S. Mottram (eds), *Flavour Science. Recent Developments*, Royal Society of Chemistry: Cambridge, pp. 126-129.

90 Kallio, H. Alhonmäki, P. and Tuomola, M. (1994) Formation of volatile sulphur compounds in cut onions, in H. Maarse and D.G. van der Heij (eds), *Trends in Flavour Research*, Elsevier: Amsterdam, pp. 463-474.

91 Mielle, P., Hivert, B. and Mauvais, G. (1995) Are gas sensors suitable for on-line monitoring and quantification of volatile compounds, in P. Étiévant and P. Schreier, *Bioflavour 95*, Institut National de la Recherche Agronomique: Dijon, pp. 81-84.

92 Rossi, V., Garcia, C., Talon, R., Denoyer, C. and Berdague, J.L. (1995) Rapid discrimination of meat products and bacterial strains using semiconductor gas sensors, in P. Étiévant and P. Schreier (eds), *Bioflavour 95*, Institut National de la Recherche Agronomique: Dijon, pp. 85-90.

93 Breheret, S., Talou, T., Bourrounet, B. and Gaset, A. (1995) On-line differentiation of mushrooms aromas by combined headspace/multi-odour gas sensors devices, in P. Étiévant and P. Schreier (eds), *Bioflavour 95*, Institut National de la Recherche Agronomique: Dijon, pp. 103-111.

INDEX